Conformational Analysis of Cyclohexenes, Cyclohexadienes, and Related Hydroaromatic Compounds

Methods in Stereochemical Analysis

Series Editor: Alan P. Marchand
Department of Chemistry
North Texas State University
Denton, Texas 76203

Conformational Analysis of Cyclohexenes, Cyclohexadienes, and Related Hydroaromatic Compounds

Edited by

PETER W. RABIDEAU

VCH
Publishers

Peter W. Rabideau
Purdue University
Department of Chemistry
1125 East 38th Street
P.O. Box 649
Indianapolis, Indiana 46223

Library of Congress Cataloging-in-Publication Data

Conformational analysis of cyclohexenes, cyclohexadienes, and related hydroaromatic compounds.

Includes bibliographies and index.
1. Aromatic compounds—Analysis. 2. Conformational analysis. I. Rabideau, Peter W.
QD335.C66 1989 547′.6 88-33970
ISBN 0-89573-702-7

British Library Cataloguing in Publication Data

The conformational analysis of cyclophexenes, cyclophexadienes and related hydroaromatic compounds.

1. Cyclic compounds. Conformational analysis
I. Rabideau, Peter W.
547′.5

ISBN 0-89573-702-7 U.S.
ISBN 3-527-26860-X W. Germany

Printed in the United States of America.

ISBN 0-89573-702-7 VCH Publishers
ISBN 3-527-26860-X VCH Verlagsgesellschaft

Printing History:
10 9 8 7 6 5 4 3 2 1

Published jointly by:

VCH Publishers, Inc.
220 East 23rd Street
Suite 909
New York, New York 10010

VCH Verlagsgesellschaft mbH
P.O. Box 1-11 16
D-6940 Weinheim
Federal Republic of Germany

VCH Publishers (UK) Ltd.
8 Wellington Court
Cambridge CB1 1HW
United Kingdom

Contents

Preface

The conformational analysis of cyclic systems is a topic of great importance for both students and researchers due in part to the rather widespread occurrence of six-membered carbocyclic and heterocyclic rings in natural and synthetic products. Much of our knowledge on this topic has been founded on the extensive studies of cyclohexane and its derivatives, however, and discussions on the conformational analysis of cyclic systems generally focus on cyclohexanes with only cursory attention given to other ring systems.

In this work, we hope to provide a broader base for the approach to conformational problems in cyclic systems by consideration of the diverse nature of partially unsaturated six-membered rings. For example, these compounds exhibit quite different geometries from cyclohexane including planar, boat, and twist conformations. Moreover, the barriers separating these structures are often quite small and overall geometry may be controlled by substituent pattern. The "effective" size of substituents can also be quite different for these systems. For example, phenyl behaves as a relatively small substituent (smaller than methyl) in dihydroanthracene, dihydronaphthalene, and dihydrophenanthrene.

The initial chapters of this book are organized according to structural type: cyclohexenes, 1,3-cyclohexadienes, and 1,4-cyclohexadienes together with the related hydroaromatics, and six-membered rings containing exocyclic double bonds. The later chapters are more specialized in their treatment of the title ring systems. They include the application of ^{13}C-NMR to the conformational analysis of hydroaromatics, an important class of compounds that serve as hydrogen donors during coal liquefaction, as well as the conformation and chemistry of biologically important diols and epoxides considered to be metabolites associated with the carcinogenic activity of polynuclear aromatic hydrocarbons. Although complete coverage of hydroaromatics containing heteroatoms would be too extensive for this book, a chapter is included that focuses primarily on sulfur-containing compounds; this chapter is especially relevant to this work due to the comparisons with 9,10-dihydroanthracenes. In view of the growing importance of computational methods as a guide to the experimentalists, a chapter is included on the application of empirical force-field calculations to the title compounds. For those readers who may be unfamiliar with the technique, a brief tutorial is provided in the appendix.

Peter W. Rabideau

Contributors

Frank A. L. Anet, Department of Chemistry and Biochemistry, University of California, Los Angeles, California 90024, U.S.A.

David M. Grant, Department of Chemistry, University of Utah, Salt Lake City, Utah 84112

Ronald G. Harvey, Ben May Institute, University of Chicago, Chicago, Illinois 60637, U.S.A.

Joseph B. Lambert, Department of Chemistry, Northwestern University, Evanston, Illinois, 60208, U.S.A.

Kenny B. Lipkowitz, Department of Chemistry, Purdue University, Indianapolis, Indiana 46223, U.S.A.

Frederick G. Morin, Department of Chemistry, McGill University, Montreal, Quebec, Canada

Peter W. Rabideau, Department of Chemistry, Purdue University, Indianapolis, Indiana 46223, U.S.A.

Andrzej Syguła, Department of Chemistry, Jagiellonian University, 30-060 Krakow, Poland

1

Conformational Analysis of Cyclohexenes

Frank A. L. Anet

1.0 INTRODUCTION

Before discussing the structures and conformational properties of cyclohexene and its derivatives, it is useful to consider the geometrical relationships of unsaturated to saturated six-membered rings in a general way. This will set the stage for the main body of the chapter and provide a quick overview of the conformations of cyclohexene. Thus, we start with cyclohexane and consider the geometric and energy consequences of introducing approximately planar cis double bonds into various forms of this molecule. After discussing the conformational properties of cyclohexene and its simple derivatives, the conformations of cyclohexene fused to other rings are presented. A brief section on the relationship of the conformations of the above molecules to some important heterocyclic analogs and to six-membered rings fused to cyclopropane or epoxide units completes the chapter.

The preference for staggered over eclipsed single bonds ensures that the lowest-energy conformation of cyclohexane is the chair, **1**.[1,2] Because the absolute value of all the ring torsional angles is 55°, the insertion of an endocyclic cis double bond into this conformation cannot be achieved without major conformational changes.

The twist-boat form of cyclohexane (**2**, D_2 point group), which is a local energy minimum situated about 5.5 kcal/mol above the chair, has no torsional angle less than about 30°,[3] preventing the introduction of a planar double bond. The boat form in cyclohexane (**3**, C_{2v} point group), which is a transition state in the pseudorotation of the twist-boat, has two 0° torsional angles and can easily accommodate up to two cis double bonds. However, the cyclohexene boat form (**4**) has not only an eclipsed CH_2–CH_2 bond but also two allylic bonds with unfavorable torsional angles (see below), and molecular mechanics calculations show it to be a conformational transition state. Nevertheless, the cyclohexene ring can be a boat in a bridged or fused ring system.[4,5]

Proceeding to still higher energies in cyclohexane, we find the half-chair transition state for the chair to twist-boat process. The half-chair (**5**, C_2 point group) pseudorotates almost freely via a "sofa" form (**6**, C_s point group),[3,6] and both forms have at least one torsional angle close to 0° and can therefore easily incorporate a double bond. Indeed, the lowest-energy conformation of cyclohexene is the half-chair (**7**, C_2 point group). It should be noted that the half-chair has much lower bond and torsional angle strains in cyclohexene than in cyclohexane. The sofa (**8**) and related arrangements (see below) have no symmetry and do not

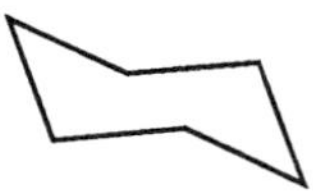

1

2

3

4

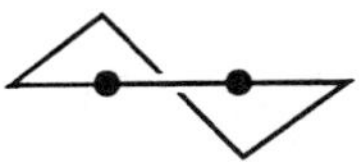

5

6

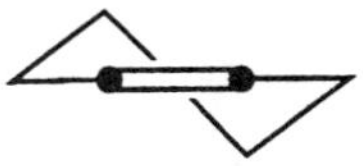

7

8

appear to be local energy minima in cyclohexene itself but are important in derivatives of this compound.

Finally, the most strained form of cyclohexane has all the carbons in one plane and thus can accommodate up to three double bonds. The planar geometry in cyclohexane itself is a high-energy two-dimensional potential energy hill that has no kinetic or thermodynamic significance, and this is also the case for planar cyclohexene.[7] However, in the case of 1,3-cyclohexadiene, which is discussed by Rabideau and Syguła in Chapter 3, the planar arrangement is a transition state for ring inversion in the lowest-energy conformation (C_2 point group) of that molecule,[3] and for 1,4-cyclohexadiene, which is discussed by Rabideau in Chapter 4, the arrangement is actually planar.

The terminology used above for the cyclohexene forms follows general usage. The boat is called a half-boat in a few publications, including two standard reference books on conformational analysis.[1,2] Since the shape of the boat is almost the same in cyclohexene as in cyclohexane, there seems to be no need for the term half-boat, as pointed out previously by Bucourt.[8] The half-chair can be considered to be intermediate in geometry between the chair and the planar form, but from another point of view, it is also intermediate between the chair and boat.

A systematic terminology to describe cyclohexene conformations, which has gained only limited acceptance, has been introduced by Bucourt and Hainaut[9,10] and is based on the relationships of eclipsed bonds, that is, bonds with 0° torsional angles. In the boat there are two such torsional angles in a 1,4 arrangement, and this geometry can be called 1,4-diplanar. The other possibilities are the 1,2-diplanar or sofa[11] (**8**) and the 1,3-diplanar forms. These are not energy extrema in cyclohexene itself, although they have well-defined geometries that correspond to (arbitrary) points on the potential energy path for the ring inversion of the half-chair and thus help to define the shape of that path. The presence of substituents or fused rings, as in steroid molecules,[12,13] can lead to conformations that are close to the ideal sofa arrangement. The boat, it should be noted, is strictly 1,4-diplanar by symmetry, whereas the half-chair, which does not have an exactly 0° torsional angle at the double bond, is only approximately monoplanar, and thus the latter term is not totally justified as an alternative name for the half-chair.

A more fundamental, although sometimes less convenient, description of puckering in rings has been proposed by Cremer and Pople,[14] who describe any puckered conformation in terms of a set of "basis" geometries, with the proportions of the bases being given by appropriate coefficients, but this and other related approaches[15] do not seem to have been used with cyclohexenes. In steroid molecules, which often contain unsaturated six-membered rings that are half-chairs or sofas (or that are intermediate between these two forms), an asymmetry parameter has been used to describe distortions from ideal torsional angle (*not* structural) symmetries. The asymmetry parameter is equal to the average distortion of sums (for ideal C_s symmetry) or differences (for ideal C_2 symmetry) of torsional angles.[12,13] Asymmetry parameters, together with the position of the corresponding (ideal) *torsional* symmetry elements, which are specified by one atom (for C_2 symmetry) or one bond (for C_s symmetry), have been used to compare the conformations of saturated and unsaturated six-membered rings in numerous steroid molecules.[12,13]

As a result of the strong torsional constraint introduced by the double bond in cyclohexene, the two unsaturated carbons can be considered to be a single pseudoatom,[15] so that cyclohexane becomes analogous to cyclopentane. These two ring systems both have C_s and C_2 forms, the chief difference being that the pseudorotation between these forms is hindered by a barrier of the order of 5–6 kcal/mol in cyclohexene as compared to the virtually free pseudorotation in cyclopentane. Nevertheless, the conformational analogy between these molecules is in some ways closer than that between cyclohexene and cyclohexane; certainly the former analogy helps to explain the much lower conformational barrier (5.3 kcal/mol) in the cyclohexene half-chair as compared to the much higher ring inversion barrier (10–11 kcal/mol) in the cyclohexane chair, as well as explaining why cyclohexene exists in two conformational families each having a local energy minimum, whereas cyclohexane, like cyclopentane, has only one conformational family and one local energy minimum.

From unsuccessful attempts made in 1937 to resolve cyclohexene-*cis*-4,5-dicarboxylic acid (**9**) on the presumption that it had a half-chair conformation and was therefore chiral,[16] it was concluded that the molecule must racemize rapidly by a ring inversion process involving vibrational motions of the ring atoms. This conclusion has turned out to be correct, although it is based on negative evidence. The same conformation was also proposed for cyclohexene in a later paper (1941) by Lister,[17] who is often considered to be the originator of the half-chair. However, it was the theoretical work of Pitzer and co-workers[18] (see below) that first provided strong evidence for the half-chair form of cyclohexene. The term half-chair, which was introduced in 1954 by Barton et al.,[19] was not used in these early papers, even though this conformation was clearly described. Corey and Sneen,[20] in another early paper (1955), used vector analysis to define the geometry of the half-chair and showed the potential use of such an approach to explain the regioselectivity of bromination in steroid ketones (see below). Their cyclohexene geometry is too puckered because of the use of ideal bond angles. Bucourt later used a similar vector approach with more realistic bond angles and applied it to explain conformational transmission effects in steroid molecules.[21]

The protons on C-4 and C-5 in the cyclohexene half-chair have nearly ideal axial (a) and equatorial (e) orientations, but those on C-3 and C-6 are less different from one another and are generally labeled pseudoaxial (a′) and pseudoequatorial (e′) as shown in **10**,[19] although some authors omit the "pseudo" qualification. These effects reflect the ring torsional angles in the half-chair, which are, in absolute values, about 62°, 46°, and 15° for the 4–5, 3–4, and 1–2 bonds, respectively. The 15° ring torsional angle at the allylic bond corresponds to a 43° torsional angle at that bond with respect to the ethylenic hydrogen and the adjacent pseudoequatorial allylic hydrogen. This leads to a rather large gauche repulsion when both of these hydrogens are replaced by bulky substituents, as shown in **11** (Johnson's $A^{(1,2)}$ allylic strain).[22] In contrast, the pseudoaxial allylic hydrogen has a torsional angle of 76° with respect to the adjacent ethylenic hydrogen, so that substituents at these positions have much smaller gauche nonbonded repulsions (**12**), although this is partly compensated

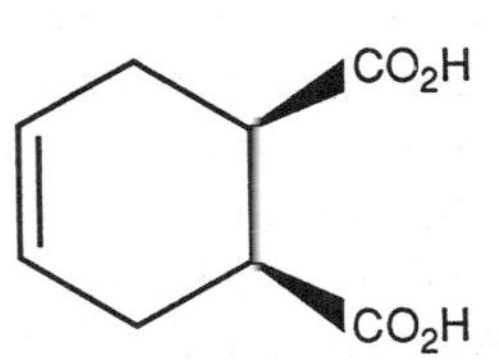

9

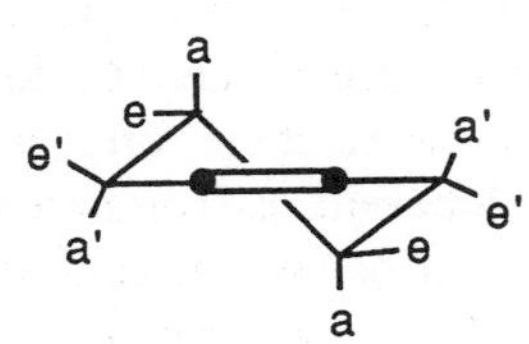

10

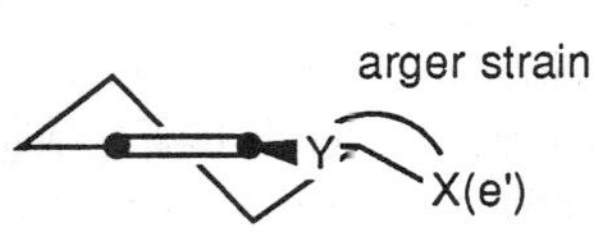

11

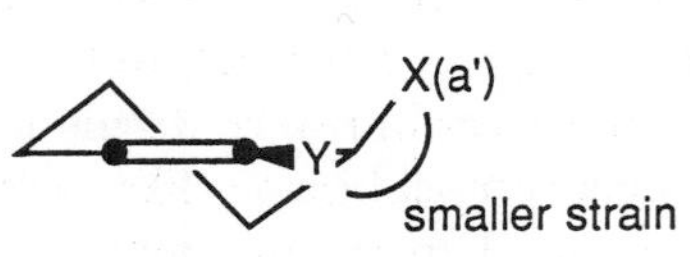

12

Johnson's $A^{(1,2)}$ Allylic Strain

by a 1,3 interaction of the pseudoaxial substituent. Molecular mechanics calculations (see below) show no $A^{(1,2)}$ strain in 1,6-dimethylcyclohexene, but the presence of a larger substituent than methyl at C-1 undoubtedly changes the picture.

Although *trans*-cyclohexene and its derivatives do not occur in compounds that are stable at room temperature, these molecules, which must have very distorted double bonds, have been detected as transitory and reactive intermediates in the photochemical additions of alcohols to cyclohexenes.[23,24] The only structural information on *trans*-cyclohexene comes from molecular mechanics calculations (see below).

1.1 CONFORMATIONAL PROPERTIES OF CYCLOHEXENES WITHOUT FUSED RINGS

1.1.1 Thermodynamic Data

Statistical mechanics can be used to calculate entropies and specific heats from vibrational data and these parameters can then be compared with experimental thermodynamic data. This approach provided some of the earliest conformational information on the nonplanarity and free pseudorotation in cyclopentane, the chair form of cyclohexane, the equatorial–axial equilibrium in methylcyclohexane, and the barrier to rotation in ethane.[1,2,25] Its application to cyclohexene by Pitzer and co-workers in 1948 led to the conclusion that the half-chair was the lowest-energy conformation, with the boat being 2.7 kcal/mol less stable than the half-chair.[18] This energy difference was required to fit the specific heat data, although the entropy fits equally well with this conclusion or with no boat form being present. However, the discrepancy between the experimental specific heat (34.65 cal/deg·mol at 400 K) and the value (33.92 cal/deg·mol) calculated on the basis of only the half-chair being populated is only 0.73 cal/deg·mol. Since the quoted errors are of the order of "several tenths of a unit," the discrepancy could well have been due to errors rather than to the presence of the boat conformation. The discrepancy cannot be blamed on a possible flexibility of the half-chair as suggested by Bucourt and Hainaut,[26] since the vibrational analysis takes all the fundamental vibrations of the molecules into account (the distinction between fundamental vibrations and overtones or combination bands, however, may possibly not have been made correctly). In retrospect, it is clear that the evidence for the boat form of cyclohexene from the thermodynamic and spectroscopic data was weak at best. Pitzer and co-workers carried out a calculation based on the known methyl rotational barriers in ethane and propene (2.8 and 1.95 kcal/mol, respectively) and also included bond angle strains. They found that the half-chair was more stable than the boat by 2.2 kcal/mol, in apparent support of the 2.7 kcal/mol value. The 2.2 kcal/mol calculated value, however, seems to be in error, because the above parameters lead to an energy difference of at least 5 kcal/mol.

1.1.2 Theoretical Calculations

Only a few quantum mechanical calculations of sufficient accuracy to be of conformational interest have been carried out on cyclohexene. *Ab initio* calculations, with geometric optimization and a double zeta basis set, but no configuration interaction, show that the half-chair is of lowest energy but unexpectedly give the planar form with an energy lower than that of a (flattened) boat (10.0 and 10.4 kcal/mol above the half-chair, respectively).[27] No results were reported for a more puckered boat with less strained bond angles. Calculations with the GAUSSIAN80 program and with optimized geometries, but only at the minimum basis set level (STO-3G), show the boat as a transition state, 5.3 kcal/mol above the half-chair,[28] in agreement with the experimental NMR data given below. At this level, the olefinic carbons are slightly pyramidalized.[28,29]

Molecular mechanics calculations based on empirically determined but transferable force constants have proved to be useful in conformational analysis and are discussed by Lipkowitz in Chapter 7. Before discussing the results of such calculations for cyclohexene, the nature of the torsional barrier at an allylic bond needs to be considered. For unsaturated molecules, for example, propene, the preferred rotamer (**13**) about the allylic single bond has the C(sp^2)—H bond staggered with respect to the methyl group.[30] This means that the sigma bond in the C(sp^2) = C(sp^2) moiety is eclipsed with the methyl group and this rotamer is therefore often referred to in the literature as being eclipsed. In a bent bond description of the double bond, the preferred rotamer is staggered[31] (e.g., **14**)

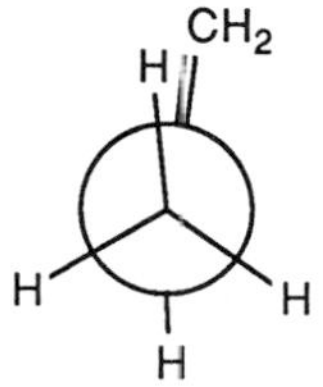

13

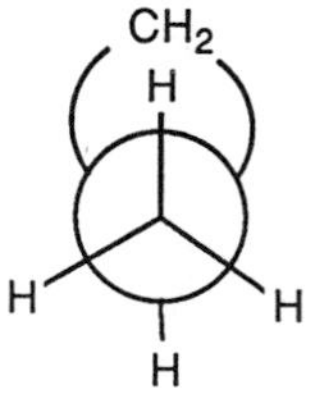

14

without any ambiguity, but this description is less suitable for calculating torsional strain energies as a function of torsional angles in molecular mechanics, since bent bonds have a curvature that is not easily defined. The barrier to internal rotation of the methyl group in propene is 2.0 kcal/mol, which is somewhat lower than the barrier in propane (3.6 kcal/mol).[32]

In the cyclohexene half-chair, the allylic bonds have favorable small ring torsional angles, whereas in the boat these torsional angles are large and unfavorable. Additionally, the boat has a fully eclipsed C(sp^3)—C(sp^3) single bond and thus the strain in this form is largely the result of torsional strains. The fusion of a cyclohexene to a benzene ring to give tetralin changes the picture significantly, because the barrier to rotation about the benzylic bond in toluene is virtually zero, in contrast to the situation described above for the allylic bond in propene.

Semiquantitative aspects of the conformational energies of cyclohexene rings bearing substituents or fused rings have been developed by Corey and Feiner for the purpose of computer-based synthetic planning in organic chemistry.[33] Allinger et al.[34] have calculated conformational equilibria of 3-methyl- and 1,6-dimethylcyclohexene by molecular mechanics and find similar preferences (0.76 and 0.84 kcal/mol) for the equatorial conformer in both cases, in disagreement with Johnson's qualitative allylic strain theory (see above),[22] at least when the group at C-1 is no larger than methyl. The above calculations were done with an early version of Allinger's alkene force field; later versions apparently have not been applied to this problem.[35] As expected, 3,3-dimethylcyclohexene is more stable than the 4,4 isomer by 0.84 kcal/mol,[34] and the stabilities of the four isomers of 3,4-dimethylcyclohexene show the following order: 3e′,4e > 3a′,4e > 3e′,4a > 3a′,4a, with relative energies of 0.00, 0.25, 0.46, and 0.71 kcal/mol, respectively.

The six isomeric benzene tetrachlorides (3,4,5,6-tetrachlorocyclohexenes) have been studied by de la Mare et al.[36] using the MM2 force field. The energies do not fit well and the chlorines tend to favor the axial positions too much unless the π electrons of the double bond are represented by two dipoles, each with a dipole moment of 1.6 D, centered 0.5 Å above and below the plane of the double bond. With this change the dipole moments also fit well, except for that of the β isomer. There is a certain amount of arbitrariness in these changes of the force field in order to satisfy a rather small group of molecules; a similar approach has been used for the naphthalene tetrachlorides, which are discussed below.

Molecular mechanics calculations on *cis*-cyclohexene (except for one early result[34]) show that the half-chair is the only local energy minimum on the potential energy surface and that the boat form of cyclohexene is a transition state.[3,7,8,26,37–40] The calculated barriers to ring inversion for the half-chair are in the range of 5–7 kcal/mol (Table 1.1), in reasonable agreement with the experimental value of 5.3 kcal/mol. The ring torsional angles of the half-chair obtained in different investigations are in good agreement with one another (Table 1.2) as well as with the experimental data on cyclohexene and its simple derivatives. The inherently twisted geometry of the half-chair (C_2 symmetry) gives rise to a slightly twisted double bond and to sp^2 carbons that are very slightly pyramidalized. The sofa form of cyclohexene is calculated to be only about 1–1.5 kcal/mol higher in energy than the half-chair, but it is not an energy minimum, at least in cyclohexene itself. The 1,3-diplanar form is somewhat lower in energy than the boat.

Table 1.1 Ring Inversion Barriers in the Cyclohexene Half-Chair Calculated by Molecular Mechanics

Force Field	Strain Energy Barrier (kcal/mol)	Reference
Bucourt	6.0	9
Bucourt	7.0	10
Bucourt	6.9	26
Favini	6.0	37
Allinger	6.4	35
Dasheva	4.2, 5.4, 6.4[a]	39
Modified Boyd	6.6	40
Ermer–Lifson	7.9 (7.3)[b]	41
Bovill–White	5.2	7

[a] Values dependent on force constant used for the barrier to rotation about the allylic bond (0, 1, and 2 kcal/mol, respectively).
[b] Free-energy barrier ($\Delta G^{\ddagger}$).

The ring inversion process in cyclohexene can be considered to be a rather hindered pseudorotation of the chiral half-chair to its mirror image via either one of two achiral boats, as shown in Figure 1.1. All the torsional angles in the half-chair change their signs but maintain their absolute magnitude upon ring inversion. The two boats also have opposite signs of torsional angles about any given bond.

The conformational properties of *trans*-cyclohexene have been investigated by molecular mechanics, although the large distortions give rise to problems in the selection of an appropriate force field.[35,41] Two conformations (chair and twist), both with C_2 symmetry, are found.[41] They have similar distortions at the double bond and have almost the same strain energies (ca. 53 kcal/mol above *cis*-

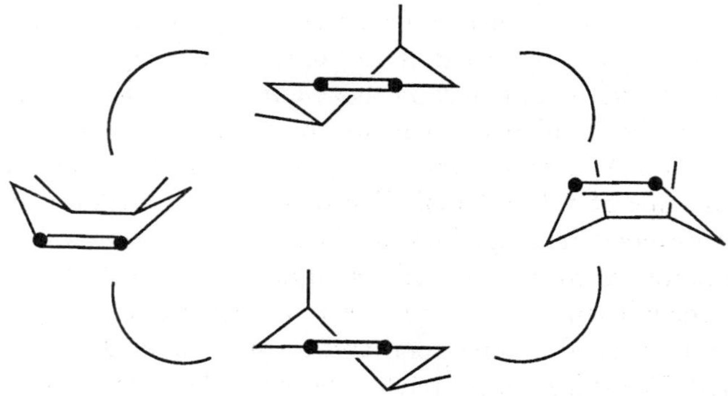

Figure 1.1 Pathways for ring inversion in the (chiral) half-chair conformation of cyclohexene. The (achiral) boats are transition states. The *cis*-4,5 bonds shown illustrate the exchanges that take place among the various forms.

Table 1.2 Ring Torsional Angles (in degrees) in the Cyclohexene Half-Chair

Ring Torsional Angle	Molecular Mechanics						Diffraction Methods		Microwave Spectroscopy[i]
	Bucourt[a]	Anet[b]	Anet[c]	Baas[d]	Favini[e]	Allinger[f]	Electron[g]	X-Ray[h]	
6,1,2,3	0	1.6	0.5	0	0	0	0	1.4	0
1,2,3,4	15.0	14.3	14	15	17.6	16	15.3	14.5	15.7
2,3,4,5	−44.4	−45.2	−44	−45	−51.1	−47	−44.9	−45.5	−46.0
3,4,5,6	60.6	62.0	60	62	67.6	64	60.7	62.2	62.6

[a] Bucourt force field.[26].
[b] Ermer–Lifson[48] force field.[15]
[c] Modified Boyd force field.[40]
[d] Bovill–White force field.[7]
[e] From limited geometric optimization, Favini force field.[37]
[f] Allinger force field.[34]
[g] Mean value of two investigations, symmetry assumed.[42,43]
[h] Mean value for three simple derivatives of cyclohexene.[44–46]
[i] Symmetry assumed.[47]

cyclohexene), but they differ in the amount of ring puckering. The ring torsional angles at the double bond are calculated to be only about 95° and the unsaturated carbons are highly pyramidalized, so that the H—C=C—H torsional angles are about −166°. The conversion of *trans*- to *cis*-cyclohexene has also been investigated and the calculated $\Delta H^{\ddagger}$ is only 9.5 kcal/mol,[41] corresponding to a lifetime of a few microseconds if entropy effects are neglected. Experimentally, it is known that the lifetimes of *trans*-cyclohexenes are about 10 μs at room temperature.[23,24] By comparison, *trans*-cycloheptene, which is still extremely strained,[3] has a lifetime at room temperature of about 1 min,[49] and *trans*-cyclooctene, although significantly strained, is indefinitely stable at room temperature.[3]

1.1.3 X-Ray Crystal Data

Early X-ray diffraction studies clearly showed twist-chair conformations for a variety of heavily substituted or ring fused cyclohexenes,[50–53] but these papers do not provide very accurate geometric parameters. Unfortunately, X-ray crystal data for cyclohexene itself or for simple cyclohexene derivatives, where the effects of substituents can be expected to be very small, do not appear to exist. The ideal molecule, of course, is cyclohexene itself, but as this compound is a liquid at room temperature, the (difficult) preparation of a single crystal at low temperatures is required before an X-ray crystal study can begin.

The Cambridge Structural Database,[54] which is an excellent source of crystal structures, contains numerous molecules with cyclohexene rings. Convenient search programs are available to search the Database and were used during the preparation of this chapter. Tabulations (atlases) of steroid structures[12,13] provide ring torsional angles as well as quantitative characterizations of the conformations of five- and six-membered rings in hundred of molecules, many with cyclohexene rings, albeit always fused to other rings. It should be noted that X-ray structure papers of ring compounds, although giving lists of bond angles and bond lengths, do not always give ring torsional angles, even though this information is critical to defining the geometry of rings and is far more useful than other structural data, at least from a conformational point of view. Fortunately, there seems to be an increasing trend in recent X-ray papers to give ring torsional angles, and, of course, these data can always be calculated from the fractional coordinates given in the papers, or more conveniently, by using the Cambridge Structural Database and its associated programs.[54]

Geometrical parameters of some relatively simple cyclohexenes whose crystal structures are accurately known are given in the partial structures: **15**,[46] **16**,[44], **17**,[45] **18**,[55] **19**,[56] **20**,[57] and **21**.[58] One of the simplest cyclohexene derivatives is a copper(I) complex (**15**) that has only a weak and labile π bond from the metal to the double bond of the cyclohexene, as shown by the near planarity of the coordinated double bond system.[46] With the exceptions of **19**, **20**, and **21**, all these molecules have half-chair conformations (in **18**, which is fairly heavily

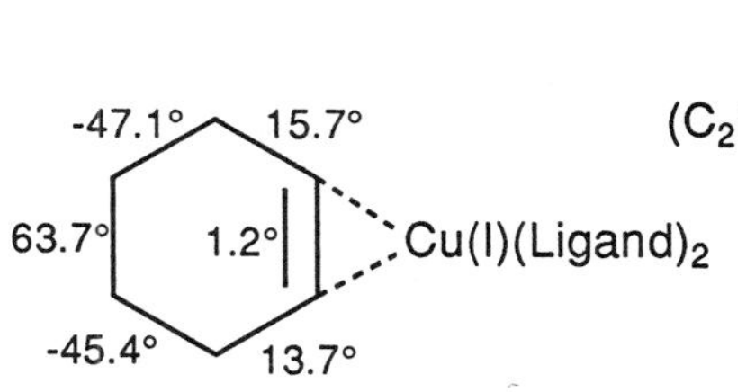

15

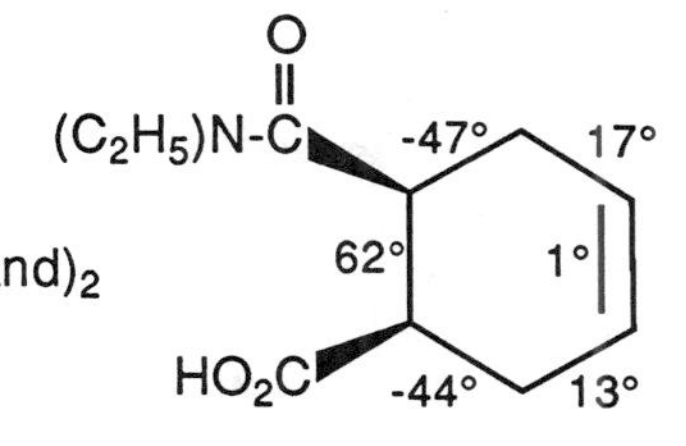

16

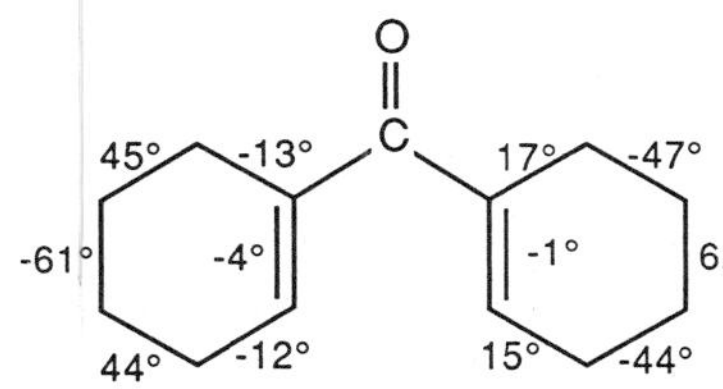

17

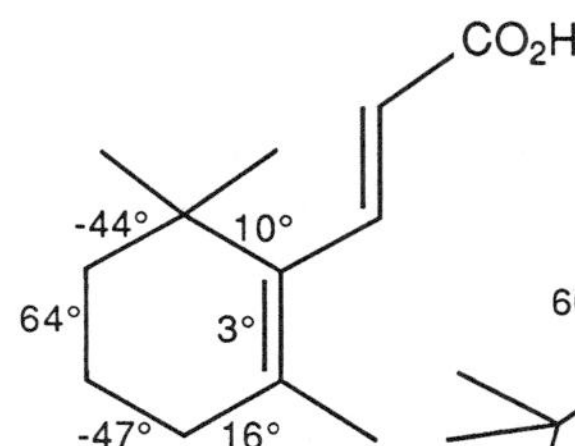

18

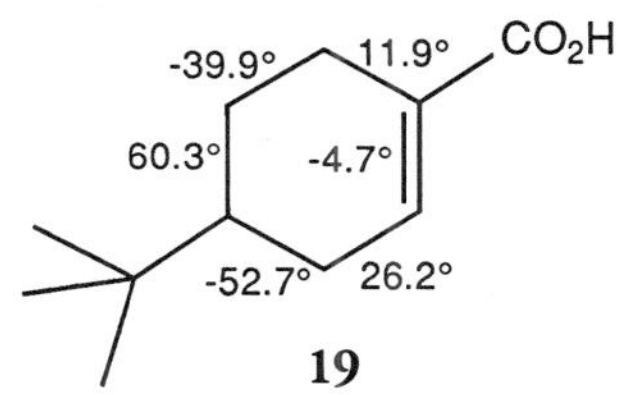

19

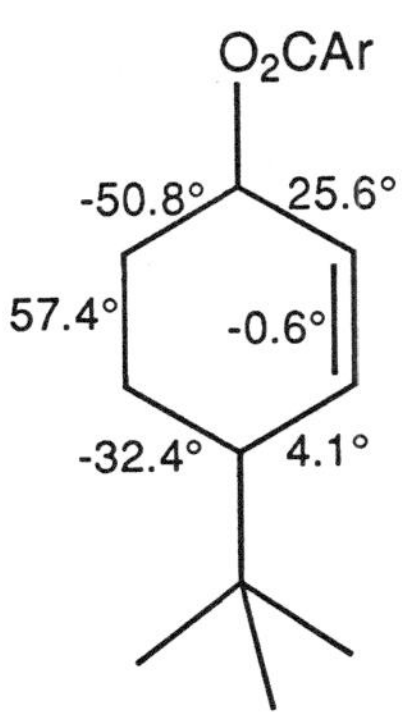

20

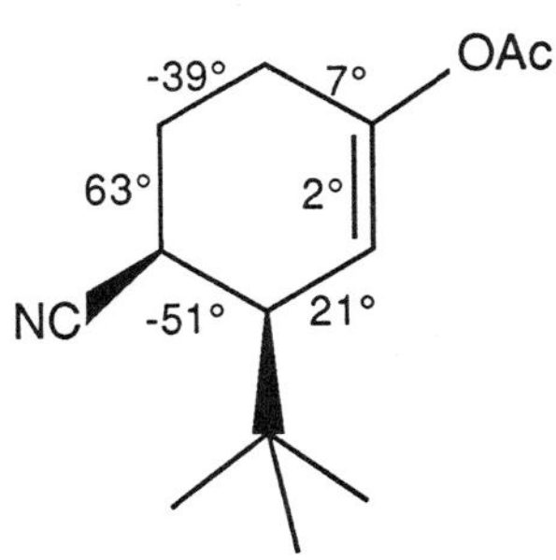

21

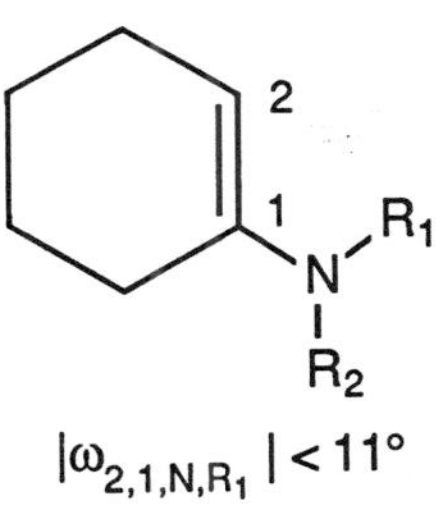

22

Ring torsional angles

substituted, the half-chair is somewhat distorted), and the average ring torsional angles are in good agreement with those obtained by other methods. The cyclohexene rings in **19**, **20**, and **21** exist as sofas rather than as half-chairs, but this seems to be the result of lattice forces rather than the presence of a *tert*-butyl substituent, because molecular mechanics calculations on 3-*tert*-butylcyclohexene show only a slightly distorted half-chair (F. A. L. Anet, unpublished results). The bond lengths and bond angles are close to standard values for alkenes and are not discussed here.

The crystal structures of a number of enamines (**22**) derived from various cyclohexanones have been obtained, and most of the interest in these molecules resides in the geometry of the side chain nitrogen atom, which is more or less pyramidalized.[59] The cyclohexene rings are half-chairs, although in some cases disorder limits the accuracy of the results. One interesting feature is the tendency for one of the piperidine, morpholine, or pyrrolidine N—C bonds to eclipse the double bond in the cyclohexene moiety. The torsional angle ω is less than 11° in 11 crystallographically or chemically different molecules. Also, the pyrrolidine enamines show the least pyramidalization at the nitrogen and therefore the greatest amount of n-π overlap.

1.1.4 Electron, Microwave, Infrared, and Photoelectron Gas-Phase Data

Several independent electron diffraction studies on cyclohexene appeared in the late 1960s and early 1970s,[38,42,43,60] and they are in substantial agreement with one another and clearly show that the major conformation in the gas phase is the half-chair (see Table 1.2 for ring torsional angles). Stereoisomers of 3,4,5,6-tetrachlorocyclohexene have also been shown to have half-chair conformations by electron diffraction, and the structures are supported by the dipole moments of the isomers.[52,61]

Microwave studies of cyclohexene and two deuterated derivatives are consistent with the half-chair but not with the boat and provide an accurate value (0.3 D) for the dipole moment of cyclohexene in the gas phase (less accurate solution values of μ for cyclohexene are in the range of 0.61–0.76 D).[47] Both the electron and microwave studies of cyclohexene gave insufficient data for a complete structural analysis, so that assumptions had to be made about the symmetry and the placement of hydrogen atoms. (X-ray crystal structures also are poor for locating precisely the positions of hydrogen atoms whereas neutron diffraction can give accurate hydrogen positions for molecules in the crystalline phase.) Microwave, infrared, and Raman data on 1-fluorocyclohexene show that it has a structure very similar to that of cyclohexene,[62] as expected.

Low-frequency infrared and Raman absorption bands give information on out-of-plane and twisting vibrations of the molecules, and from these data calculations of the energy required for ring inversion or even to completely flatten the ring can be carried out. In the case of cyclohexene, these energies are calculated to be 7.6 ± 0.6 and 25.4 ± 5 kcal/mol.[63] The former value is higher than that obtained by NMR, although it is in agreement with some of the higher molecular mechanics values mentioned above. The energy required to flatten the ring appears to be much too high, since a molecular mechanics calculation gives a value of only 8 kcal/mol.[7] In a later paper on 1-fluorocyclohexene, Lord et al. have pointed out that the calculations of high barriers from low frequency vibrational data (the 163 cm^{-1} ring bending mode in cyclohexene corresponds to an energy of only 0.47 kcal/mol) is prone to large errors.[62]

Microwave and low-frequency Raman and infrared spectra of 2-cyclohexenone in the gas phase show that this molecule adopts a sofa conformation with C-5 being the out-of-plane atom (**23**).[64–66]

Photoelectron spectra of the *cis*- and *trans*-3-methoxy-6-*tert*-butylcyclohexenes (**24** and **25**) show that the cis isomer, where the methoxy group is pseudo-

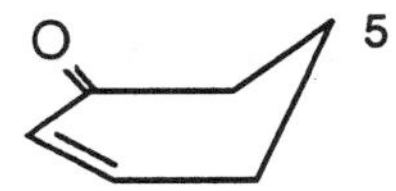

23

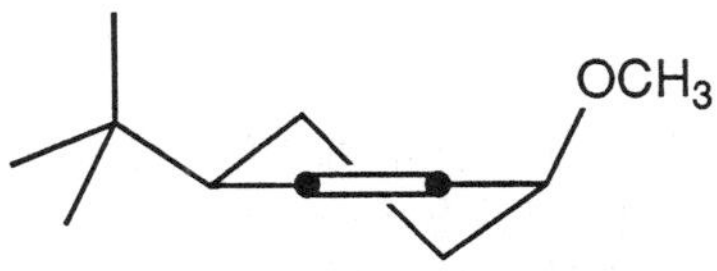

24 (cis)

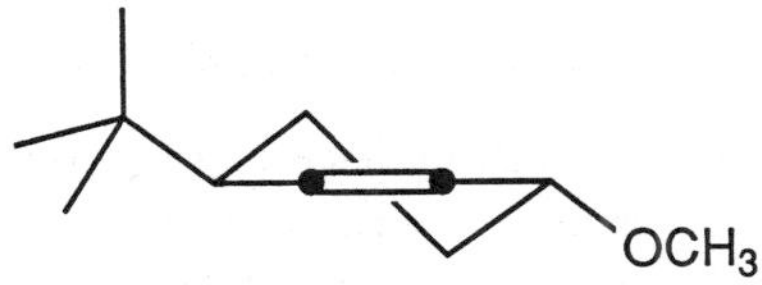

25 (trans)

axial on the half-chair, has the higher ionization energy because of the stabilizing effect of the near-parallel relationship of the π orbital and the σ^* orbital of the C(3)-methoxy bond.[67] Analogous effects are found in related octalin derivatives (see below), but the allylic ether group in acyclic systems without steric hindrance tends to be eclipsed with the double bond, especially when there is an electron attracting group trans to the alkoxy group.[68]

1.1.5 Solution Infrared Data

In monosubstituted cyclohexenes, particularly when the substituent is a halogen, infrared and Raman bands in the liquid or solution state can often be assigned either to the axial or the equatorial half-chair conformer, and the temperature dependence of the intensities of these bands allow $\Delta H°$ for the equatorial–axial equilibrium to be calculated. The equilibrium constant (or $\Delta G°$) is much more difficult to obtain, because the absolute intensities of the infrared bands in the two forms are generally not known and may even vary somewhat with temperature.[69] Only one conformer is generally present in the crystalline state, and this can be useful in the assignments of bands.

Sakashita has found by variable-temperature infrared and Raman spectroscopy that the chlorine substituent in 3-chlorocyclohexene prefers the pseudoaxial (**26**) over the pseudoequatorial (**27**) position with a $\Delta H°$ of 0.64 kcal/mol.[70] The analogous 3-bromo compound has a $\Delta H°$ of 0.70 kcal/mol, whereas 4-bromocyclohexene has only a slight enthalpic preference for the axial form ($\Delta H°$

= 0.05 kcal/mol).[71] *Trans*-4,5-dichlorocyclohexene has also been studied by the same methods and shows a preference for the diaxial form, **28**.[72]

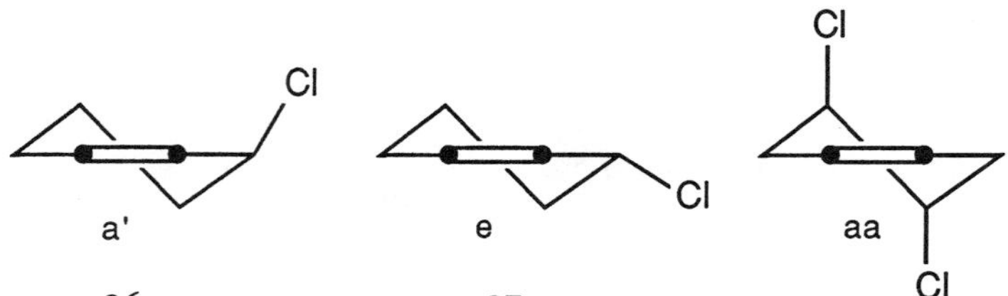

26 **27** **28**

1.1.6 Solution NMR Data

Nuclear magnetic resonance (NMR) is a powerful tool for conformational studies of ring molecules in solution,[73,74] and it has been extensively applied to cyclohexene and its derivatives. Dynamic NMR can give energy barriers,[75] provided that these are not too low and that the equilibrium is not too one-sided, as is indeed the case in many but not all monosubstituted cyclohexenes. Furthermore, slow-exchange spectra provide essential information on symmetry and on the number and populations of conformations present. Room-temperature measurements of vicinal $^1H—^1H$ coupling constants provide data on (averaged) torsional angles via the Karplus relationship, and the temperature dependence of these coupling constants can also be measured to give values of $\Delta H°$ and $\Delta S°$, although not without problems of reliability.[73–75] Most of the existing NMR data dealing with conformational aspects in cyclohexenes are for protons, but ^{13}C-NMR has also been used more recently.

1.1.6.1 Coupling Constants

The ratio of J_{trans}/J_{cis} for protons in a $CH_2—CH_2$ ring moiety has been called the *R* value by Lambert,[76] who has shown that it depends in a sensitive manner on the ring torsional angle of the C—C bond: it is thus a valuable tool for the determination of ring puckering. Buys has tabulated *R* values for a large number of six-membered rings.[77] For torsional angles of 47–65°, the *R* values change from 1.12 to 3.62, respectively. In cyclohexene-1,2,3,3,6,6-d_6, J_{trans} = 8.94 Hz and J_{cis} = 2.95 Hz, giving R = 3.03 and a torsional angle of 63° for the 4,5 bond.[78] More recent work on other cyclohexene isotopomers has confirmed these coupling constants and has also given values for other coupling constants, which are in agreement with the half-chair: J_{23} (3.73 Hz), J_{34} (5.67 and 6.77 Hz for cis and trans, respectively), and J_{13} (−2.2 Hz).[79] The *R* value (1.19) for the 3,4 bond corresponds to a torsional angle of 48°.

The conformational equilibria in 1-phenylcyclohexenes with substituents in the 6 positions have been examined by Garbisch using the widths of the C-6 proton signals and the torsional dependence of vicinal coupling constants.[80] Large C-6 substituents, for example, phenyl, *tert*-butyl, or even nitro, take up the pseudoaxial position (**29**) on the half-chair because of nonbonded repulsions with the phenyl group at C-1 in the alternate pseudoequatorial orientation, and the proton halfwidth is about 10 Hz. However, in the analogous series with geminal dimethyl groups at the 4 position, the 1,3-diaxial interaction forces the substituent at C-6 to be pseudoequatorial (**30**) and the proton halfwidth then rises to about 20 Hz. The preference for the axial position of the nitro group in 6-nitro-1-phenylcyclohexene is also supported by equilibrium data on the *trans*- and *cis*-4-*tert*-butyl derivatives (**31** and **32**, respectively), where the trans:cis ratio is about

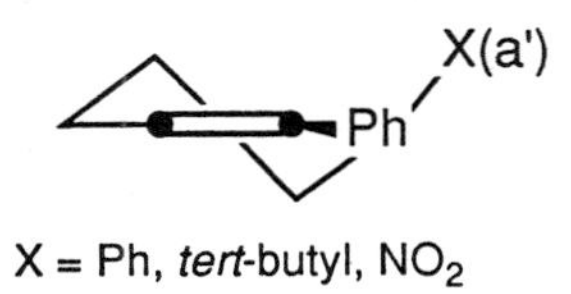

29

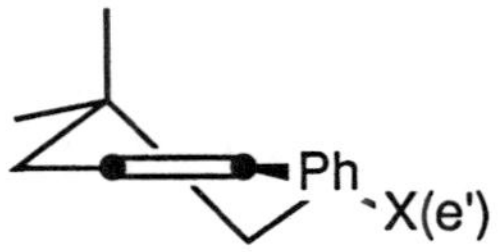

30

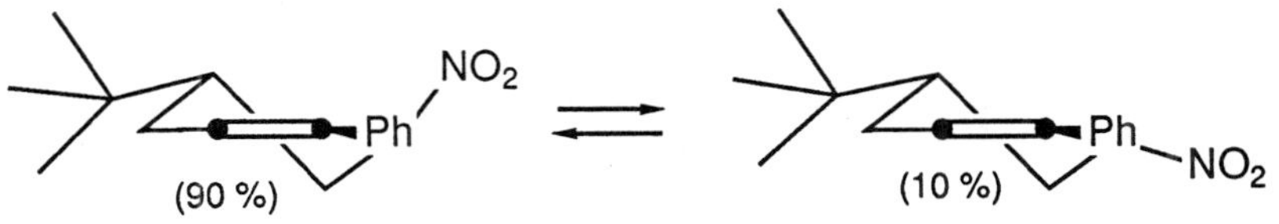

31 **32**

10:1. The high energy of the pseudoequatorial nitro group is an example of the $A^{(1,2)}$ allylic strain principle proposed by Johnson.[22] Garbisch has also used vicinal coupling constants to establish that the conformational equilibrium in 4-phenylcyclohexene favors the equatorial over the axial position by 1 kcal/mol.[78]

Zefirov et al. have examined the NMR spectra of cyclohexenes deuterated at the 3 and 6 positions (initially without decoupling the deuterons[81]) and measured conformational equilibria between **33** and **34** or between **35** and **36** for

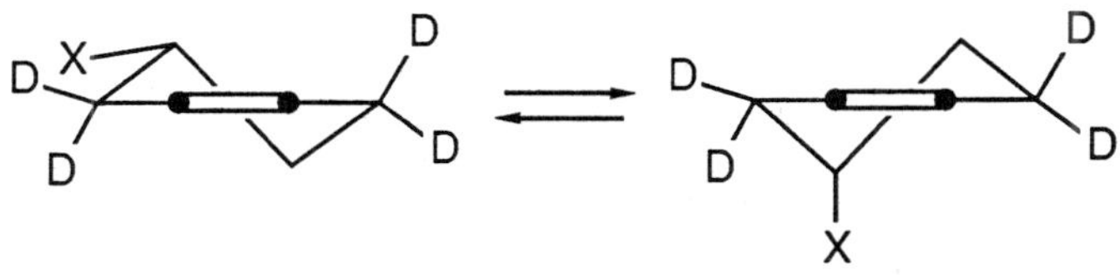

33 **34**

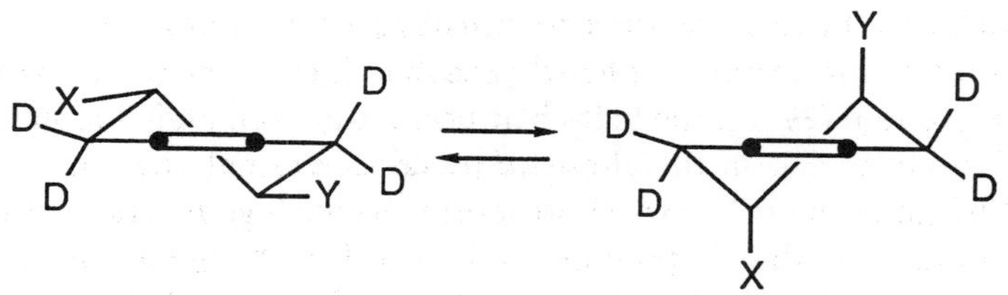

35 **36**

substituents such as CO_2Me, CHO, CN, NO_2, CCl_3, CF_3, and COPh (Table 1.3).[81–84] These workers used standard vicinal coupling constants derived from cyclohexane derivatives ($J_{aa} = 12, J_{ae} = 3$, and $J_{ee} = 2$ Hz). They showed that a previous suggestion[85] that such substituents in the axial orientation have a strong attractive interaction with the double bond is untenable. When the substituents at the 4,5 positions are large, for example, CCl_3 at C-4 and CO_2H at C-5, the conformational equilibrium between diequatorial and diaxial is about 1:1, whereas with smaller substituents, for example, CH_3 at C-4 and NO_2 at C-5, the equilibrium is completely on the side of the diequatorial form.[83] Zefirov et al. have measured peak widths and chemical shifts in 4-substituted cyclohexenes as a

Table 1.3 Conformational Equilibria in 4- and 4,5-Disubstituted Cyclohexenes

Substituents 4	Substituents 5	Solvent	$J_{45\text{-cis}} + J_{43\text{-trans}}$ (Hz)	Percentage Equatorial	Reference
CO_2H	H	None	14.1	85	81
CN	H	$CHCl_3$	11.3	55	81
NO_2	H	None	11.9	60	81
COPh	H	C_6H_6	12.7	70	81
CHO	H	$CHCl_3$	13.0	70	82
CO_2CH_3	H	$CHCl_3$	14.3	85	82
			J_{45}		
CHO	Ph	CH_2Cl_2	10.2	—	82
CN	NO_2	C_6H_6	10.2	—	82
CCl_3	COCl	[a]	5.9	46	83
CCl_3	CO_2CH_3	[a]	7.2	61	83
CCl_3	NO_2	[a]	6.3	51	83
$CHCl_2$	CO_2H	[a]	9.95	93	83
CH_3	NO_2	[a]	10.7	(100)	83
CF_3	NO_2	CH_2CHCl	—	72	84
CCl_3	NO_2	CH_2CHCl	—	60	84
CCl_3	NO_2	CS_2	—	59	84
CCl_3	CN	CS_2	—	74	84
CH_2Br	CN	CH_2CHCl	—	83	84
CH_3	CN	CH_2CHCl	—	82	84

[a] Ten percent in an unspecified solvent.

function of solvent polarity, with dielectric constants ranging from 2 to 50.[86] The changes in linewidths are quite small (<2 Hz in lines that are 20–25 Hz wide) and show some scatter with the dielectric constant, but there seems to be a clear, although small, stabilization of the axial over the equatorial form as the solvent polarity increases. Deuterated molecules would provide more reliable data, as coupling constants would then be available rather than just linewidths. In other work on trans 4,5-disubstituted cyclohexenes, Zefirov and co-workers made use of partially deuterated molecules and measured coupling constants as a function of temperature to obtain diaxial–diequatorial equilibrium constants.[84] They fitted four parameters (J_{aa}, J_{ee}, $\Delta H°$, and $\Delta S°$) to their data and calculated $\Delta G°$ at 20 °C. Such an analysis is prone to correlated errors, particularly in $\Delta H°$ and $\Delta S°$. The diequatorial form is always the more populated, the range being 60–85%, but the molecules with relatively large substituents, for example, CH_3 and NO_2, have the least percentage (60%) of this form.

Aycard, Bodot, and co-workers, using the ratio ($R_{trans/cis}$) of the J_{trans} to J_{cis} to calculate the fraction of the equatorial form, have found only a small solvent dependence of the conformational equilibrium in 4-cyanocyclohexene-3,3,6,6-d_4.[87] In later work on a series of partially deuterated 1,3-, *cis*- and *trans*-3,4-, and *cis*- and *trans*-3,5-disubstituted cyclohexenes,[88–90] they measured coupling constants and analyzed the data in terms of the magnitudes of gauche interactions. They point out that the e′a′ energy difference for a 3-*tert*-butyl group in cyclohexene is much less than the ea energy difference of a *tert*-butyl group in cyclohexane. A selection of their NMR and equilibration data is given in Table 1.4.

Lessard et al. have used the linewidth method to determine the conformational equilibria in a variety of 3-substituted cyclohexenes and have been especially interested in the effect of an electron-donating group at the 1 position.[91] Such a group can stabilize the no-bond resonance structure, **37**, and thereby increase the preference of an electronegative group at C-3 for the pseudoaxial position, which is ascribed by these authors partly to electrostatic interactions and partly to no-bond resonance, that is, to an anomeric effect. In a later paper,[92] Lessard and co-workers investigated the relative preferences of acetoxy and methoxy groups for the (pseudo)axial versus the (pseudo)equatorial positions when next to unsaturation in cyclohexenes, tetralins, and methylenecyclohexanes. They also report that 3-methoxycyclohexene, but not 3-acetoxycyclohexene, gives a dynamic NMR effect in the vicinity of -150 °C at 100 MHz and that the equilibrium is largely on the side of the pseudo-axial conformer, but they were unable to obtain quantitative data. The lack of a dynamic NMR effect in the acetoxy compound is ascribed to an insufficient concentration of the pseudoequatorial form, which is therefore less populated than that of the methoxy derivative. At room temperature, the average percentages of pseudoaxial forms in $CDCl_3$ solution are 57 (41 and 65 as extreme values) for 3-methoxycyclohexene and 63 (47 and 84 as extreme values) for 3-acetoxycyclohexene. The data were obtained from ^{13}C and ^{1}H chemical shifts, coupling constants, and linewidths

Table 1.4 Chemical and Conformational Equilibria in Substituted Cyclohexenes

Substituents		Solvent	Temperature (°C)	Percentage cis	Percentage 3e, 4e′ (trans)[a]	Reference
3-CH_3	4-CO_2CH_3	CH_3OH	65	20	~100	88
3-$C(CH_3)_3$	4-$CO_2C_2H_5$	C_2H_5OH	78.5	1		89
3-$C(CH_3)_3$	4-CN	$(CH_3)_3OH$	82	34	50	89
3-CH_3	4-CN				82	90
4-$C(CH_3)_3$[b]	5-CO_2CH_3	CH_3OH	65	9		89

[a] In CS_2 at room temperature.
[b] 1,2-Dimethyl substituents also present.

using *tert*-butyl-substituted molecules as models. The authors conclude that the 3-acetoxy group prefers the pseudo-axial position more than does the 3-methoxy group, because of the former's greater stability as an anion in the π-σ^*_{C-O} interaction, but the large scattering of the reported data makes this conclusion rather uncertain. They point out that the use of a *tert*-butyl group in model molecules is known to give geometric distortions and to lead to incorrect results, and that while a 4-*tert*-butyl group is entirely equatorial, a 3-*tert*-butyl group is not entirely (pseudo)equatorial. In the case of tetralin and methylenecyclohexane derivatives with the methoxy or acetoxy substituent in the benzylic or allylic position, Lessard et al. find that the methoxy prefers the pseudoaxial or the axial position, but the acetoxy surprisingly greatly prefers the pseudoequatorial or the equatorial position. The equatorial preference in these systems seems to be general for unsaturated substituents and is termed the "unsaturation effect," but its origin is obscure.

A tetramethylcarvone has been shown to exist in two conformations separated by a free-energy barrier of 14 kcal/mol.[31] The six-membered rings are sofas in both cases (**38** and **39**), and there is some evidence to support the axial disposition of the unsaturated side chain, with **38** being favored over **39** by 0.2 kcal/mol. A long range coupling (^4J = 1.2 Hz) between the olefinic proton at C-3 and the (equatorial) methine proton at C-5 occurs in **38**, but these protons are not coupled in **39**. This is ascribed to small distortions in the ring of **39** that are not

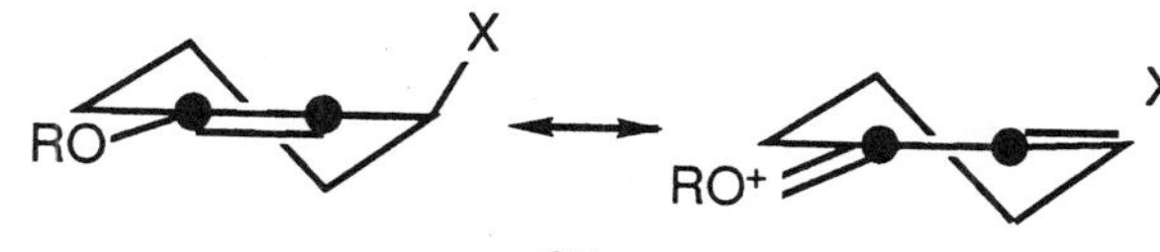

37

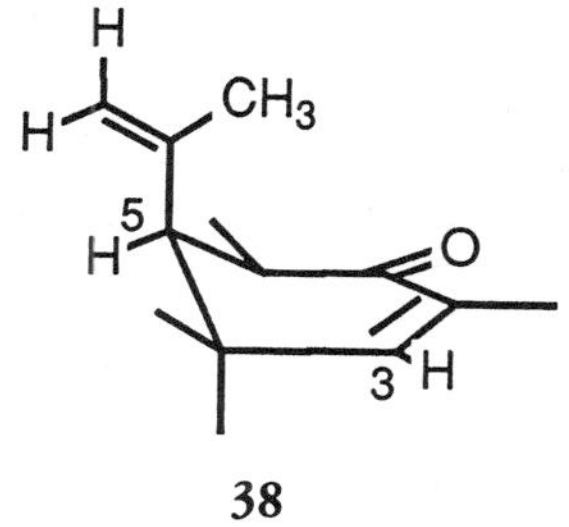

38

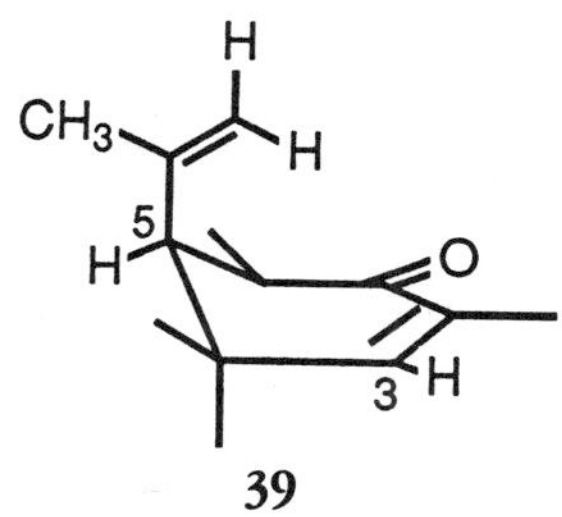

39

present in **38** and to the sensitivity of long-range couplings to geometric factors. However, this argument is not entirely convincing because of the nearness in energy of the two forms. The dynamic NMR effect observed in this system results from strongly hindered rotation about the bond linking the unsaturated side chain to the ring.

1.1.6.2 Low-Temperature NMR Data on Barriers and Populations

In 1965, Anet and Haq reported a dynamic NMR study in cyclohexene-3,3,4,5,6,6-d_6 in $CBrF_3$ solution.[93] The 4,5 protons exhibit two chemical shifts below about −160 °C, which is the coalescence temperature, corresponding to a free-energy barrier of 5.3 kcal/mol. Because these workers accepted Pitzer's value for the half-chair/boat energy difference (2.7 kcal/mol, see above)[18] and were unaware of Bucourt's 1964 calculation showing that the boat form is 7.4 kcal/mol higher in energy than the half-chair (see below),[9] they interpreted their results in terms of a half-chair inverting via the boat as an intermediate. Simultaneously with the Anet–Haq report, Jensen and Bushweller, who also do not quote Bucourt's work, reported preliminary dynamic NMR data on 4-chlorocyclohexane,[94] and in 1969 they published an extensive investigation of the dynamic NMR properties of a series of cyclohexene derivatives.[69]

A single halogen substituent at the 4 position is found to raise the ring inversion barrier in cyclohexene by more than 1 kcal/mol. These increased barriers, as well as the even larger increases observed in the *cis*-4,5-disubstituted cyclohexene, **40**, are good evidence for eclipsing of the 4,5 bond in the transition state (Table 1.5). In contrast, cyclohexane and the monohalogenated cyclohexanes have essentially the same inversion barriers because in these cases the eclipsing takes place in the boat/twist-boat pseudorotation and the energies of these forms

Table 1.5 Ring Inversion (Free-Energy Barrier, $\Delta G^‡$ in Cyclohexenes Without Fused Rings from Low-Temperature NMR or ESR Data

Substituent	Solvent	$\Delta G^‡$	Reference
1,2,3,3,4,5,6,6-d_8	$CBrF_3$	5.3	93
2,3,3,4,5,5,6,6-d_8	Vinyl chloride-d_3	5.4	69
3-CH_2·	[a]	6.1	95
4-Fluoro	Vinyl chloride-d_3	5.3	69
4-Chloro	Vinyl chloride-d_3	6.3	69
4-Bromo	Vinyl chloride-d_3	6.3	69
4-Iodo	Vinyl chloride-d_3	6.5	69
4-CH_2·	[a]	6.2	95
3,3-Dimethyl,1,2,5,5-d_4	$CHCl_2F$	6.3	96
4,4-Dimethyl	$CHCl_2F$	6.1	96
cis-4,5-Dicarbomethoxy	Vinyl chloride-d_3	7.4	69
3,3,6,6-Tetramethyl	Vinyl chloride-$CHCl_2F$	8.4	96
Perfluoro	CCl_2F_2	6.6	97
1,2-Dichloro-3,3,4,4,5,5,6,6-octafluoro	CCl_2F_2	6.0	97
Cylohexanesemidiones (**46**)			
Unsubstituted	DMF	4.2	98
3,3-Dimethyl	DMF	5.2	98
4,4-Dimethyl	DMF	5.1	98
3,3,5,5-Tetramethyl	DMF	4.1	98

[a] At −50 °C in nonpolar solvents.

are much less than those of the transition states.[73] St.-Jacques and co-workers have shown that the effect of gem dimethyl groups on the ring inversion barrier is much greater in cyclohexene than in cyclohexane.[96] The especially high barrier (8.4 kcal/mol) in 3,3,6,6-tetramethylcyclohexene (**41**) provides strong support for a boat transition state, as the inward-pointing methyls should suffer strong nonbonded repulsions in that form. It has been reported that the 60 MHz ^{1}H-NMR spectrum of *cis*-4,5-dimethylcyclohexene does not give a dynamic NMR effect down to −161°C,[99] but this is inconsistent with data on related compounds discussed above and probably results from the low observation frequency and the complexity of the spectrum.

Anderson and Roberts have studied ring inversion in two polyfluorinated cyclohexenes (**42** and **43**) by ^{19}F dynamic NMR at low temperatures and find free-

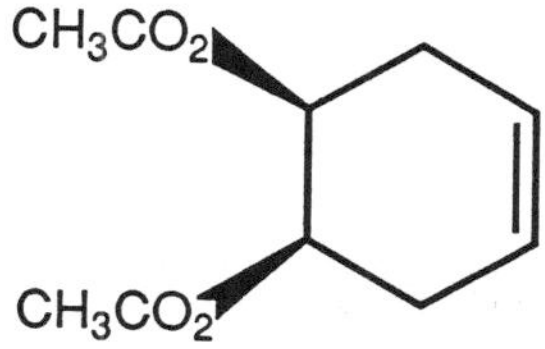

40

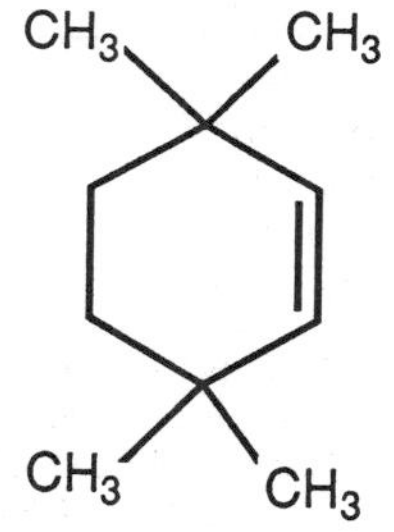

41

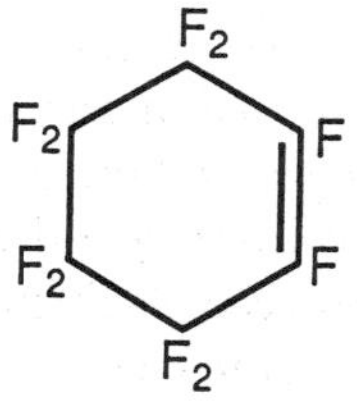

42

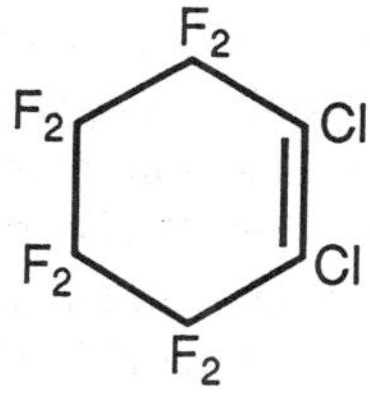

43

energy barriers of 6.3 and 6.8 kcal/mol, respectively.[97] The large ^{19}F shifts allow rate constants to be measured over a wide temperature range and accurate entropies of activation to be obtained; the latter are not significantly different from zero. The similarity of the conformational barriers in corresponding perfluoro and perhydro series found in cyclohexene also occurs in other ring systems.[73]

Conformational equilibria in 4-halogenated cyclohexenes have been measured by direct integration at low temperatures and present an interesting contrast with the corresponding cyclohexane analogs (Table 1.6).[69,100] The lack of

Table 1.6 Conformational Equilibria in 4-Substitued Cyclohexenes and Monosubstituted Cyclohexanes from Low-Temperature NMR Data

	$\Delta G°$(eq $\rightleftharpoons$ ax) (kcal/mol)	
Substituent	4-Substituted Cyclohexene	Substituted Cyclohexane
Fluorine	0.014[a]	0.276[b]
Chlorine	0.20,[a] 0.31,[c] −0.02[d]	0.528[b]
Bromine	0.077,[a] 0.2[c]	0.476[b]
Iodine	−0.016,[a] 0.16[c]	0.468[b]
Cyano	0.15,[c] 0.02[d]	0.24[b]
Hydroxy	0.22[c]	1.04[e]
OTMS	0.31[c]	
Subtituent	**4-Substituted 1,2-Dimethylcyclohexene**	
Chlorine	0.38,[c] 0.06[d]	
Bromine	0.30[c]	
Iodine	0.20[c]	
Cyano	0.14,[c] 0.14[d]	
Hydroxy	0.70[c]	

[a] At about −160°C in deuterated vinyl chloride.[69]
[b] At about −80°C in CS_2.[100]
[c] At −145°C in CF_2Cl_2.[101]
[d] At 145°C in CHF_2Cl.[101]
[e] At −83°C in CD_3OD.[100]

preference for the equatorial position in the cyclohexene series has been explained on the basis of only a single 1,3-diaxial interaction and of a significant dipole-induced dipole attraction involving the halogen and the π electrons in the double bond. Lambert and Marko report some interesting differences in the solvent dependence of conformational equilibria between 4-substituted cyclohexenes and 4-substituted 1,2-dimethylcyclohexenes, obtained at −150 °C and 500 MHz (Table 1.6).[101] Whereas the conformational equilibrium in 4-cyano-1,2-dimethylcyclohexene is independent of solvent polarity, that in 4-cyanocyclohexene increasingly favors the equatorial form in the more polar solvent (CHF_2Cl). This has been ascribed to the expected lower dipole moment in 1,2-dimethylcyclohexene as compared to cyclohexene, since the double bond in the former molecule is nearly symmetrically substituted. Thus, a significant dipole–dipole interaction, which results in an electrostatic energy that is dependent on solvent polarity, should only occur in 4-cyanocyclohexene, as observed. These workers therefore conclude that the equilibrium data (Table 1.6) on 4-halogenated cyclohexenes reported by Jensen and Bushweller,[69] who employed the rather polar vinyl chloride as a solvent, are strongly influenced by solvent polarity. The effects of substituents on the conformational equilibria of cyclohexenes and methylenecylohexanes has been reviewed recently by Lambert et al.[103] (see also Chapter 2 for the conformational analysis of methylenecylohexanes).

Lambert and Marko did not obtain coalescence temperatures during their low-temperature 500 MHz NMR investigation of the above compounds.[101] Such measurements could have revealed the effects of substituents such as 4-cyano, 4-hydroxy, and 1,2-dimethyl-4-X (X = halogen, cyano, and hydroxy) on the ring inversion barrier in cyclohexene. The presence of the 1,2-dimethyl groups might be expected to increase the inversion barrier because eclipsing effects of the allyl bond are greater in the boat than in the half-chair. Also, it would be valuable to know the dependence of the barriers in the above substituted cyclohexenes on solvent polarity. Thus, much remains to be learned about the factors determining the size of ring inversion barriers in substituted cyclohexenes, despite the firm foundation laid down by Jensen and Bushweller[69] and by Bernard and St.-Jacques.[96]

1.1.7 ESR Data

Ring inversion equilibria and kinetics of the isomeric cyclohexenylmethylene free radicals, **44** and **45**, have been recently measured by ESR in nonpolar solvents.[95] The low barriers to ring inversion in cyclohexenes are favorable for ESR measurements because destruction of the short-lived radicals can be slow compared to the ring inversion process, which simplifies the analysis, and because the lineshape changes occur in a convenient temperature range. The $CH_2\cdot$ group, which is presumably planar, shows no conformational preference when in the 3 position and only 0.17 ± 0.03 kcal/mol preference ($\Delta G°$) for the equatorial position when in the 4 position. The activation energies for ring inversion are 5.5 and 5.7 kcal/mol and are similar to the free-energy barriers in cyclohexanes determined by NMR (Table 1.5).

Russell et al. have measured ring inversion barriers in cyclohexanesemi-diones (**46**).[98] The values of $\Delta G^‡$ (Table 1.5) are similar to those of cyclohexenes,

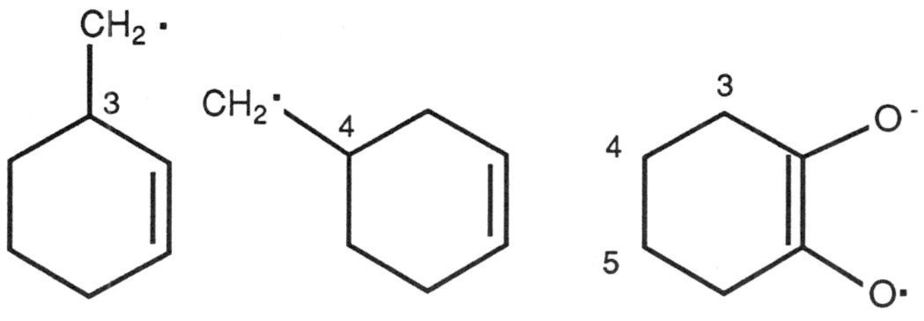

44 **45** **46**

except for the 3,3,5,5-tetramethyl derivative of **46**, where the 1,3-diaxial interaction in the half-chair raises its energy and thereby lowers the barrier to inversion.

1.1.8 Chemical Methods and Equilibrium of Diastereomers

Rickborn and Lwo have measured rates and product ratios for the epoxidation of *cis*-4,5-dimethylcyclohexene from which they deduced by comparison with data on related compounds that a 4-methyl group prefers the equatorial over the axial orientation by 1 kcal/mol.[103] This value is nearly one-half of the *A* value of the methyl group in methylcyclohexane and is in line with conformational preferences found subsequently in other 4-substituted cyclohexenes by direct NMR integration of peaks at low temperatures.[101] Unfortunately, the NMR integration method fails with 4-methylcyclohexene, presumably because of the low percentage of the axial form (estimated to be about 1%) at low temperatures. Aycard and Bodot have measured base-catalyzed chemical equilibria in diastereomeric cyclohexenes containing CN or CO_2R groups in connection with their NMR work (Table 1.4).[88,89]

1.2 FUSED CYCLOHEXENES

Many steroids, terpenoids, alkaloids, and other natural products contain fused cyclohexene rings in a diversity of environments. The greatest amount of structural as well as conformational information on such molecules is to be found in steroids.[8–10,12,13,26] Because of this diversity and the large literature on these compounds, the coverage in the following sections is only representative.

1.2.1 Theoretical Calculations for Fused Cyclohexenes

The *cis*- and *trans*-octalin systems have attracted much attention in attempts to explain the regiochemistry of bromination and other electrophilic substitution reactions of steroid ketones.[8] These reactions proceed by enol or enolate intermediates that contain cyclohexene moieties, and the rate of formation of these intermediates can determine product ratios in reactions such as bromination. In other reactions, such as enol acetate formation from ketones under acidic conditions, thermodynamic equilibria betweem regio-isomers can occur.

Bucourt[8,21,104] and Bucourt and Hainaut[10] have investigated conformational equilibria and strain energies in unsaturated steroids and in bicyclic models, initially by a simple examination of the distortions induced by unsaturation in the ring torsional angles of the molecule, and later by increasingly more quantitative molecular mechanics calculations. The difference in the energies between Δ^2-*trans*-octalin (**47**) and Δ^3-*trans*-octalin (**48**) in the absence of substituents is calculated to be 0.8 kcal/mol, with the Δ^2 isomer favored, and the cyclohexene moiety in the Δ^3 isomer is in a conformation intermediate between the half-chair and the sofa. A methyl group at C-9 increases the energy of the Δ^2 isomer. This effect, which had previously been deduced by Corey and Sneen using less appropriate molecular geometries,[20] explains the direction of enolization of steroid ketones, which appear to be thermodynamically controlled under acidic conditions. For example, Djerassi and co-workers have observed that treatment of 3-keto A/B trans fused steroids with acetic amhydride yields exclusively the Δ^2 enol acetate, whereas the analogous 19-normethyl series gives a mixture of the Δ^2 and Δ^3 enol acetates.[105]

When the A/B rings are cis fused, the B/C trans ring fusion in steroids prevents ring inversion, and enolization of the 3-keto group takes place toward C-4 and not C-2. This is supported by calculations that show that the presence of a 19-methyl group is an important factor in this preference.[10] However, if the C/D fusion is changed to cis, enolization occurs toward C-2. The situation is more complex in the simple bicyclic series, because the presence of two possible conformations, but enolization toward C-2 seems more common than toward C-4.

Allinger et al. have calculated heats of formations of all the isomeric octalins (**47**, **48**, **49**, **50**, **51**, and **52**).[34] Entropies arising from symmetry differences were then used to calculate free energies and percentage compositions. The Δ^9 and $\Delta^{1(9)}$ isomers are more stable than those containing only disubstituted double bonds,

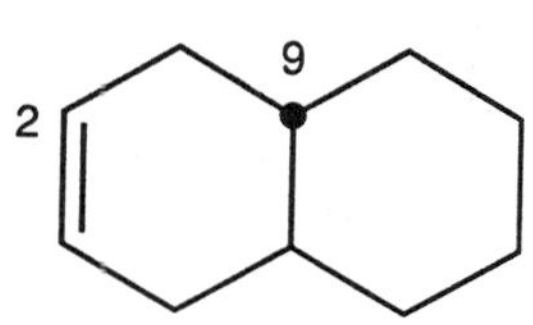

47

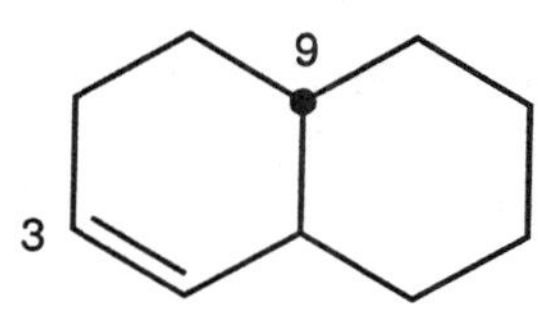

48

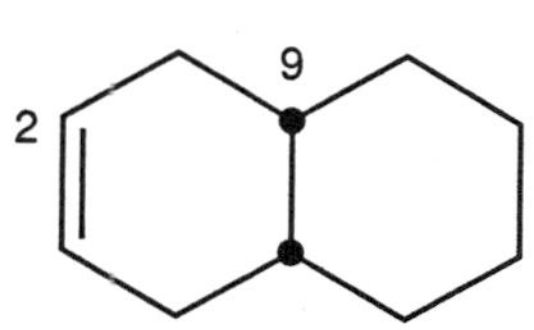

49

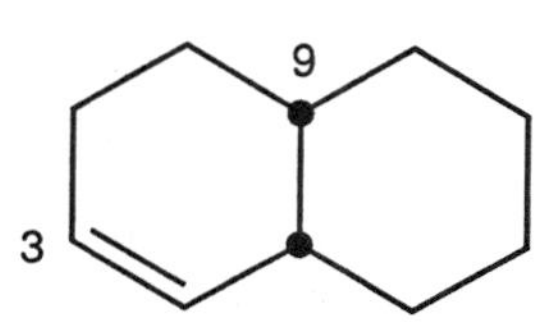

50

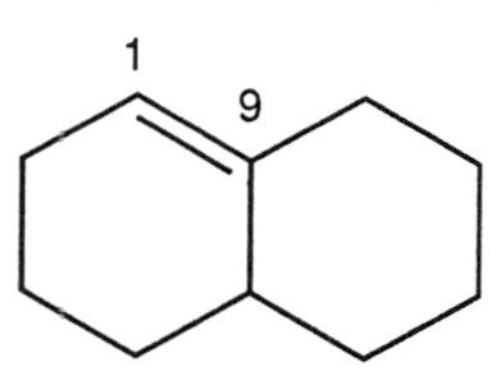

51

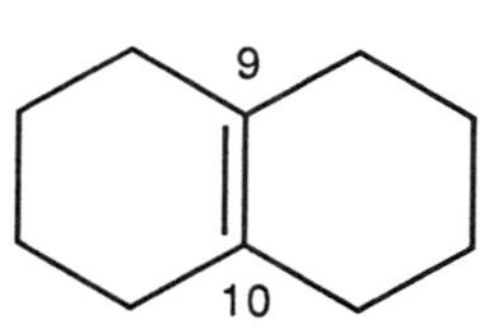

52

and in the latter cases, the trans fused isomers are more stable than the cis fused ones. The order of stabilities of the Δ^1 and Δ^2 isomers are opposite in the cis and trans fused series as discussed above.

Molecular mechanics calculations on the ring inversion process in the *cis*-Δ^2-octalin, **49**, have been carried out with the White–Bovill force field.[7] The lowest-energy conformation (**53**) is slightly distorted from C_2 symmetry. The transition state for ring inversion, **54**, has a half-chair in the saturated ring with the eclipsing remote from the ring fusion, so that the cyclohexene moiety is still a half-chair. The difference in strain energy between the ground and transition states in **54** is 8.2 kcal/mol. Following this transition state are some pseudorotating forms of relatively high energies that have boat cyclohexene and cyclohexane rings. Passage over a second transition state, which is the mirror image of **54**, finally gives the mirror image of **53**, that is, ring inverted **53**. The calculated strain energy difference between **53** and **54** is 8.2 kcal/mol, which is somewhat higher than the experimental free-energy barrier (9.9 kcal/mol) in **53**, but this has been ascribed to the White–Bovill force field giving too low a ring inversion barrier for cyclohexane, although the barrier for cyclohexene is well reproduced (Table 1.1). Thus, ring inversion in **53** is dominated by the saturated ring. Calculations for the

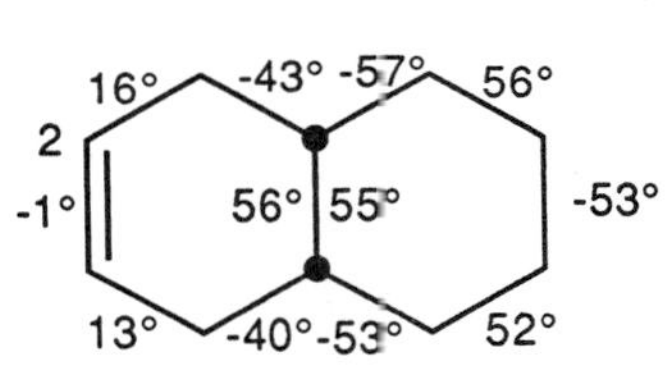

53

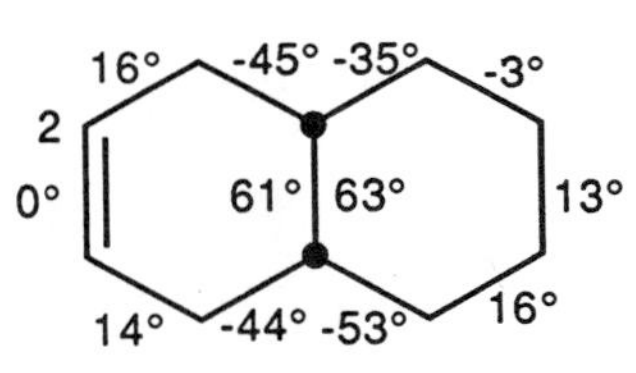

54

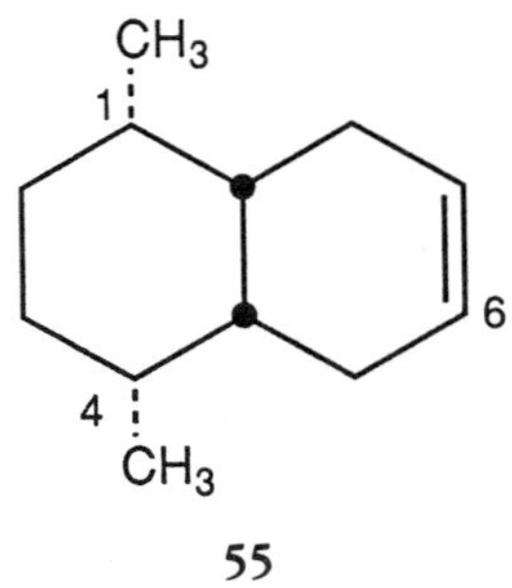

55

cis-1,4-dimethyl-*cis*-Δ^6-octalin **55**, on the other hand, show that the transition state for ring inversion has a twist-boat cyclohexane and an approximately 1,3-diplanar cyclohexene ring.

Bucourt has investigated the hexahydronaphthalenes, which are the appropriate models for understanding the direction of enolization in 3-keto steroids containing unsaturation in ring B.[8,21] Enolization occurs toward C-2 or C-4 depending on the position of the double bond at 7,8 (**56**) or at 6,7 (**57**), respectively. In steroids with an aromatic A ring, a B/C cis fusion (i.e. 9α) is

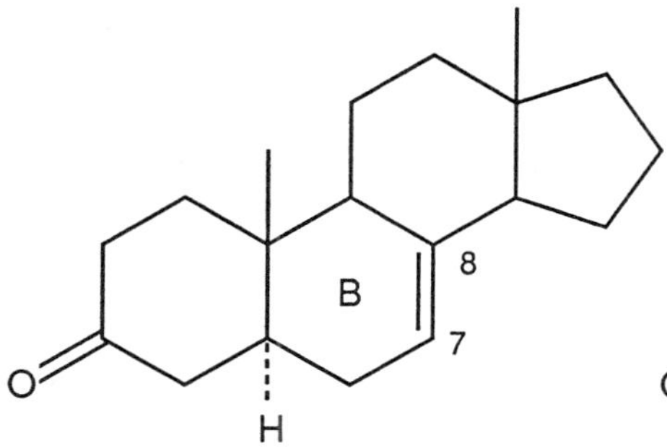

56

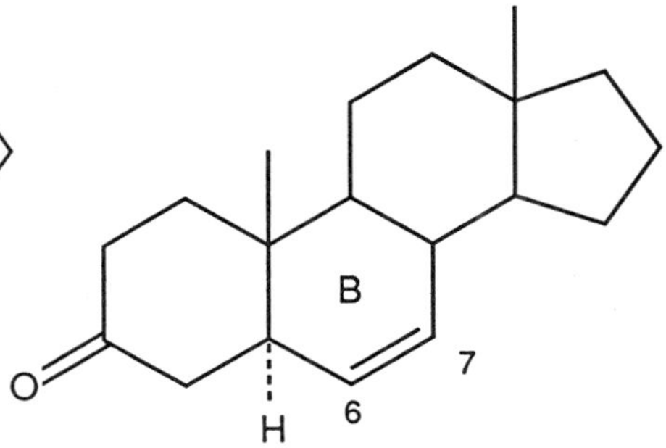

57

not as unfavorable as in trans A/B steroids; nevertheless, the 9β configuration is always favored.

Molecular mechanics calculations with a more or less modified MM2 force field on the isomers of naphthalene tetrachloride have been reported by de la Mare.[36] Although geometries are well reproduced, the energies and dipole moments are in poor agreement with experiment unless account is taken of the polarization of the aromatic ring by the chlorine atoms. The conformations of these compounds are discussed in the NMR section below.

1.2.2 X-Ray Crystal Data for Fused Cyclohexenes

The most common conformation of fused cyclohexene rings is the half-chair. However, there are many examples of sofas, as well as of forms intermediate between the half-chair and the sofa. Fusion to a ring that causes eclipsing at the 4,5 position, or certain steric constraints, especially in tetralin derivatives, can result in a boat form.

Some crystals contain more than one molecule in the unit cell and the cyclohexene rings in these crystallographically different molecules sometimes have rather different conformations.[12,13] Thus, crystal lattice effects must be important, at least in these cases, and it can also be deduced that the half-chair is easily distorted in the direction of the sofa, in agreement with molecular mechanics calculations on cyclohexene.

The Diels–Alder adduct **58**, which has a cyclohexene fused to a cyclohexenedione has a crystal structure showing two half-chairs.[106,107] Related compounds

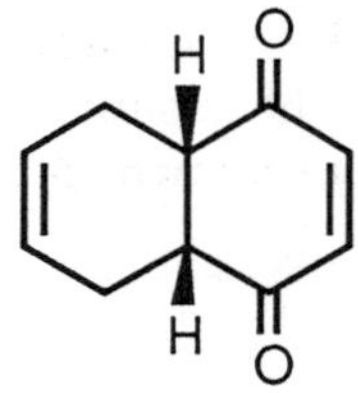

58

and their photodimers also have half-chair conformations, sometimes appreciably distorted toward sofa forms.[106,108]

For double bonds in ring A or ring B of steroids with no other unsaturation in the cyclohexene moiety, the conformations are most often half-chairs or half-chairs more or less distorted toward sofas.[12,13] All the steroid structures discussed here have the normal trans B/C ring fusion. Ring A in $\Delta^{2,3}$ and $\Delta^{5,10}$ steroids generally have rather undistorted half-chairs, as expected from the position of ring fusion at the double bond (for $\Delta^{5,10}$), or trans diequatorial at the 4,5 position (cyclohexene numbering) (for $\Delta^{2,3}$), whereas $\Delta^{4,5}$ steroids have this ring more or less distorted from the half-chair. Ring B in $\Delta^{5,6}$ steroids, which are of common occurrence, are half-chairs with distortions toward either of the two closed possible sofas, and the maximum torsional angle distortion with respect to the ideal C_2 axis is about 10–15°. Conformations ranging from half-chairs to sofas are found in ring A of $\Delta^{5,6}$-3-keto steroids; average torsional angles are shown in **59** and **60**.

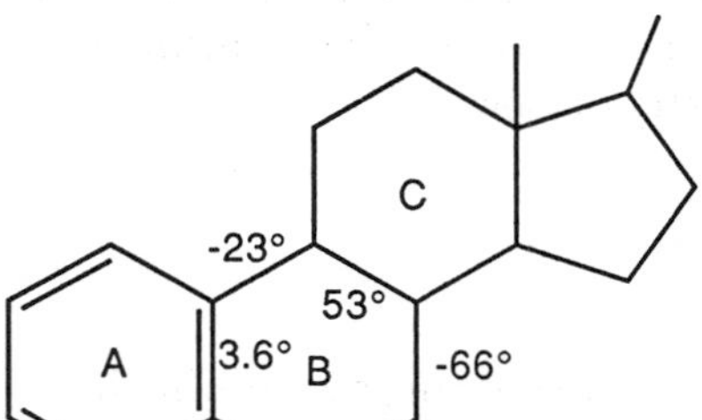

Half-chair forms

59

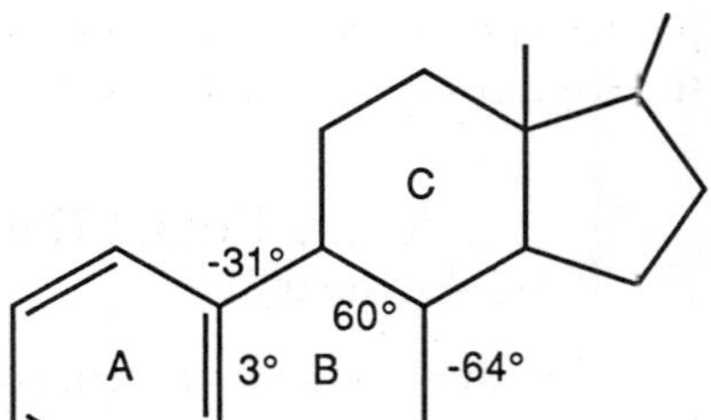

Sofa forms

60

Average ring torsional angles in ring B

Steroids with ring A aromatic are about equally distributed between half-chairs and intermediate (half-chair/sofa) forms for ring B, with only a few molecules containing forms close to the sofa.[12,13] Average torsional angles for ring

B for the half-chairs and sofas are shown in **61** and **62** respectively. One interesting exception to the above generalization is a molecule having two substituents at the 11 position (11β-hydroxy-11α-methyl), where the half-chair would have large nonbonded repulsions with the 1 position in the benzene ring.[13] Ring B in this case is an almost symmetrical boat and the torsional angles, which are shown in **63**, correspond to a highly puckered ring. The average of the absolute values of the

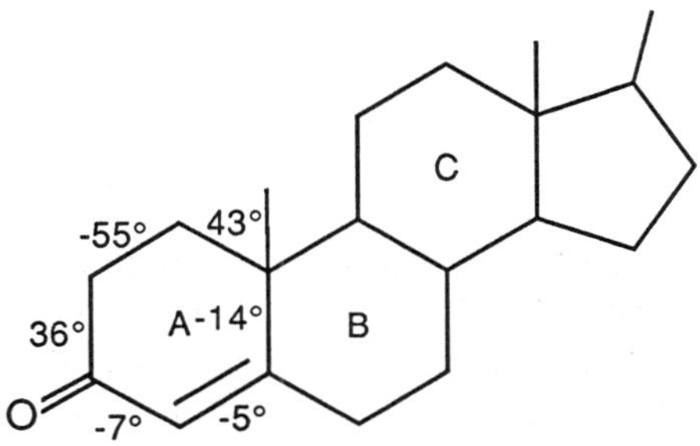

Half-chair forms

61

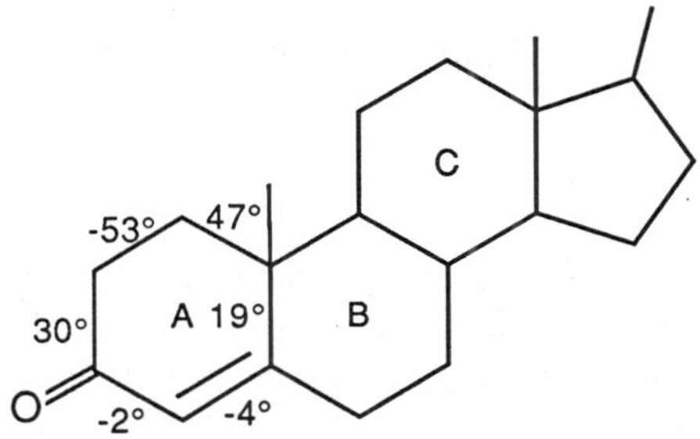

Sofa forms

62

Average torsional angles in ring A

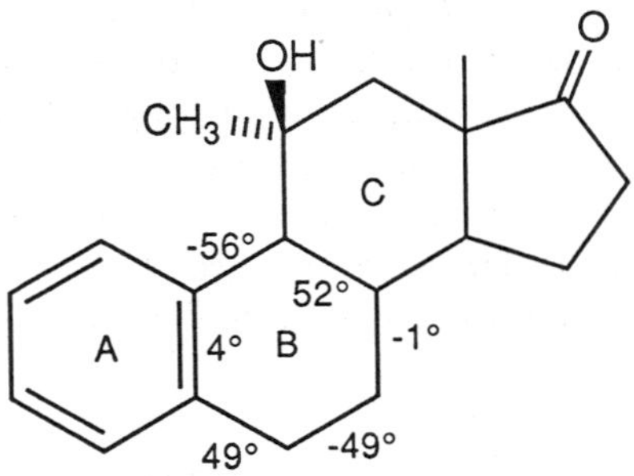

Ring torsional angles in ring B (boat)

63

torsional angles is 51°, as compared to the 39° value calculated by molecular mechanics for boat cyclohexene.[7]

The α, ε, and the 5,6-dichloro derivative of the γ isomers of naphthalene tetrachloride have half-chair conformations in the crystalline states,[51,109,110] and their conformations in solution are discussed below in connection with their NMR spectra.

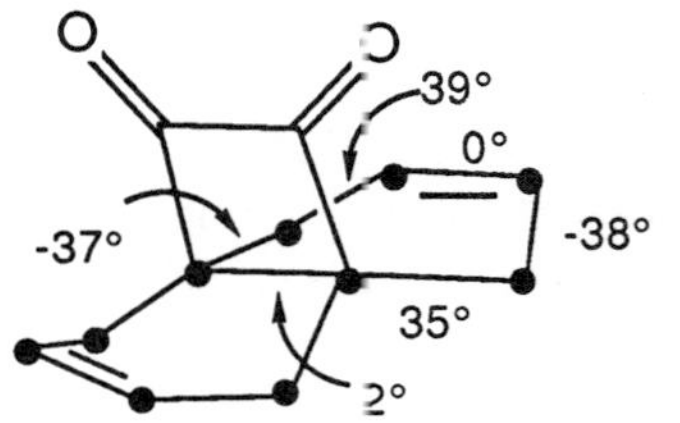

Ring torsional angles (ca. C_{2v})

64

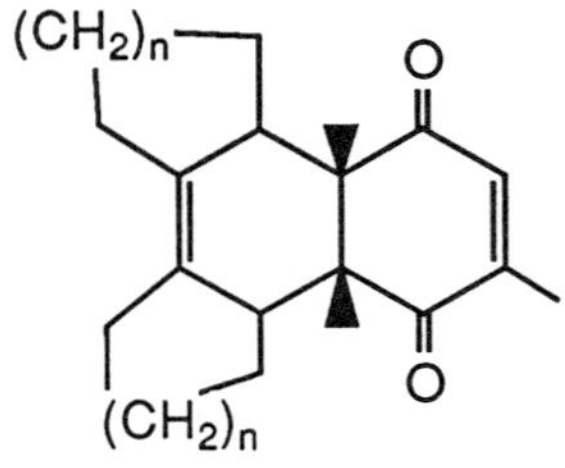

65

The crystal structures of [4,4,2]propella-3,8-dienes show two cyclohexene moieties in boat forms that are oriented as shown in **64**.[4,5] The four-membered ring causes an enlargement of the endocyclic bond angles in the cyclohexene ring at the position of fusion (θ = ~115°). The average of the absolute values of the ring torsional angles in the six-membered rings is 37°, which is close to the 39° in boat cyclohexene.

Diels–Alder adducts of 1,1′-bi(cycloalkenyl)s with quinones have a cyclohexene ring fused to three other rings (e.g., **65**). The crystal structures of six such molecules show half-chair conformations that are more or less distorted toward the sofa, just as in simpler systems.[111] However, in two cases, both with two seven-membered rings fused to the cyclohexene moiety, the distortions in the latter rings are much larger and the conformations are well on the way to being twisted boats, with 4,5 torsional angles (cyclohexene numbering) of 25–30°. The seven-membered rings are apparently not responsible for this effect, and it is probable that the half-chair/twist-boat energy differences in these Diels–Alder adducts are much lower than in simple cyclohexenes because of the highly unsaturated six-membered ring that is fused to the cyclohexene 4,5 position.

1.2.3 NMR Data for Fused Cyclohexenes

Modern one- and two-dimensional high-field ^{1}H and ^{13}C-NMR experiments, such as NOE difference spectroscopy and COSY, are particularly valuable for investigating the structures and conformations of complex molecules, because many values of the $^3J_{H-H}$ can be obtained and related to torsional angles. One recent example, which involves a cyclohexene ring fused to a seven-membered ring (**66**), has given several vicinal coupling constants in the cyclohexene ring.[112] The vicinal coupling constants between protons on C-1 and C-11 are 3.4 and 7.8 Hz and show that the C-10 substituent is predominantly equatorial on the half-chair in **66**.

The coupling constants in the ^{1}H-NMR spectra of four isomeric naphthalene

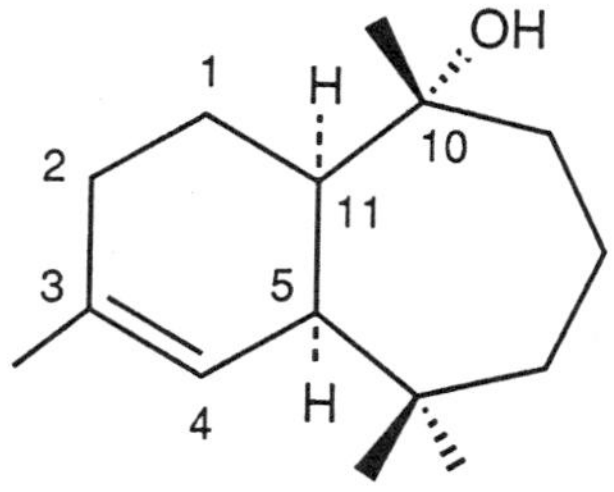

66

tetrachlorides have been determined.[113,114] In the α isomer (**67**), whose configuration and conformation (a′eea′) are known from an X-ray crystal structure,[51] the large coupling constant (11.0 Hz) between H-2 and H-3 in the AA′BB′ spectrum clearly shows that these protons are also axial in solution. The small coupling constant (3.5 Hz) between H-1 and H-2 is consistent with these protons being in the e′ and a positions, respectively. In the δ isomer, which shows four different chemical shifts, the three vicinal coupling constants (3.2, 10.9, and 7.6 Hz) are consistent only with the a′eee′ conformation, **68**. All the J's are 8.0 Hz in the AA′BB′ spectrum of the γ isomer, and the all-trans configuration has been assigned with the predominant conformation (**69**) having all the chlorines equatorial or pseudo equatorial. However, the value of J_{23} does not fit well with this suggested conformation, and it appears that some of the a′aaa′ form, or possibly a boat form, must also be present in equilibrium; on the other hand, the value of J_{12} does not indicate any of the a′aaa′ conformation, unless the reference values used for the coupling constants are not appropriate for this isomer. With J_{12} and J_{23} values of 8.0 and 11.0 Hz for the e′eee′ form and 4.0 and 2.0 Hz for the a′aaa′ form, respectively, which are conceivable values, the calculated coupling constants for a 3:1 mixture of the e′eee′ and a′aaa′ (**70**) forms are 7.0 and 8.7

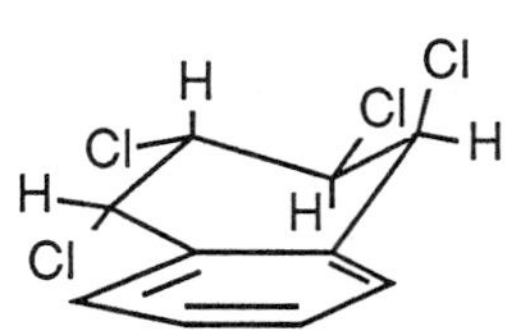

67

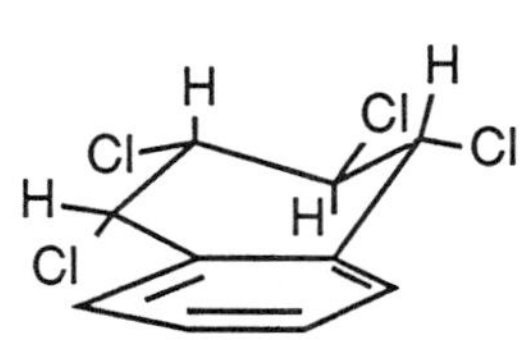

68

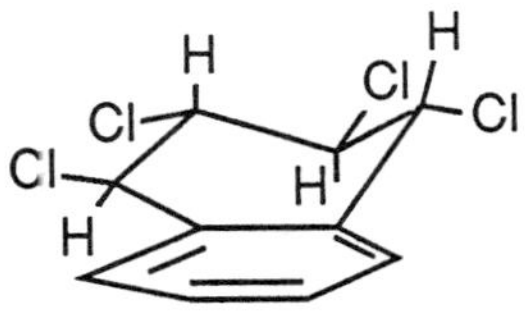

69

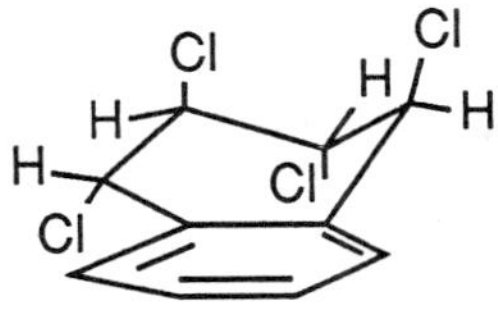

70

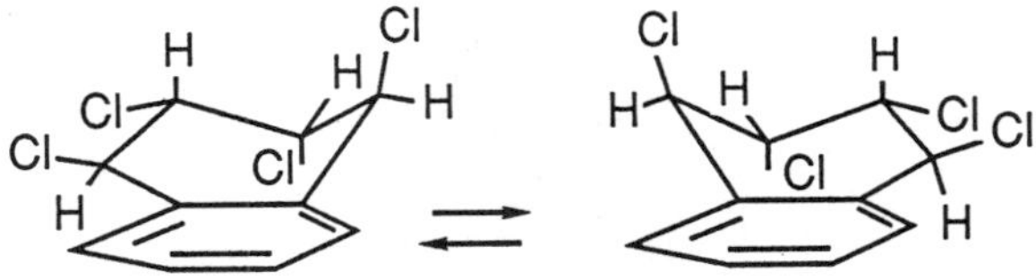

71 **72**

Hz, in reasonable agreement with experiment. Recent molecular mechanics calculations,[36] described above, show that the γ isomer probably exists as a mixture of conformations, in agreement with the above analysis. The 5,6-dichloro derivative of the γ isomer has an NMR spectrum that is consistent with a predominant boat conformation, even though the crystal contains only the half-chair, with all the chlorines axial.[110] The suggested boat forms in these tetralins require confirmation before being fully accepted.

The ϵ isomer of naphthalene tetrachloride shows an AA′BB′ spectrum with coupling constants of 5.8 and 3.5 Hz and it has been assigned to the degenerate equilibrium between the a′aee′ (**71**) and e′eaa′ (**72**) forms. The average of the expected $J_{ee'}$ and $J_{aa'}$ for H-1 and H-2 (ca. 2 and 7.6 Hz) is 4.8 Hz in fair agreement with the observed value, and J_{ee} between H-2 and H-3 should be small, as found. Subsequently, the crystal structure of this isomer has been determined and is in agreement with the configuration determined by NMR.[109]

Alkaline dehydrochlorination reactions of the above four isomeric naphthalene tetrachlorides have been shown to take place in two stages and to give dichloronaphthalene isomers in various proportions and to take place at quite different rates.[115] These are complex reactions because it is not necessarily the most populated conformation that gives rise to the products. The slowest rate is found in the γ isomer, which has all its chlorine atoms equatorial or pseudoequatorial, and thus cannot undergo a 1,2 anti-elimination.

The methine protons in *trans*-1,3-dimethyl-1,2,3,4-tetrahydronaphthalene (**73**) are 0.5 ppm more shielded than those of the cis isomer (**74**), and this is

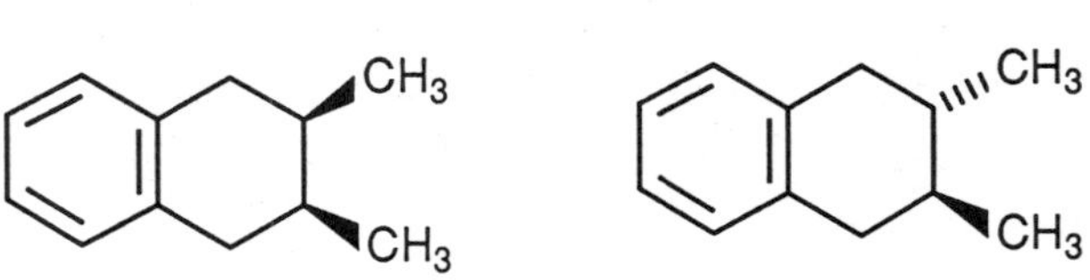

73 **74**

consistent with half-chair conformations with the trans isomer having these protons in the axial positions, whereas the cis isomer has one methine proton axial and the other equatorial.[99] The cis isomer does not give a dynamic NMR at 60 MHz down to − 169 °C, so that ring inversion seems to be significantly faster in tetralins than in cyclohexenes. However, the spectra are complex at the low ^{1}H frequency used and this compound needs to be reinvestigated at the much higher proton frequencies available on current NMR spectrometers.

No ring inversion barriers appear to have been determined by NMR in tetralin and its derivatives. Unsuccessful attempts in 1-acetoxy- and 1-methoxytetralin,[92] whose equilibria were discussed previously together with those of related cyclohexenes, may be caused by enancomeric equilibria at low temperatures, or possibly by very low barriers. The ring inversion barrier ($\Delta G^{\ddagger}$) in *cis*-$\Delta^{6,7}$-octalin, which involves inversion of a cyclohexene as well as a cyclohexane ring, is 9.9 kcal/mol (F. De Pessemier, P. Vanhee, M. J. O. Anteunis, and D. Tavernier, unpublished results cited in Ref. 7). This conformational process has been examined by molecular mechanics (see above).[7] Brown et al. have shown that the preferred conformation in the *cis*-octalin, **75**, has the methoxy group pseudoaxial

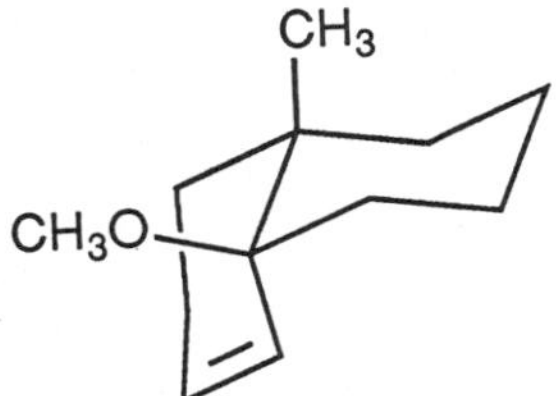

75

on the cyclohexene moiety to the extent of 85%, as shown by low-temperature ^{13}C-NMR.[67] The photoelectron spectra of **75** and related molecules are consistent with an anomeric effect when the methoxy group is perpendicular to the plane of the double bond.

Ring inversion in some radical cations (**76**, **77**, and **78**) derived from cyclohexene-1,2-dithiol and heterocyclic derivatives have been measured by dynamic ESR.[116] These barriers are comparable to those measured by NMR in simple cyclohexenes. The radical anion (**79**) derived from 1,2-diphenylcyclohex-

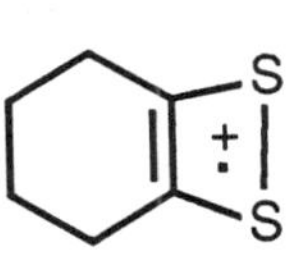

76

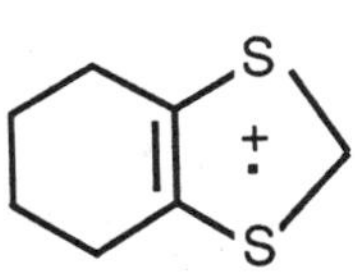

77

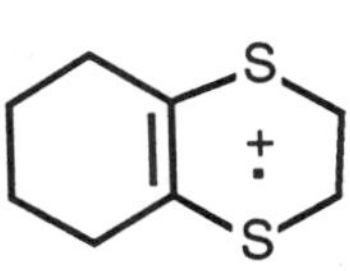

78

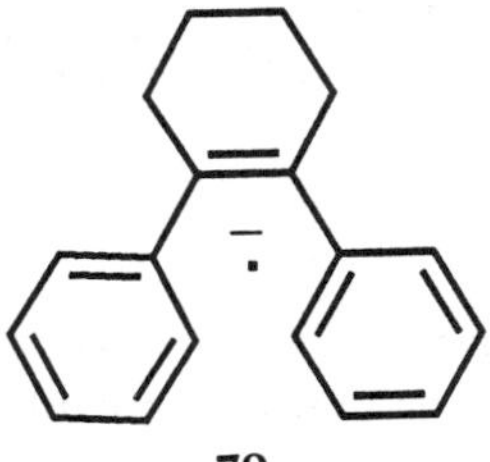

79

ene seems to exist in a half-chair but does not undergo ring inversion on the ESR time scale even at room temperature.[117] Possibly, the double bond in the radical anion is somewhat twisted to relieve the steric repulsion between the two phenyl rings; such an effect should increase the ring puckering and presumably also the ring inversion barrier.

McDowell et al. have examined a series of methyl-substituted tetrahydronaphthoquinones in the crystalline phase by ^{13}C-NMR, with cross-polarization and magic angle spinning.[118] They find twice as many peaks as in solution and interpret this as indicating that the lattice prevents inversion of the two half-chairs (**80** and **81**) in the crystalline phase. The average of the two chemical shifts of a

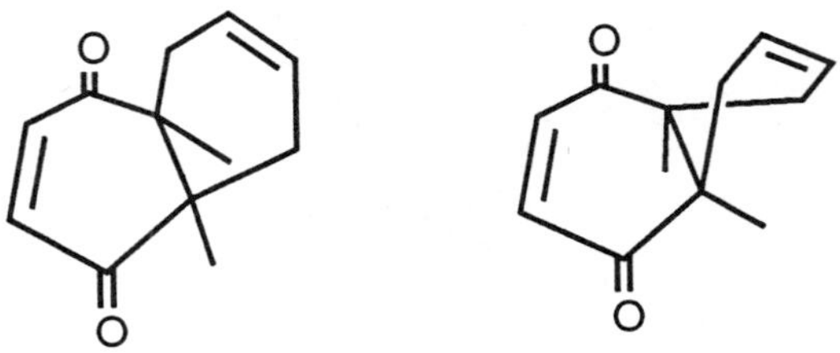

80 **81**

carbon in the solid state is nearly, but not quite, equal to the shift in solution as a result of the different environments of the molecules. Nevertheless, the magnitudes of the chemical shift splittings in the solid state are useful from a structural point of view. In solution, these compounds give dynamic NMR effects between 25 and -90 °C, corresponding to free-energy barriers of about 9 kcal/mol.[119]

The barriers to ring inversion in three derivatives of *cis*-$\Delta^{2,6}$-hexalin with CH_2R substituents at the ring junction (R = CN, Br, CO_2CH_3) are between 12 and 14 kcal/mol and are much higher than those in monocyclic cyclohexenes.[120] The unsaturated propellane, **82**, which has three mutually fused cyclohexene

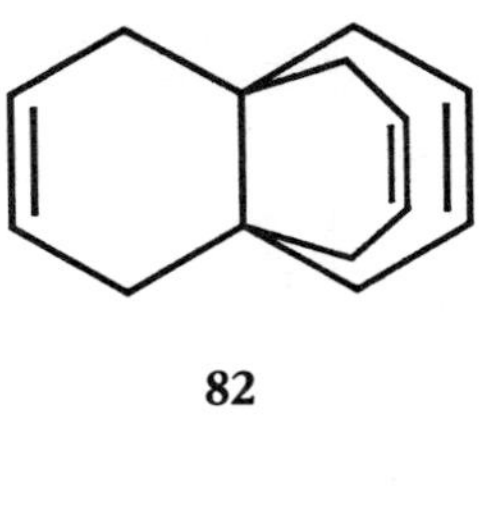

82

rings, has an even larger barrier to ring inversion (16.7 kcal/mol). The large barriers in these molecules show that the individual cyclohexene moieties, which presumably have half-chair conformations, must invert in a concerted manner.

This is consistent with the ring inversion process in cyclohexene having a single transition state with no intermediate that is local energy minimum.

1.3 MISCELLANEOUS APPLICATIONS OF CONFORMATIONAL ANALYSIS IN CYCLOHEXENES

The addition of molecular chlorine to ethers and esters derived from 4-hydroxycyclohexene (**83**) in $CHCl_3$ at -50 °C takes place slightly more slowly for

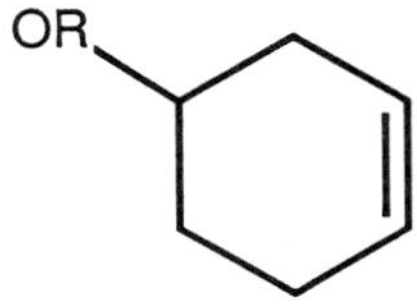

83

the equatorial than for the axial conformer, as shown by the ratio of the products formed in the reaction.[121] The ratios of k_a/k_e for OCD_3, O_2CCH_3, and O_2CCF_3 as the 4-substituents are 1.5, 1.2, and 1.5, respectively. This has been taken to confirm MO predictions of the deactivating effect of an equatorial electronegative substituent in the 4 position,[122] but the above ratios do not follow the increasing order of inductive effects of the substituents. For bromination the reaction is faster for the equatorial than for the axial conformer, the k_a/k_e ratios being 0.54, 0.67, and 0.57, respectively, for the same three substituents. The differences in reactivity between the axial and equatorial forms of these 4-substituted cyclohexenes are so small and devoid of trends that assigning them to any single factor, such as the electronic effect proposed by Zefirov,[122] does not seem to be warranted.

Nakanishi and co-workers have shown that the absolute configurations of 2-cyclohexen-1-yl benzoates (or *p*-substituted benzoates) can be obtained by circular dichroism measurements and a consideration of the chiral exciton coupling between nondegenerate chromophores, irrespective of whether the substituent is pseudoequatorial (**84**) or pseudoaxial (**85**).[123] Thus, (R)-2-cyclohexen-1-yl ben-

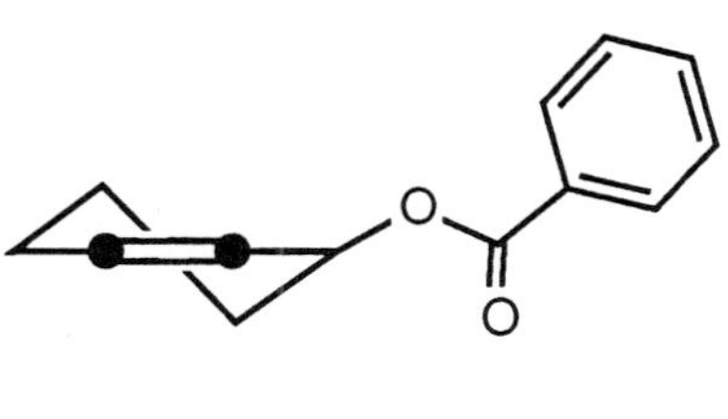

84

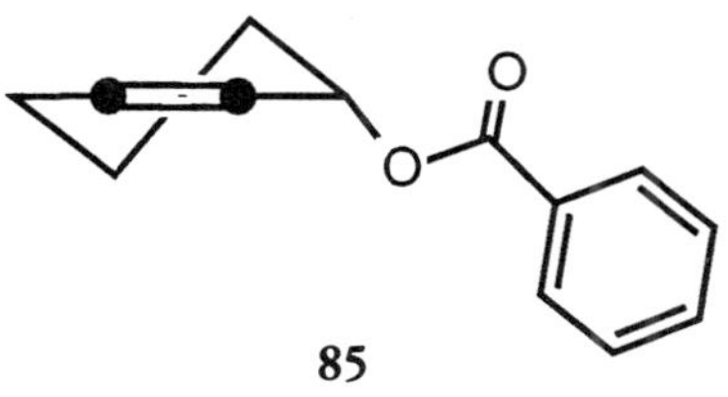

85

zoate gives a positive benzoate Cotton effect. The method has been applied to cyclohexenol moieties in terpenes and steroids. When the chromophore in the six-membered ring is a conjugated enone rather than a double bond, the sign of the Cotton effect changes when the half-chair inverts.[124]

Scheffer and co-workers have studied the photochemistry of molecules related to tetrahydronaphthoquinone.[125] The ring system consists of a cyclohexene fused to a 4-hydroxycyclohexenone ring. They find that the products formed in solution and in the crystalline phase can be understood in conformational terms and that the photochemical effects of the introduction of methyl groups also have a conformational basis. The cyclohexene ring can exist in the two possible half-chairs (**86** and **87**). Photochemistry in the solid state proceeds only by intramolecular hydrogen abstraction and the product depends on the particular half-chair present, which depends on the substitution pattern. In solution, by contrast, there appears to be a fast equilibrium between the two half-chairs and a higher energy boat (**88**), not only in the ground state but also in the excited state. The product **89**, which has a cyclobutane ring and is formed by a 2 + 2 cycloaddition, is derived from the boat. The lattice constraints present in the crystalline phase prevent the formation of the boat and of the photochemical products derived from this conformation. Figure 1.2 summarizes these photochemical results.

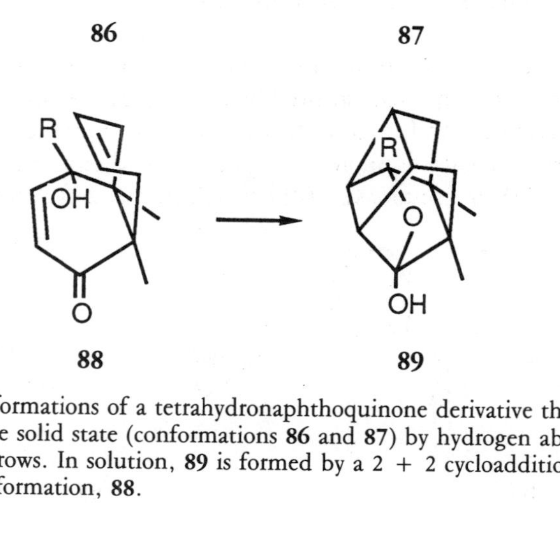

Figure 1.2 Conformations of a tetrahydronaphthoquinone derivative that leads to photoproducts in the solid state (conformations **86** and **87**) by hydrogen abstraction as shown by the arrows. In solution, **89** is formed by a 2 + 2 cycloaddition from the boat cyclohexene conformation, **88**.

Enols and enolates derived from six-membered ketones have a cyclohexene moiety, and their reactions are dependent on conformational effects, particularly when one of the product suffers from Johnson's allylic strain ($A^{(1,2)}$ strain).[22] Solvent effects on the conformational equilibrium (**90**) of the potassium enolate of

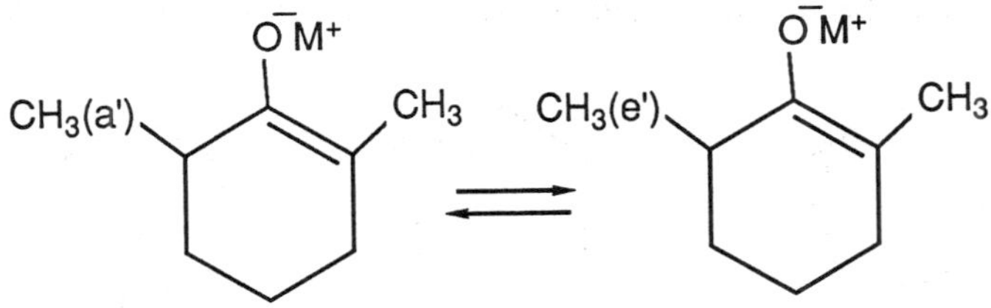

90

2,6-dimethylcyclohexanone have been interpreted in terms of $A^{(1,2)}$ strain, but the effect is relatively small, and an alternative explanation has been given that also helps to explain the observed equilibria in six-membered ring enolates.[126]

Enamines derived from cyclohexanone are important in organic synthesis, and reference has already been made to the conformation in the crystal state of some enamines derived from cyclohexanones. Allylic strain of the $A^{(1,2)}$ kind comes into play in enamines even more than in enolates and compounds such as the enamine of 2,4-dimethylcyclohexanone has the trans configuration with a pseudoaxial methyl group even when prepared from the cis isomer of the ketone.[22] The cis enamine, unlike the trans isomer, suffers from allylic strain. Other examples of $A^{(1,2)}$ strain on the stereochemistry of enamines are given in Johnson's review.[22]

1.4 HETEROCYCLIC, CYCLOPROPANE, AND EPOXIDE ANALOGS OF CYCLOHEXENE

This section presents some highlights of carbocyclic and oxygen-containing heterocyclic molecules whose conformations have close relationships to those of the cyclohexenes (unsaturated heterocyclic six-membered rings are discussed by Harvey in Chapter 8). Dihydropyran (**91**), which is an interesting heterocyclic of cyclohexene, has a half-chair conformation with a free-energy barrier of 6.6 kcal/mol, which is more than 1 kcal/mol greater than the corresponding barrier in cyclohexene.[127] The increased barrier has been ascribed to changes in the n-π interaction, which depends on the O—C(sp^2) torsional angle and is more favorable in the half-chair than in the boat. In fact, the conformation of dihydropyran should be intermediate between a half-chair and a sofa as a result of this interaction, which can also be described as being due to resonance (**92**). The conformations of some substituted dihydropyrans have been studied by ^{1}H-NMR chemical shifts and coupling constants and interpreted on the basis of half-chairs.[128]

The conformations and barriers to inversion in 2,3-dihydropyran and 1,4-dioxene (**93**) have been investigated by far-infrared spectroscopy by Lord et al.,

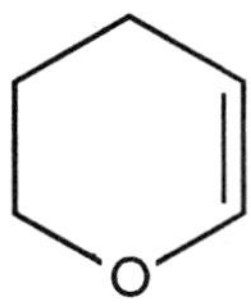

91

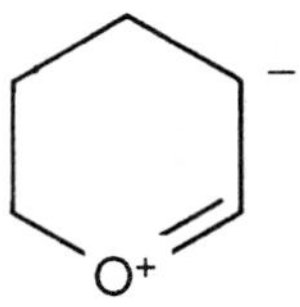

92

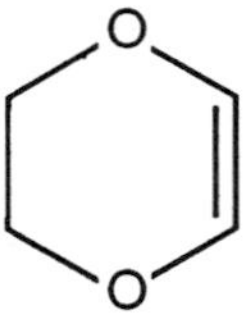

93

who find half-chairs with ring inversion barriers of 8.4 and 8.3 kcal/mol, respectively, and deduce that boats are metastable intermediates that are 1.0 and 1.5 kcal/mol, respectively, above the twist forms. Durig et al. have measured Raman spectra of the same two molecules and have changed some of the band assignments made on the basis of the infrared work, but the calculated barriers to ring inversion change only slightly (7.6 kcal/mol in both molecules).[63]

The UV spectrum of dihydropyran immediately after irradiation by a pulse of laser IR light shows the presence of two species, which supports the existence of two local energy minima, that is, the half-chair and the boat.[129] The interaction of the two lone pairs of electrons on oxygen with the π orbitals of the double bond does not give a simple torsional potential, according to quantum mechanical calculations,[130] but the reason why these molecules, unlike cyclohexene, have two local energy minima is not obvious.

The conformations of δ-valerolactone (**94**) and δ-valerolactam (**95**), which

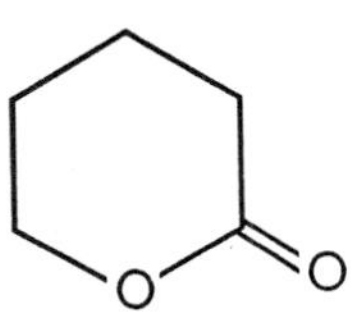

94

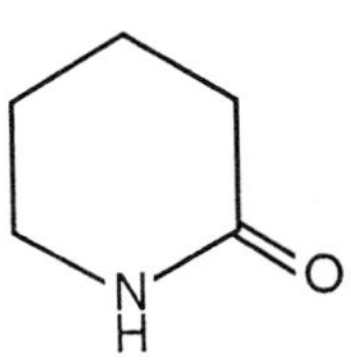

95

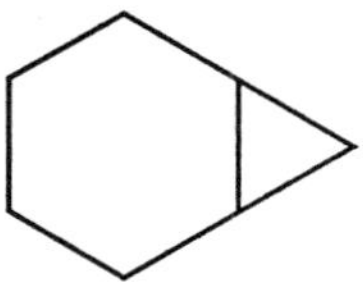

96

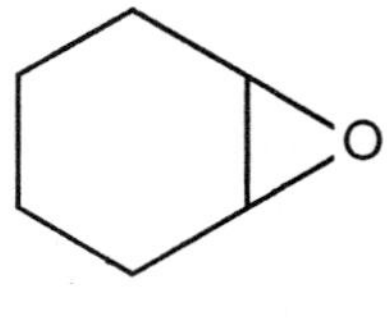

97

have a partial double bond in an endocyclic position of their six-membered rings because of resonance, appear to have a variety of conformations. Six-membered lactones appear to be very flexible,[131] and a survey of X-ray data of molecules with δ-lactones shows that many have nonplanar lactone groups and that half-chairs, boats, and twist-boats occur.[132] Much less work has been done on δ-valerolactam, but molecular mechanics calculations show that the half-chair is more stable than the boat by at least 3 kcal/mol.[133]

The fusion of a three-membered ring to cyclohexane leads to the same strong torsional constraint as in cyclohexene, although the torsional potentials about a bond to a three-membered ring and a bond to a double bond have different shapes.[3] Molecular mechanics and quantum mechanical calculations on noncarane (**96**) and cyclohexene oxide (**97**) show that the half-chair is the stable conformation.[134] Electron diffraction results on cyclohexene oxide show a half-chair,[60] as do X-ray crystal structures of derivatives of cyclohexene oxide.[135] Half-chairs have also been found in noncarane by electron diffraction[136] and in crystalline noncarane derivatives by X-ray diffraction.[13]

1.5 CONCLUSION

Cyclohexenes and related molecules, unless strongly constrained, have half-chair conformations, which may be more or less distorted toward the sofa. Boat forms are in general transition states for the ring inversion of the half-chair, except when constraints are present,[4,5] or possibly in some heterocyclic analogs of cyclohexene. Most of the effects of substituents on conformational equilibria and barriers appear to be fairly well understood, and molecular mechanics calculations are eminently well suited to this ring system, although some questions remain concerning the interaction of polar groups with one another and the possible importance of induced dipoles—even saturated hydrocarbons are polarizable![3,137]

REFERENCES

1. Eliel, E. L.; Allinger, N. L.; Angyal, S. J.; Morrison, G. A. *Conformational Analysis*; Wiley-Interscience: New York, 1965.
2. Hanack, H. *Conformational Theory*; Academic Press: New York, 1965.
3. Burkert, U.; Allinger, N. L. *Molecular Mechanics*. American Chemical Society: Washington, DC, 1982.
4. Newkome, G. R.; Fronczek, F. R.; Baker, G. R. *Tetrahedron Lett.* **1982**, *23*, 2725.
5. Fink, R.; Van der Helm, D.; Neely, S. C. *Acta Crystallogr., Sect. B* **1975**, *31*, 1299.
6. Pickett, H. M.; Strauss, H. L. *J. Am. Chem. Soc.* **1970**, *92*, 7281.
7. Vanhee, P.; Tavernier, D.; Baass, J. M. A.; van der Graaf, B. *Bull. Soc. Chim. Belg.* **1981**, *90*, 697.

8. Bucourt, R. *Top. Stereochem.* **1974**, *8*, 159.
9. Bucourt, R.; Hainaut, D. *C. R. Acad. Sci. Paris, Sect. C.* **1964**, *258*, 3305.
10. Bucourt, R.; Hainaut, D. *Bull. Soc. Chim. Fr.* **1965**, 1366.
11. Philbin, E. M.; Wheeler, T. S. *Proc. Chem. Soc.* **1958**, 167.
12. *Atlas of Steroid Structure*, Duax, W. L.; Norton, N. A. Eds.; IF/Plenum, New York, 1975; Vol. 1.
13. *Atlas of Steroid Structure*, Griffin, J. F.; Duax, W. L.; Weeks, C. M. Eds.; IF/Plenum, New York, 1984; Vol. 2.
14. Cremer, D.; Pople, J. A. *J. Am. Chem. Soc.* **1975**, *97*, 1354.
15. Anet, F. A. L. In *Conformational Analysis in Medium-Sized Ring Heterocycles*; Glass, R. Ed.; VCH Publishers: Deerfield Beach, FL 1988.
16. Boëseken, J.; van der Gracht, W. J. F. de Rijck *Recl. Trav. Chim. Pays-Bas* **1937**, *56*, 1203.
17. Lister, M. W. *J. Am. Chem. Soc.* **1941**, *63*, 143.
18. Beckett, C. W.; Freeman, N. K.; Pitzer, K. S. *J. Am. Chem. Soc.* **1948**, *70*, 4227.
19. Barton, D. H. R.; Cookson, R. C.; Klyne, W.; Shoppee, C. W. *Chem. Ind. (London)* **1954**, 21.
20. Corey, E. J.; Sneen, R. A. *J. Am. Chem. Soc.* **1955**, *77*, 2505.
21. Bucourt, R. *Bull. Soc. Chim. Fr.* **1963**, 1262.
22. Johnson, F. *Chem. Rev.* **1968**, *68*, 375.
23. Bonneau, R.; Joussot-Dubien, J.; Salem, L.; Yarwood, A. J. *J. Am. Chem. Soc.* **1976**, *98*, 4329.
24. Goodman, J. L.; Peters, K. S.; Misawa, H.; Caldwell, R. A. *J. Am. Chem. Soc.* **1986**, *108*, 6803.
25. Aston, J. G. In *Determination of Organic Structures by Physical Methods*; Braude, E. A.; Nachod, F. C. Eds.; Academic Press: New York, 1955; Vol. 1, p. 515.
26. Bucourt, R.; Hainaut, D. *Bull. Soc. Chim. Fr.* **1967**, 4562.
27. Saebø, S.; Boggs, J. E. *J. Mol. Struct.* **1981**, *73*, 137.
28. Burke, L. A. *Theor. Chim. Acta* **1985**, *68*, 101.
29. Rondan, N. G.; Paddon-Row, M. N.; Caramella, P.; Houk, K. N. *J. Am. Chem. Soc.* **1981**, *103*, 2436.
30. Dorigo, A.; Pratt, D. W.; Houk, K. N. *J. Am. Chem. Soc.* **1987**, *109*, 6591.
31. Hoffmann, H. M. R.; Giguere, R. J.; Pauluth, D.; Hofer, E. *J. Org. Chem.* **1983**, *48*, 1155.
32. Veillard, A. In *Internal Rotation in Molecules*; Orville-Thomas, W. J. Ed.; Wiley: London, 1974; pp. 385–424.
33. Corey, E. J.; Feiner, N. F. *J. Org. Chem.* **1980**, *45*, 765.
34. Allinger, N. L.; Hirsch, J. A.; Miller, M. A.; Tyminski, I. J. *J. Am. Chem. Soc.* **1968**, *90*, 5773.
35. Allinger, N. L.; Sprague, J. T. *J. Am. Chem. Soc.* **1972**, *94*, 5734.
36. de la Mare, P. B. D.; Hall, D.; Pavitt, N. *J. Comput. Chem.* **1983**, *4*, 114.
37. Favini, G.; Buemi, G.; Raimondi, M. *J. Mol. Struct.* **1968**, *2*, 137.
38. Naumov, V. A.; Dashevskii, V. G.; Zaripov, N. M. *J. Struct. Chem. USSR* **1971**, *11*, 736.
39. Dashevsky, V. G.; Lugovskoy, A. A. *J. Mol. Struct.* **1972**, *12*, 39.
40. Anet, F. A. L.; Yavari, I. *Tetrahedron* **1978**, *34*, 2879.
41. Ermer, O. *Aspecte von Kraftfeldrechnungen*; W. Baur Verlag: München, 1981.
42. Chiang, J. F.; Bauer, S. H. *J. Am. Chem. Soc.* **1969**, *91*, 1898.
43. Geise, H. J.; Buys, H. R. *Recl. Trav. Chim. Pays-Bas* **1970**, *89*, 1147.
44. Pedone, C.; Benedetti, E.; Immirzi, A.; Allegra, G. *J. Am. Chem. Soc.* **1970**, *92*, 3549.
45. Holbrook, S. R.; van der Helm, D. *Acta Crystallogr., Sect. B* **1975**, *31*, 1689.
46. Thompson, J. S.; Whitney, J. F. *Acta Crystallogr., Sect. C* **1984**, *40*, 756.
47. Scharpen, L. H.; Wollrab, J. E.; Ames, D. P. *J. Chem. Phys.* **1968**, *49*, 2368.

48. Ermer, O.; Lifson, S. *J. Am. Chem. Soc.* **1973**, *95*, 4121.

49. Inoue, Y.; Ueoka, T.; Kuroda, T.; Hakushi, T. *J. Chem. Soc., Chem. Commun.* **1981**, 1031.

50. Pasternak, R. S. *Acta Cystallog.* **1951**, *4*, 316.

51. Lasheen, M. A. *Acta Crystallogr.* **1952**, *5*, 593.

52. Bastiansen, O.; Markali, J. *Acta Chem. Scand.* **1952**, *6*, 442.

53. Lindsey, J. M.; Barnes, W. H. *Acta Crystallogr.* **1955**, *8*, 227.

54. Taylor, R.; Kennard, O. *J. Chem. Inf. Comput. Sci.* **1986**, *26*, 28.

55. Schenk, H. *Cryst. Struct. Commun.* **1972**, *1*, 143.

56. Całdeki, Z.; Głowka, M. L. *Cryst. Struct. Commun.* **1978**, *7*, 317.

57. Bellucci, G.; Colapietro, M.; Ingrosso, G.; Spagna, R.; Zambonelli, L. *J. Chem. Res.* **1978**, 2417.

58. Viani, R.; Lapasset, J. *Acta Crystallogr., Sect. B* **1978**, *34*, 1195.

59. Brown, K. L.; Damm, L.; Dunitz, J. D.; Eschenmoser, A.; Hobi, R.; Kratky, C. *Helv. Chim. Acta* **1978**, *61*, 3108.

60. Naumov, V. A.; Bezzubov, V. M. *J. Struct. Chem.* **1967**, *8*, 466.

61. Bastiansen, O. *Acta Chem. Scand.* **1952**, *6*, 975.

62. Lord, R. C.; Lee, G. D.; Stanley, A. E.; Groner, P.; Li, Y. S.; Durig, J. R. *Spectrochim. Acta* **1985**, *41A*, 115.

63. Durig, J. R.; Carter, R. O.; Carreira, L. A. *J. Chem. Phys.* **1974**, *60*, 3098.

64. Manley, S. A.; Tyler, J. K. *J. Chem. Soc., Chem. Commun.* **1970**, 382.

65. Carreira, L. A.; Towns, T. G.; Malloy, T. B., Jr. *J. Chem. Phys.* **1979**, *70*, 2273.

66. Smithson, T. L.; Wieser, H. *J. Chem. Phys.* **1980**, *73*, 2518.

67. Brown, R. S.; Marcinko, R. W.; Tse, A. *Can. J. Chem.* **1979**, *57*, 1890.

68. Tronchet, J. M. J.; Xuan, T. N. *Carbohydr. Res.* **1978**, *67*, 469.

69. Jensen, F. R.; Bushweller, C. H. *J. Am. Chem. Soc.* **1969**, *91*, 5774.

70. Sakashita, K. *Nippon Kagaku Zasshi* **1960**, *81*, 49; *Chem. Abstr.* **1959**, *54*, 12015.

71. Sakashita, K. *Nippon Kagaku Zasshi* **1959**, *80*, 972; *Chem. Abstr.* **1959**, *54*, 2008.

72. Sakashita, K. J. *Chem. Soc. Jpn., Pure Chem. Sect.* **1953**, *74*, 315; *Chem. Abstr.* **1953**, *48*, 1087.

73. Anet, F. A. L.; Anet, R. In *Dynamic Nuclear Magnetic Resonance Spectroscopy*; Jackman, L. M.; Cotton, F. A. Eds.; Academic Press: New York, 1975; pp. 543–619.

74. Ōki, M. *Applications of Dynamic NMR Spectroscopy to Organic Chemistry*; VCH Publishers: Deerfield Beach, FL, 1985.

75. Sandström, J. *Dynamic NMR Spectroscopy*; Academic Press: London, 1982.

76. Lambert, J. B. *J. Am. Chem. Soc.* **1967**, *89*, 1836.

77. Buys, H. R. *Recl. Trav. Chim. Pays-Bas* **1969**, *88*, 1003.

78. Garbisch, E. W., Jr.; MacKay, K. D. Abstracts of the 155th National Meeting of the American Chemical Society, San Francisco, CA, March–April 1968, Abstract No. PO65.

79. der Hyde, W. A.; Lüttke, W. *Chem. Ber.* **1978**, *111*, 2384.

80. Garbisch, E. W., Jr. *J. Org. Chem.* **1962**, *27*, 4249.

81. Zefirov, N. S.; Chekulaeva, V. N.; Belozerov, A. I. *Tetrahedron* **1969**, *25*, 1997.

82. Zefirov, N. S.; Sergeev, N. M.; Chekulaeva, V. N.; Gurvich, L. G. *Proc. Acad. Sci. USSR (Engl. Transl.)* **1970**, *190*, 42.

83. Zefirov, N. S.; Victorova, N. M.; Knyazev, S. P.; Sergeyev, N. M. *Tetrahedron Lett.* **1972**, 1091.

84. Viktorova, N. M.; Knyazev, S. P.; Zefirov, N. S.; Gavrilov, Yu. D.; Nikollaev, G. M.; Bystrov, V. F. *Org. Magn. Reson.* **1974**, *6*, 236.

85. Kugatova-Shemyakina, G. P.; Ovchinnikov, Yu. A. *Tetrahedron* **1962**, *18*, 697.

86. Zefirov, N. S.; Samoshin, V. V.; Akhmetova, G. M. *J. Org. Chem. USSR* **1985**, *21*, 203.
87. Aycard, J.-P.; Bodot, H.; Garnier, R.; Lauricella, R.; Pouzard, G. *Org. Magn. Reson.* **1970**, *2*, 7.
88. Aycard, J.-P.; Geuss, R.; Berger, J.; Bodot, H. *Org. Magn. Reson.* **1973**, *5*, 473.
89. Aycard, J.-P.; Bodot, H. *Can. J. Chem.* **1973**, *51*, 741.
90. Aycard, J.-P.; Bodot, H. *Org. Magn. Reson.* **1975**, *7*, 226.
91. Lessard, J.; Tan, P. V. M.; Martino, R.; Saunders, J. K. *Can. J. Chem.* **1977**, *55*, 1015.
92. Ouédargo, A.; Viet, M. T. P.; Saunders, J. K.; Lessard, J. *Can. J. Chem.* **1987**, *65*, 1761.
93. Anet, F. A. L.; Haq, M. Z. *J. Am. Chem. Soc.* **1965**, *87*, 3147.
94. Jensen, F. R.; Bushweller, C. H. *J. Am. Chem. Soc.* **1965**, *87*, 3285.
95. Walton, J. C. *J. Chem. Soc., Perkin Trans. Q.* **1986**, 1641.
96. Bernard, M.; St.-Jacques, M. *Tetrahedron* **1973**, *29*, 2539.
97. Anderson, J. E.; Roberts, J. D. *J. Am. Chem. Soc.* **1970**, *92*, 97.
98. Russell, G. A.; Underwood, G. R.; Lini, D. C. *J. Am. Chem. Soc.* **1967**, *89*, 6636.
99. Peters, H.; Archer, R. A.; Mosher, H. S. *J. Org. Chem.* **1967**, *32*, 1382.
100. Jensen, F. R.; Bushweller, H. C. *Adv. Alicyclic Chem.* **1971**, *3*, 139.
101. Lambert, J. B.; Marko, D. E. *J. Am. Chem. Soc.* **1985**, *107*, 7978.
102. Lambert, J. B.; Clikeman, R. R.; Taba, K. M.; Marko, D. E.; Bosch, R. J.; Xue, L. *Acc. Chem. Res.* **1987**, *20*, 454.
103. Rickborn, B.; Lwo, S.-Y. *J. Org. Chem.* **1965**, *30*, 2212.
104. Bucourt, R. *Bull. Soc. Chim. Fr.* **1964**, 2080.
105. Berkoz, B.; Chavez, E. P.; Djerassi, C. *J. Chem. Soc.* **1962**, 1323.
106. Phillips, S. E. V.; Trotter, J. *Acta Crystallogr., Sect. B* **1977**, *33*, 984.
107. Phillips, S. E. V.; Trotter, J. *Acta Crystallogr., Sect. B* **1977**, *33*, 996.
108. Phillips, S. E. V.; Trotter, J. *Acta Crystallogr., Sect. B* **1977**, *33*, 991.
109. Godfrey, J. E.; Waters, J. M. *Cryst. Struct. Commun.* **1975**, *4*, 45.
110. Burton, G. W.; de la Mare, P. B. D.; Sibley, L. D.; Waters, J. M. *J. Chem. Res. (S)* **1980**, 132.
111. Kaftory, M.; Weisz, A. *Acta. Crystallogr., Sect. C* **1984**, *40*, 456.
112. Prakash, O.; Roy, R.; Kulshreshtha, D. K. *Magn. Reson. Chem.* **1988**, *26*, 47.
113. de la Mare, P. B. D.; Johnson, M. D.; Lomas, J. S.; Sanchez del Olmo, V. *Chem. Commun.* **1965**, 483.
114. de la Mare, P. B. D.; Johnson, M. D.; Lomas, J. S.; Sanchez del Olmo, V. *J. Chem. Soc. B* **1966**, 827.
115. de la Mare, P. B. D.; Koenisgsberger, R.; Lomas, J. S. *J. Chem. Soc. B* **1966**, 834.
116. Russell, G. A.; Law, W. C. *Heterocycles* **1986**, *24*, 321.
117. Gerson, F.; Martin, W. B. *Helv. Chim. Acta* **1987**, *70*, 1558.
118. McDowell, C. A.; Naito, A.; Scheffer, J. R.; Wong, Y.-F. *Tetrahedron Lett.* **1981**, *22*, 4779.
119. Ariel, S.; Scheffer, J. R.; Trotter, J.; Wong, Y.-F. *Tetrahedron Lett.* **1983**, *24*, 4555.
120. Gilbao, H.; Altman, J.; Loewenstein, A. *J. Am. Chem. Soc.* **1969**, *91*, 6062.
121. Zefirov, N. S.; Samoshin, V. V.; Zemlyanova, T. G. *Tetrahedron Lett.* **1983**, *24*, 5133.
122. Zefirov, N. S. *Tetrahedron Lett.* **1975**, 1087.
123. Harada, N.; Iwabuchi, J.; Yokota, Y.; Uda, H.; Nakanishi, K. *J. Am. Chem. Soc.* **1981**, *103*, 5590.
124. Koreeda, M.; Harada, N.; Nakanishi, K. *J. Am. Chem. Soc.* **1974**, *96*, 266.
125. Appel, W. K.; Jiang, Z. Q.; Scheffer, J. R.; Walsh, L. *J. Am. Chem. Soc.* **1983**, *105*, 5354.
126. House, H. O.; Tefertiller, B. A.; Olmstead, H. D. *J. Org. Chem.* **1968**, *33*, 935.

127. Bushweller, C. H.; O'Neil, J. W. *Tetrahedron Lett.* **1969**, 4713.

128. Achmatowicz, O., Jr.; Jurczak, J.; Konowal, A.; Zamojski, A. *Org. Magn. Reson.* **1970**, *2*, 55.

129. Garcia, D.; Grunwald, E. *J. Am. Chem. Soc.* **1980**, *102*, 6407.

130. Anderson, G. M., III; Kollman, P. A.; Domelsmith, L. N.; Houk, K. N. *J. Am. Chem. Soc.* **1979**, *101*, 2344.

131. Philip, T.; Cook, R. L.; Malloy, T. B., Jr.; Allinger, N. L.; Chang, S.; Yuh, Y. *J. Am. Chem. Soc.* **1981**, *103*, 2151.

132. Thomas, S. A. *J. Cryst. Spectr. Res.* **1985**, *15*, 115.

133. Warshel, A.; Levitt, M.; Lifson, S. *J. Mol. Spectrosc.* **1970**, *33*, 84.

134. Todeschini, R.; Pitea, D.; Favini, G. *J. Mol. Struct.* **1981**, *71*, 279.

135. Domiano, P.; Macchia, F. *Acta. Crystallogr., Sect. B* **1980**, *36*, 3067.

136. Naumov, V. A.; Bezzubov, V. M. *Proc. Acad. Sci. USSR, Chem. Sect. (Engl. Transl.)* **1970**, *193*, 477.

137. Boyd, R. H.; Kesner, L. *J. Chem. Phys.* **1980**, *72*, 2179.

2

Conformational Analysis of Six-Membered Carbocyclic Rings with Exocyclic Double Bonds

Joseph B. Lambert

2.1 CONFORMATIONAL EFFECTS OF THE EXOCYCLIC DOUBLE BOND ON THE RING

Introduction of an exocyclic double bond on a six-membered ring alters both the static and the dynamic properties of the ring. These properties have been studied in solution by nuclear magnetic resonance (NMR) spectroscopy. Static properties also have been studied in the solid state by X-ray diffraction. Vibrational spectroscopy has provided information about the overall symmetry of these molecules.

Cyclohexanone (**1**) and methylenecyclohexane (**2**) are parent molecules that possess a single exocyclic double bond. The simplest such molecule for which an X-ray structure is available is 4,4-diphenylcyclohexanone (**3**).[1] The conformational

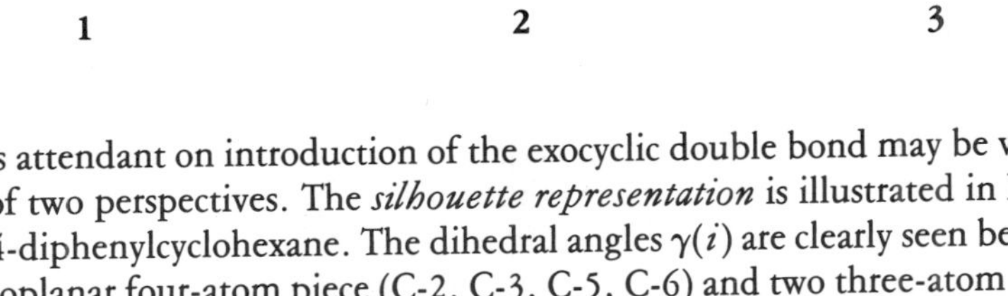

1 **2** **3**

distortions attendant on introduction of the exocyclic double bond may be viewed in either of two perspectives. The *silhouette representation* is illustrated in Figure 2.1 for 4,4-diphenylcyclohexane. The dihedral angles $\gamma(i)$ are clearly seen between a central coplanar four-atom piece (C-2, C-3, C-5, C-6) and two three-atom pieces (either C-6, C-1, C-2 or C-3, C-4, C-5). The variable i in $\gamma(i)$ refers to the middle atom of the three-atom piece. In Figure 2.1, $\gamma(1)$ is seen to be $38 \pm 2°$ and $\gamma(4)$ to be $52 \pm 2°$. These numbers may be compared with that in the parent

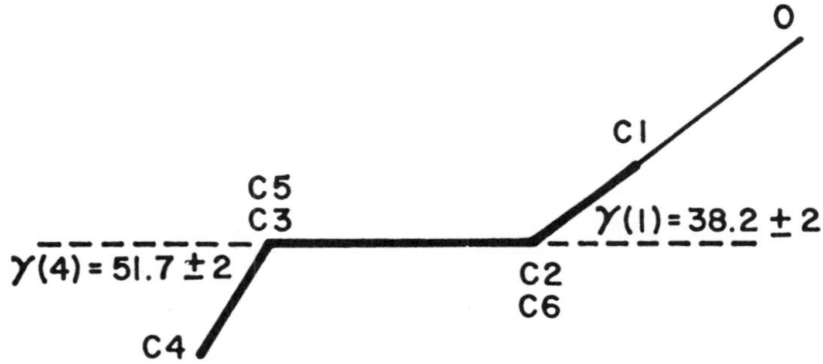

Figure 2.1 A silhouette representation of 4,4-diphenylcyclohexanone: projection onto the plane passing through C-1 and C-4 and perpendicular to the plane of C-2—C-3—C-5—C-6. Reprinted with permission from J. B. Lambert, R. E. Carhart, and P. W. R. Corfield, *J. Am. Chem. Soc.* **1969**, *91*, 3569. Copyright © 1969 American Chemical Society.

cyclohexane, 49°.[1] Values smaller than 49° indicate flattening of the ring, and larger values indicate puckering. If cyclohexane had perfectly tetrahedral carbon centers, its γ would be 54.5°. Opening of the C—C—C angles from the tetrahedral 109.5° to 111.5°, however, causes the ring to be slightly flattened.[2] In 4,4-diphenylcyclohexanone, the remaining silhouette angles are 50 ± 4° for $\gamma(3, 5)$ and 42 ± 4° for $\gamma(2, 6)$.[1] This molecule thus shows pronounced flattening around the C-6, C-1, C-2 portion of the ring, caused by opening of the C—C—C angle from 111.5° in cyclohexane to 116.8° in **3**. The remainder of the ring is relatively undistorted from the shape of the cyclohexane.

The silhouette representation is useful only if the four basal atoms are coplanar. If the opposing bonds are skewed, $\gamma(i)$ may still be calculated from a least-squares plane, but the concept of interplanar angles becomes compromised. As additional substituents, heteroatoms, or sp^2 atoms are added to a ring, the likelihood that four atoms are coplanar decreases. Consequently, in the general case, the conformation of portions of a six-membered ring is better described by the *dihedral angle representation*.[1] Any four consecutive atoms, C(i − 1)—C(i)—C(i + 1)—C(i + 2) define a dihedral angle $\phi(i, i + 1)$. The value in cyclohexane is 54.5° (not the 60° expected for tetrahedral symmetry). Values larger than 54.5° in other systems indicate puckering in that portion of the ring, and smaller values indicate flattening. In 4,4-diphenylcyclohexanone, $\phi(1, 2)$ and $\phi(6, 1)$ have an average value of 42° because of the presence of the sp^2 carbon atom; $\phi(2, 3)$ and $\phi(5, 6)$ have an average value of about 52° and $\phi(3, 4)$ and $\phi(4, 5)$ of about 59°, indicative of slight flattening and puckering, respectively.[1]

The structure of 4,4-diphenylcyclohexanone has been described here in detail as an illustration of the basic distortion brought about by introduction of a single sp^2 atom in a carbocyclic six-membered ring. The region of the ring around the sp^2 atom is considerably flattened. The remainder of the ring is relatively undistorted (compared with cyclohexane), although there may be a slight puckering at the end of the ring opposite to the sp^2 atom that is a reflex to the flattening distortion and presumably reduces strain throughout the ring.

In solution, the primary tool for obtaining quantitative information about the shape of a six-membered ring is the *R* value method and other Karplus-derived NMR methods.[3,4] Any —CH_2—CH_2— segment in a ring should exhibit both a trans and a cis vicinal coupling constant in the ^1H spectrum. The ratio of these coupling constants ($J_{trans}/J_{cis} = R$) is independent of substituent effects and hence depends entirely on the ring geometry. The internal dihedral angle Ψ of any XCH_2CH_2Y portion of the ring then is given by Equation (2.1).[3]

$$\cos \Psi = \left(\frac{3}{2+4R} \right)^{1/2} \qquad [2.1]$$

The symbol Ψ is used for this dihedral angle to indicate that it is the angle inside the ring, **4**. The H—H dihedral angles may be calculated from Ψ.[4] The internal

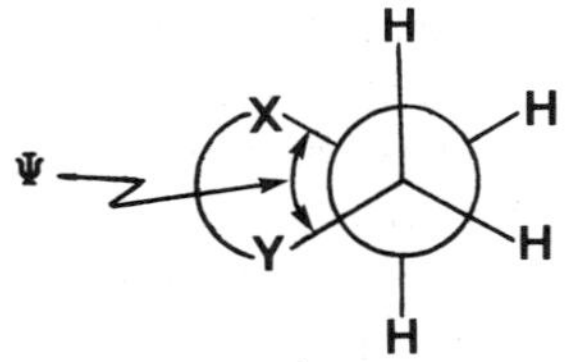

4

dihedral angle for the 2–3 segment of a number of six-membered rings containing a single sp^2 atom has been measured, including that for cyclohexanone (54°),[4] 4,4-diphenylcyclohexanone (53°),[4] and methylenecyclohexane (56°).[5] As seen from the X-ray structure, the distortion in the 2–3 segment is much reduced from that in the 6–1–2 segment. The NMR results similarly showed only slight flattening in this region of the ketone. Comparison of numerous *R*-value-based dihedral angles with those from other structural methods (primarily X-ray) indicates an average variance in Ψ between the two methods of only 1.4°.[6]

Introduction of a second sp^2 center, as in **5** and **6**, should increase flattening

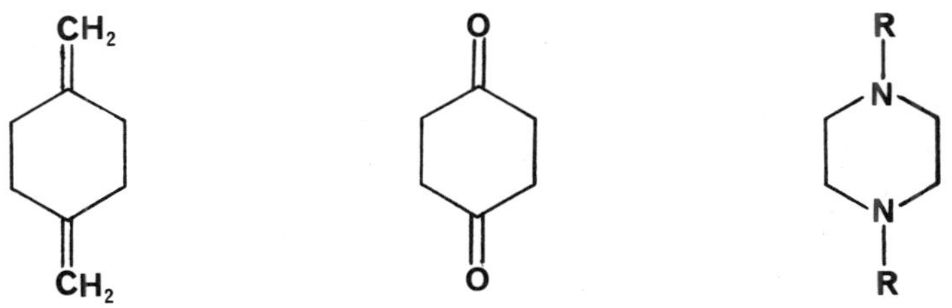

5 **6** **7**

of the ring. The dihedral angles for a number of such molecules have been measured: 1,4-dimethylenecyclohexane (**5**, 51°),[4] 1,4-cyclohexanedione (**6**, 50°),[4] 1,4-cyclohexanedione dioxime (47°),[4] *anti-N,N*-dinitrosopiperazine (**7**, R = NO, 52°),[4,7] and *anti-N,N'*-dicarbomethoxypiperazine (**7**, R = CO_2Me, 55°).[8] The piperazines are least flattened, presumably because the nitrogen center is not fully trigonal. The diene **5**, the dione **6**, and the dioxime show increasing amounts of flattening.

As flattening increases, eclipsing interactions within the —CH_2CH_2— portions of the molecule increase to the point that the chair form becomes less stable than the twist-boat form. 1,4-Cyclohexanedione, in fact, exists as a twist-boat in the solid.[9] The *R* value detects deformations in portions of a ring in solution, but it cannot provide a perspective of the entire ring that could distinguish between chair and twist forms. The selection rules of vibrational spectroscopy provide such a perspective. The chair form is centrosymmetric and should show no coincidences between the infrared and Raman spectra. The twist form is noncentrosymmetric and should show only coincident bands. In this fashion, it was demonstrated that in solution 1,4-cyclohexanedione is in the twist form but 1,4-dimethylenecyclohexane is in the chair form.[10]

These deformations have an important influence on the dynamics of chair–chair ring interconversion. There may be at least two causes of these effects. First, the flatter the ring becomes, the closer it is to the transition state to chair–chair ring reversal. Second, the barrier to ring reversal is related to the barrier to rotation about the individual bonds of the ring.[11] The sp^2–sp^3 bond of C-1—C-2,6 for either **1** or **2** rotates more rapidly than any sp^3–sp^3 bond of cyclohexane. The effect is larger in the ketone **2** than in the hydrocarbon **1**, and it is larger in rings with two exocyclic double bonds than in rings with only one. Thus, the free energy of activation to ring reversal in cyclohexanone is 4.1 kcal/mol at its coalescence temperature near −180°C,[12] and that in methylenecyclohexane is 8.4 kcal/mol,[5] both lower than that in cyclohexane, 10.8 kcal/mol.[13] In 1,4-dimethylenecyclohexane, with two sp^2 centers, the barrier is 7.5 kcal/mol.[14] 1,4-Cyclohexanedione (**6**) is in a twist-boat conformation, which pseudorotates rapidly between mirror images and is not available for dynamic NMR measurements.

Thus, the major conformational effects result from the opening of the bond angle at the sp^2 center and from the faster bond rotation around the sp^2–sp^3 bond. Variation or accumulation of these two factors can explain most of the conformational statics and dynamics in these exocyclic systems.

2.2 CONFORMATIONAL EFFECTS OF THE EXOCYCLIC DOUBLE BOND ON ADJACENT SUBSTITUENTS

In a typical open chain system, bonds to an sp^3 center have no eclipsing interactions with bonds to an adjacent sp^3 center, because of their staggered arrangement. In contrast, an sp^2 center in a six-membered ring has an eclipsing interaction between the exocyclic double bond and the adjacent equatorial bond (**8**).[6] This interaction was first recognized by Robins and Walker,[15,16] and the phenomenon was referred to later as the 2-alkyl ketone effect.[16,17] The C=O/C—R interaction should destabilize the equatorial conformer, thereby increasing the proportion of the axial conformer, which has the weaker C=O/C—H interaction.

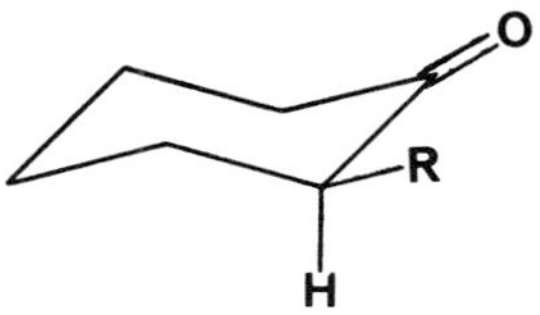

8a

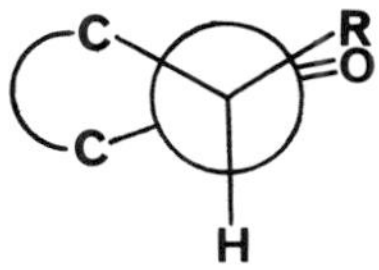

8b

The eclipsing interaction between single and double bonds is not necessarily repulsive, but the nonbonded interactions between the exocyclic oxygen and the atoms of the equatorial 2 substituent should enhance any repulsion. The distance between the carbonyl oxygen and an equatorial 2-methyl group, however, is too large to give rise to a palpable effect. Equilibration experiments showed the 2-methyl ketone effect to be negligible.[16] Larger 2 substituents, however, do exhibit such an effect. A 2-ethyl group destabilizes the equatorial conformer by about 0.7 kcal/mol and a 2-isopropyl group by about 1.7 kcal/mol.[16]

Because the major effect of an exocyclic double bond on the adjacent equatorial substituent probably derives from nonbonded interactions, the repulsion should be exaggerated when the double bond is alkenic rather than ketonic. Malhotra and Johnson[18,19] recognized this effect in 1965 and referred to it as allylic 1,3 or $A^{(1,3)}$ strain. For the general case of Equation (2.2), these authors

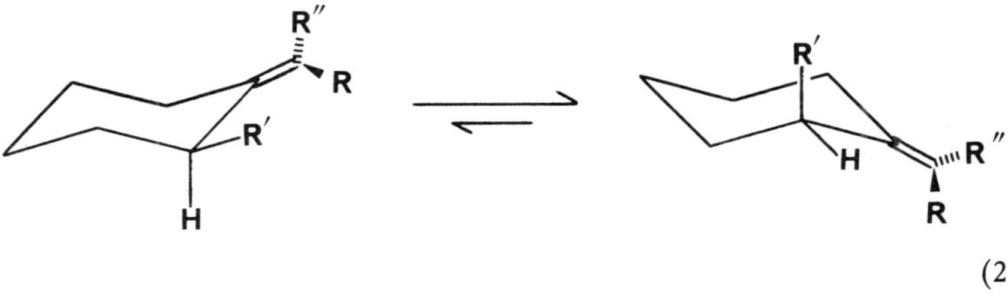

(2.2)

stated that when R and R′ are of moderate size, there is a substantial interference that results in an increase in the proportion of the 2-axial conformer. For R = R′ = CH_3, they estimated that the interaction was similar to that between two peri methyls in naphthalene, or about 7.5 kcal/mol, larger than that between 1,3-diaxial methyls. When R is H and R′ is methyl, however, the effect is much smaller.

Johnson's review[19] contains an impressive collection of examples of allylic 1,3 strain, of which he considers the 2-alkyl ketone effect to be a derivative. In an example of the exocyclic carbon–carbon bond, **9**, the axial-methyl conformer is clearly favored when the carboxyl group is syn to the 2 substituent.[20] When carboxyl is anti (not shown), the molecule prefers the equatorial form. The interaction of methyl with carboxyl is sufficient to destabilize the equatorial form, but the methyl–hydrogen interaction has little effect.

The carbon–carbon bond of the enolate ion **10** presents a similar situation.

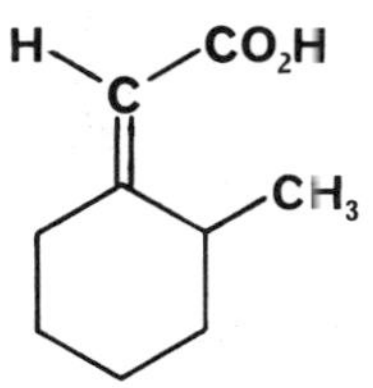

9

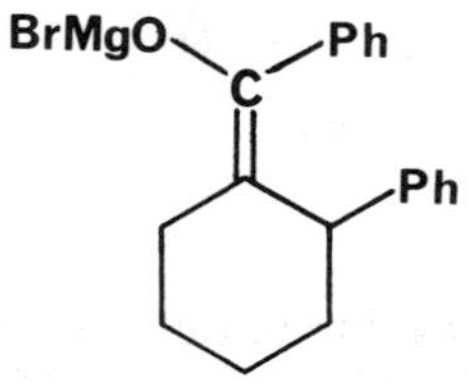

10

By NMR, the 2-phenyl group prefers the axial conformation.[18] This preference occurs in both the isomer shown and in the isomer in which the phenyl group on the double bond is anti to the 2-phenyl ring. Thus, both phenyl and the oxide are sufficiently large to destabilize the equatorial 2-phenyl conformation.

Exocyclic carbon–oxygen and carbon–nitrogen double bonds cause similar effects. In 97.7% sulfuric acid, 2,6-dimethylcyclohexanone (**11**) has 44% of the trans (axial/equatorial) isomer present (shown), compared with 8% in dilute acidic or in basic solution (the remainder is the cis or diequatorial isomer).[21] The presence of the proton on oxygen destabilizes the equatorial syn methyl group in the cis isomer, although the effect is not large. The *aci*-nitro compound **12** exists

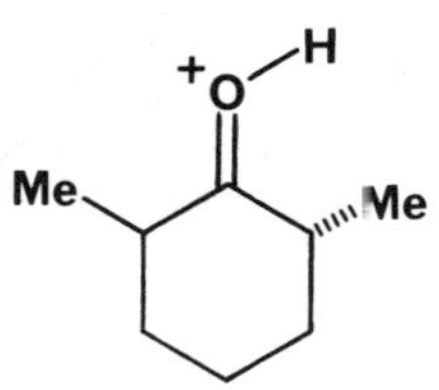

11

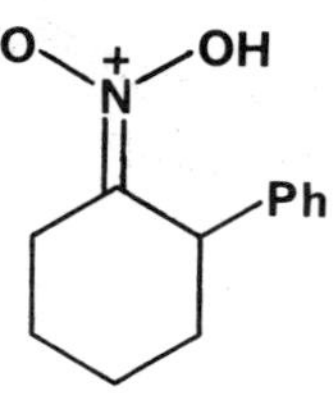

12

predominantly with the phenyl group axial, because of the allylic 1,3 interaction between the oxygens and the phenyl.[18]

The nitrogen–nitrogen bond in nitrosamines has a large degree of double-bond character. Moreover, the oxygen atom can serve as the R substituent of Equation (2.2). In a particularly extreme case, the trans, syn, trans tetramethyl derivative of *N,N*-dinitrosopiperazine (**13**) prefers the all-axial conformer to the all-equatorial conformer.[22]

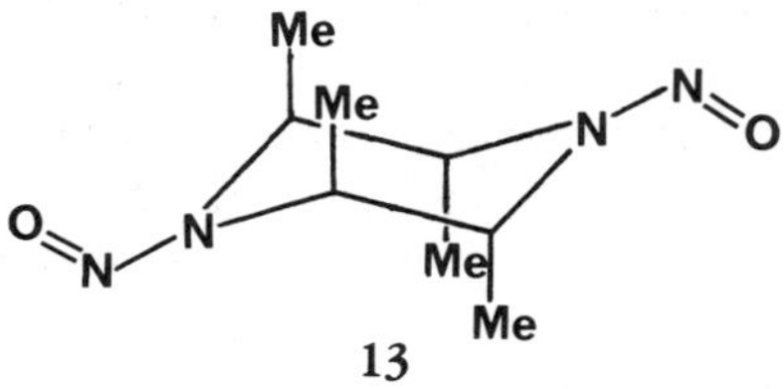

13

These selected examples demonstrate the generality of the allylic 1,3 effect in six-membered rings. Substituents on an exocyclic double bond interact repulsively with equatorial substituents at the 2 position. Depending on the size of the group on the double bond that is syn to the substituted adjacent position [R in Equation (2.2)], the 2-equatorial conformation is destabilized and substantial amounts of the axial conformer may be observed.

When the substituent is polar, there is the additional factor of possible electrostatic interactions. The situation is reasonably clear in 2-halocyclohexanones,[23] since the relative polarity of the axial and equatorial conformations is easily discernible [Equation (2.3)].

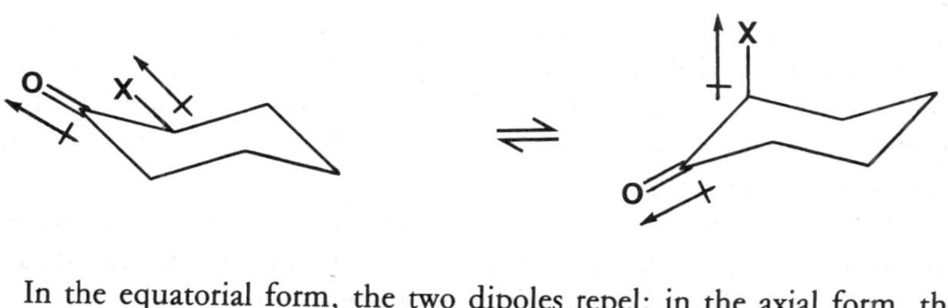

(2.3)

In the equatorial form, the two dipoles repel; in the axial form, the two dipoles cancel to some extent. Thus, the more polar equatorial form should be disfavored not only by the steric (allylic 1,3) effect but also by electrostatic repulsion. More polar solvents, however, would increasingly favor the equatorial form and reduce the amount of the axial form.

Experimental observations based on dipole moment measurements generally corroborate this line of reasoning.[23] For 2-bromocyclohexanone, there is about 85% of the axial conformer in heptane but 62% in 1,4-dioxane. For 2-chlorocyclohexanone, there is about 76% of the axial conformer in heptane but 37% in 1,4-dioxane. Dipole moment measurements unfortunately are not quantitatively reliable, but interpretation of the qualitative trends should be valid. Bromine should have a smaller electrostatic effect than chlorine but a larger steric effect. Both atoms clearly have a much destabilized equatorial conformation. In 2-fluorocyclohexanone, there is 48% of the axial conformer in heptane and 15% in 1,4-dioxane. The reduced amount of axial presumably is caused in part by the smaller size of fluorine, but it might have been thought that the larger polarity of the C—F bond could have led to a reduced equatorial population. For the case of fluorine, however, there may be an additional orbital interaction.[24] As there is a variety of such possibilities, the C—F/C=O interaction in the equatorial form of 2-fluorocyclohexanone requires further clarification.

Later studies of 2-methoxycyclohexanone discussed these conformational preferences in terms of the generalized anomeric effect, which is comprised not only of dipole–dipole interactions but also hyperconjugative interactions, the exact mix of which is not clear.[25] The σ bond of the 2-axial substituent may donate electrons to either the σ^* or the π^* orbital of the carbonyl group ($\sigma_{C-X} \rightarrow \sigma^*_{C=O}$ or $\sigma_{C-X} \rightarrow \pi^*_{C=O}$), or alternatively a lone pair on X may donate electrons to the π^* orbital. The hyperconjugative interaction, which may be depicted as **14**, occurs

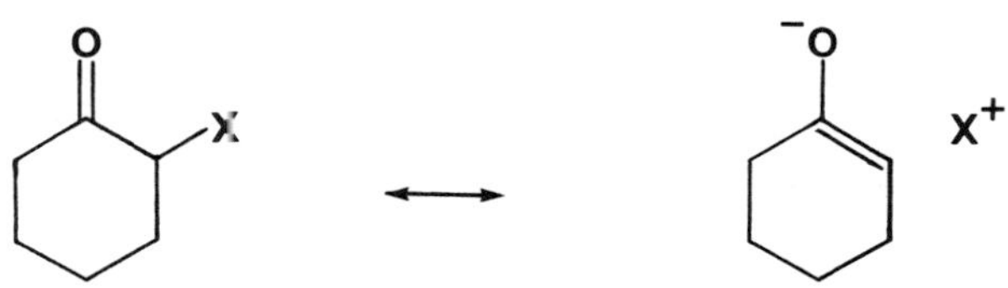

14a **14b**

optimally in the axial conformation because of favorable orbital arrangements but is stereoelectronically disfavored in the equatorial conformation. Thus, for example, 2-methoxycyclohexanone is 63% axial. Interpretation for the 2-halocyclohexanones[23] needs to be revised along these lines, although dipolar interactions should not be discounted.

Conformational interactions between polar 2 substituents and an exomethylene group have been studied extensively by the groups of Saunders and Zefirov.[26–30] At least three distinct effects have been discerned.

1. The allylic 1,3 effect has strictly a steric source and reflects the nondipolar interaction between the syn-exo group and the 2-equatorial group, R and R′ in Equation (2.2). For example, 2-methyl-1-(methylene)cyclohexane (**15**, X = CH_3, Y = H) has a free-energy difference for the 2-methyl group of 1.0 ± 0.2 kcal/mol favoring the equatorial position, compared with about 1.6 for methylcyclohexane. Thus, the $A^{(1,3)}$ effect (the decrease in the equatorial preference) in the case of H/CH_3 [R/R′ in Equation (2.2)] is about 0.6 kcal/mol. Zefirov and Barenenkov[29] have found the following 1,3 allylic effects (R/R′): CN/CH_3, 2.1 kcal/mol; CN/Br, 1.2 kcal/mol; and CN/O(CO)CH_3, 0.3 kcal/mol.

2. Hyperconjugative donation of electrons occurs from the exocyclic double bond to the σ^* orbital of the 2-axial C—X bond ($\pi_{C=C} \rightarrow \sigma^*_{C-X}$), as shown in **16a**. Hyperconjugation in **16** moves in the opposite sense to that in **14**. For 2-methoxy-1-(methylene)cyclohexane (**15**, X = OCH_3, Y = H), the conformational

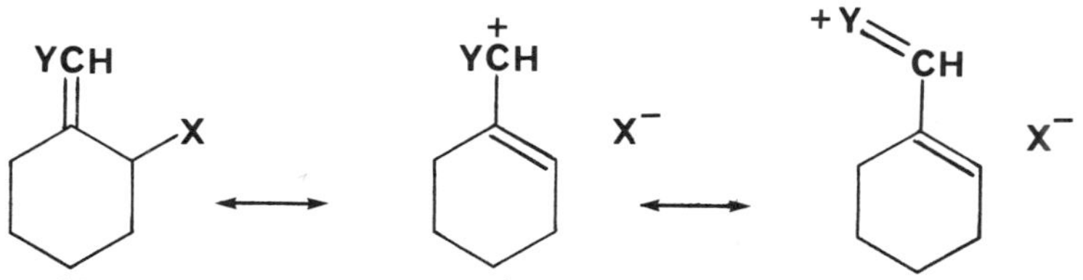

15 16a 16b

preference is 0.4 kcal/mol in favor of the axial form, compared with about 0.6 in favor of the equatorial form for methoxycyclohexane. This difference of about 1.0 kcal/mol is composed of possibly 0.3 kcal/mol of the $A^{(1,3)}$ steric effect and 0.7 kcal/mol of the hyperconjugative effect. Presence of an electron-donating substituent on the exomethylene carbon enhances hyperconjugation, so that *anti*-2,7-dimethoxy-1-(methylene)cyclohexane (**15**, X,Y = OCH_3) exists entirely in the axial conformation.[26] Electron-withdrawing substituents such as cyano (**15**, Y = CN) decrease the hyperconjugative effect and increase the equatorial proportion of the polar 2 substituent.[(28,29)]

3. It would seem that a more strongly electron-withdrawing 2 substituent (X in **15** and **16**) would enhance the hyperconjugative effect and decrease the amount of the 2-equatorial conformer. Such appears not to be the case for unsaturated 2-substituents, including X = vinyloxy, phenoxy, and acetoxy.[28–30] These substituents actually favor the equatorial position (61, 74, and 80%, respectively). Saunders has referred to this as the unsaturated effect and assigned it to an unknown source or possibly to electrostatic phenomena arising from reorientation of the side chain. The normal dipole–dipole effect should disfavor the equatorial form [see Equation (2.3)], so this effect would require an unusual arrangement of dipoles. The effect may be due to an unusual orbital effect,[24] as suggested above for 2-fluorocyclohexanone. Although the electrostatic effect was considered to be very important for cyclohexanones carrying polar 2 substituents, its role in the less polar exomethylenecyclohexanes is still not clear.

2.3 CONFORMATIONAL EFFECTS OF THE EXOCYCLIC DOUBLE BOND ON NONADJACENT SUBSTITUENTS

In monosubstituted cyclohexanes, the major determinant of the axial/equatorial equilibrium constant is the repulsive interaction between the axial substituent and the syn-axial hydrogens at the 3 and 5 positions. Robins and Walker[31] and Klyne[32] also recognized that conformational preferences could be perturbed for substituents at the 3 position of cyclohexanones. Removal of the syn-axial proton should permit a larger proportion of axial substituent, a phenomenon that was termed the 3-alkyl ketone effect [Equation (2.4), M = O]. As a first approximation, the conformational preference of about 1.6 kcal/mol for methyl should be halved, to 0.8 kcal/mol, since it would seem that half of the

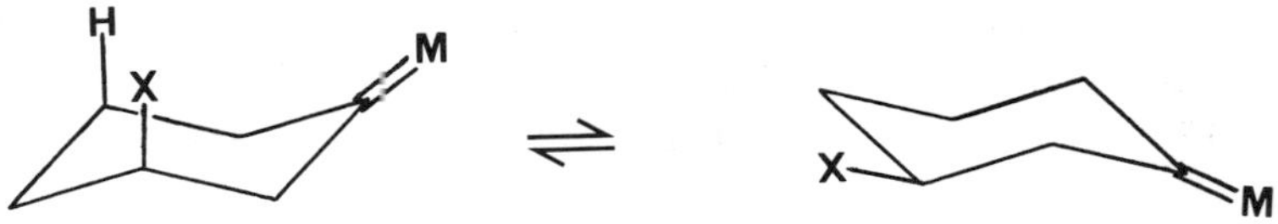

(2.4)

repulsive interactions have been removed. Closer analysis by Allinger and co-workers, however,[16] showed that the reduction in the free-energy difference may be only 0.5 rather than 0.8 kcal/mol.

Because of difficulties in carrying out the equilibration experiments and because the coalescence temperature of cyclohexanones is extremely low, experiments on the 3-alkyl ketone effect were never extended to other alkyl groups and only recently to polar groups.[33] The axial conformer population is enhanced for oxygen and fluorine substituents but decreased for less electronegative atoms such as chlorine, bromine, and sulfur. The analogous interaction between 3 substituents and an exomethylene group [Equation (2.4), M = CH_2], however, has been thoroughly explored by Lambert and co-workers. They chose to study 3-substituted methylenecyclohexanes by the direct NMR method, because these systems have a convenient coalescence temperature around −100 °C. Harsh conditions required for chemical equilibration can thus be avoided. The lower polarity of the exomethylene group compared with the ketonic group makes the mechanism of interaction between the exocyclic group and the substituent less clear.

Table 2.1 contains the free-energy differences measured for several substituents in the exomethylene system [Equation (2.4), M = CH_2].[34,35] The effect of exomethylene on 3-methyl closely resembles the expectations from the 3-alkyl ketone effect. In both CF_2Cl_2 and $CHFCl_2$, there is much more axial methyl, and the free-energy difference is approximately half that for methylcyclohexane. Thus, replacement of a ring CH_2 group by $C{=}CH_2$ substantially reduces the repulsive interactions experienced by the axial 3-methyl conformer and permits a higher proportion of the axial conformer.

The remaining substituents in Table 2.1 are polar and generally have decreased proportions of the axial conformer in the noninteracting solvent CF_2Cl_2, contrary to the expectations from the 3-alkyl ketone effect. Lambert and co-workers have described the repulsion between the exomethylene double bond and polar 3-axial substituents in terms of a classical dipole–dipole interaction. They make the following arguments to prove this point.

1. An interaction between induced dipoles (van der Waals repulsion) seems unlikely because of the large distance (>3 Å) between a 3-axial C—X bond and

Table 2.1 Free Energy Differences for 3-Substituted Methylenecyclohexanes[a]

Substituent (X)	A^b	$-\Delta G°(CF_2Cl_2)$ (kcal/mol)	$-\Delta G°(CHFCl_2)$ (kcal/mol)
CD_3	1.6	0.80 ± 0.10	0.70 ± 0.15
OH	0.61	1.12 ± 0.04	0.69 ± 0.03
OAc[c]	0.71	0.61 ± 0.03	0.38 ± 0.02
OMs[c]	0.56		0.39 ± 0.02
OBs[c]			0.36 ± 0.05
OTs[c]	0.52		0.44 ± 0.05
OPNB[c]			0.48 ± 0.05
OCD_3	0.55	0.80 ± 0.02	0.11 ± 0.01
SCH_3	1.07	1.22 ± 0.02	0.65 ± 0.04
SC_6H_5			0.36 ± 0.02
CN	0.24	0.26	0.07

[a] Data from Refs. 33 and 34.
[b] The A value in kcal/mol: $-\Delta G°$ for monosubstituted cyclohexanes, in CCl_4, CS_2, or $CS_2/CDCl_3$; see footnote *c* in Table I of Ref. 33.
[c] Symbols respectively for acetoxy, mesyloxy, brosyloxy, tosyloxy, and *p*-nitrobenzoyloxy.

the exocyclic C=C bond. Moreover, the effect does not increase when OCD_3 is replaced with SCH_3, which is more polarizable and has a larger van der Waals radius.

2. The effect appears to be related to the size of the charge on the axial oxygen in the X = OR series [Equation (2.4)]. Hydroxyl, with the highest negative charge density, has the largest $-\Delta G°$, followed by OCD_3, in the low-polarity solvent CF_2Cl_2. Ester resonance in acetoxyl considerably reduces the charge density. For this substituent, there is slightly less axial conformer in the exomethylene compound than in the monosubstituted cyclohexane. The order of effective substituent size as measured by $-\Delta G°$ of the 3-substituted methylenecyclohexanes in CF_2Cl_2 is thus OAc < OCD_3 < OH, which follows the order of charge on the substituent, whereas the order in monosubstituted cyclohexanes is OCH_3 < OH < OAc, which approximately follows the order of substituent size.

3. All the above comparisons used data from the low-polarity, non-hydrogen-bonding solvent CF_2Cl_2 in Table 2.1. In this solvent intramolecular interactions are maximized. If the equilibrium constant of the 3-substituted methylenecyclohexanes [Equation (2.4), M = CH_2] is determined primarily by repulsive dipolar interactions between the C—X and C=CH_2 bonds in the axial conformer, it should be sensitive to the polarity of the solvent. The dipoles in the axial conformer reinforce each other, so that the axial conformer is more polar than the equatorial, in which the dipoles are opposed. Therefore, a change to a more polar solvent should shift the equilibrium toward the more polar axial form. The data in Table 2.1 substantiate this argument: every polar substituent

including acetoxyl has a lower $-\Delta G°$ (higher proportion of axial) in the more polar, hydrogen-bonding $CHFCl_2$. The value for the nonpolar methyl group, however, is insensitive to the polarity of the solvent. This behavior is consistent with a dipole–dipole interaction for all the polar substituents. In contrast, the two conformers of monosubstituted cyclohexanes have essentially the same dipole moments, so that their equilibrium is nearly insensitive to the polarity of solvent.[36]

4. The dipole of the $C{=}CH_2$ bond is small but nonzero (the dipole moment of a model, isobutylene, is 0.50 D). The double bond in an isopropylidene system [Equation (2.4), M = CMe_2], however, should have essentially no dipole, since it is a nearly symmetrical, tetrasubstituted double bond. The prediction for this system is that there should be considerably more axial conformer. This hypothesis was tested by Lambert and Taba[37] in 3-methoxy-1-(isopropylidene)cyclohexane [Equation (2.4), X = OCD_3, M = CMe_2]. In this compound, $-\Delta G°$ was found to be 0.19 kcal/mol in CF_2Cl_2, compared with 0.80 in the methylenecyclohexane [Equation (2.4), X = OCD_3, M = CH_2]. Thus, removal of the dipole of the double bond indeed substantially reduces the amount of axial conformer, in agreement with a dipole–dipole repulsion in the $C{=}CH_2$ system. The experiment with the $C{=}CMe_2$ system also provides information on the possibility of an interaction between the dipole of the C—X bond and the quadrupole of the double bond. Whereas the change from $C{=}CH_2$ to $C{=}CMe_2$ alters the dipole of the bond considerably, it has little effect on the quadrupole of the bond. The observation of a substantial reduction in the amount of axial conformer when exomethylene was replaced by isopropylidene therefore is consistent with a dipole–dipole rather than a dipole–quadrupole interaction in the exomethylene system.

The cyano figure in Table 2.1 is particularly interesting.[35] By the *A* value measure, cyano is an extremely small group. When the exomethylene group is placed at a 3 position with respect to it, however, there is little or no change in the free-energy difference, from 0.24 to 0.26 kcal/mol. The latter value corresponds to the low-polarity solvent CF_2Cl_2, in which intramolecular interactions should be maximized. The most likely explanation is in terms of the disposition of charge within the cyano dipole. The negative end of the cyano dipole (nitrogen) is one bond further removed from the ring than the negative end of oxygen or sulfur substituents. The greater distance would then reduce the dipole–dipole interaction, even if the dipole itself is not smaller. The small effect of cyano on the proportion of axial conformer also indicates that a direct π-π interaction between cyano and the exo double bond does not contribute significantly to the conformational preference.

Some data have been reported on dibromomethylene and dichloromethylene systems [Equation (2.4), M = CBr_2 and CCl_2, respectively].[37] Apparently contrary to the expectations of a simple dipole–dipole explanation, these double bonds, like that of isopropylidene, permit larger amounts of axial conformer. It was suggested[30] that an attractive $n \rightarrow \pi^*$ interaction can occur in these systems because the LUMO energy is reduced in comparison with those of the M = CH_2 and CMe_2 systems. More recent studies[38] (J. B. Lambert and L. Xue, unpublished results) using *ab initio* calculations, however, indicate that the conformational populations in these systems too may be explained in terms of atomic charges and hence electrostatic considerations.

The electrostatic repulsion of an allenic group [Equation (2.4), M = $C{=}CH_2$] with methoxyl (X = OCH_3) has been found to be substantially less than that of the exomethylene group (M = CH_2).[38]

The interaction of an exomethylene double bond with a 3 substituent has also been studied by ultraviolet photoelctron spectroscopy.[39] The photoelectron spectra of **17**–**20** provided information on orbital energies. The highest occupied

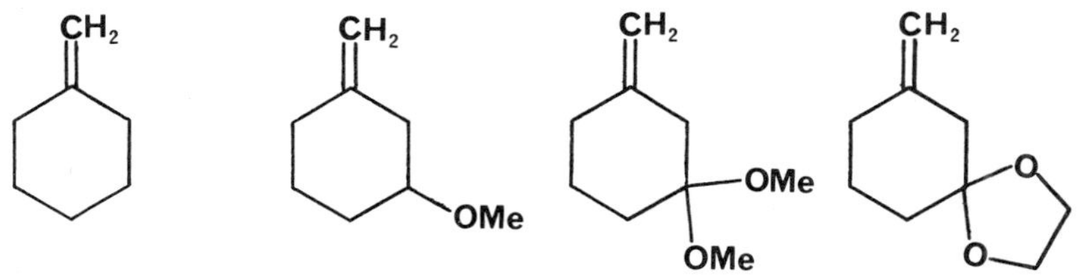

17 **18** **19** **20**

molecular orbital in each case was found to be the π orbital. Compound **17** provided the energy (9.18 eV) of an unperturbed π orbital. The equatorial methoxy group of **18** has essentially no effect on the π orbital (9.17 eV). The axial ether group of **19** or **20** destabilizes the π orbital by 0.1–0.2 eV (to 9.07 and 8.99 eV, respectively). *Ab initio* calculations on these systems reproduced these experimental observations.[39] This orbital destabilization by an axial ether group parallels the NMR observation of decreased proportions of the axial conformation for polar groups at the 3 position.

The interaction between an exomethylene double bond and a substituent at the 3 position has proved to be very complex. The magnitude of the interaction depends on several factors: (1) the angular orientation (axial or equatorial) between the double bond and the substituent, (2) the polarity of the solvent, (3) the nature of the 3 substituent (nonpolar like CH_3, polar like OCH_3, special cases like CN), and (4) substituent pattern on the double bond ($C{=}CY_2$, where Y can be H, CH_3, halogen, or $=CH_2$). The major repulsive interaction appears to result from a dipole–dipole mechanism and favors the 3-equatorial conformer. When the substituent (e.g., CH_3) or the double bond (e.g., tetraalkyl substituted) is nonpolar, the repulsive interaction disappears, and the proportion of 3-axial conformer increases.

Finally, a 4-alkyl ketone effect is conceivable, but the distances involved would suggest a small magnitude. By the same token, the effect of an exomethylene group on even a polar 4 substituent should be relatively small.

ACKNOWLEDGMENTS

Support of our own work on this subject has been provided by the National Science Foundation and by the Donors of the Petroleum Research Fund, administered by the American Chemical Society.

REFERENCES

1. Lambert, J. B.; Carhart, R. E.; Corfield, P. W. R. *J. Am. Chem. Soc.* **1969**, *91*, 3567.
2. Davis, M.; Hassel, O. *Acta Chem. Scand.* **1963**, *17*, 1181.
3. Lambert, J. B. *J. Am. Chem. Soc.* **1967**, *89*, 1836.
4. Lambert, J. B. *Acc. Chem. Res.* **1971**, *4*, 87.
5. Gerig, J. T.; Rimerman, R. A. *J. Am. Chem. Soc.* **1970**, *92*, 1219.
6. Lambert, J. B.; Sun, H.-n. *Org. Magn. Reson.* **1977**, *9*, 621.
7. Lambert, J. B.; Gosnell, J. L., Jr.; Bailey, D. S.; Henkin, B. M. *J. Org. Chem.* **1969**, *34*, 4147.
8. Lambert, J. B.; Gosnell, J. L., Jr.; Stedman, D. E. *Rec. Chem. Progr.* **1971**, *32*, 119.
9. (a) Groth, P.; Hassel, O. *Acta Chem. Scand.* **1964**, *18*, 923. (b) Mossel, A.; Romers, C. *Acta Crystallogr.* **1964**, *17*, 1217.
10. (a) Allinger, N. L.; Blatter, H. M.; Freiberg, L. A.; Karkowski, F. M. *J. Am. Chem. Soc.* **1966**, *88*, 2999. (b) Bhatt, M. V.; Srinivasan, G.; Neelakantan, P. *Tetrahedron* **1965**, *21*, 291. (c) Bailey, D. S.; Lambert, J. B. *J. Org. Chem.* **1973**, *38*, 134.
11. Lambert, J. B.; Featherman, S. I. *Chem. Rev.* **1975**, *75*, 611.
12. Anet, F. A. L., Chmurny, G. N.; Krane, J. *J. Am. Chem. Soc.* **1973**, *95*, 4423.
13. Anet, F. A. L.; Bourn, A. J. R. *J. Am. Chem. Soc.* **1967**, *89*, 760.
14. St.-Jacques, M.; Bernard, M. *Can. J. Chem.* **1969**, *47*, 2911.
15. Robins, P. A.; Walker, J. *J. Chem. Soc.* **1955**, 1789.
16. For reviews, see Allinger et al. in Ref. 10 and Eliel, E. L.; Allinger, N. L.; Angyal, S. J.; Morrison, G. A. *Conformational Analysis*; Interscience: New York, 1965; pp. 112–113.
17. Rickborn, B. *J. Am. Chem. Soc.* **1962**, *84*, 2414.
18. Malhotra, S. K.; Johnson, F. *J. Am. Chem. Soc.* **1965**, *87*, 5493.
19. Johnson, F. *Chem. Rev.* **1968**, *68*, 375.
20. Hauth, H.; Stauffacher, D.; Niklaus, P.; Melera, A. *Helv. Chem. Acta* **1965**, *48*, 1087.
21. D'Silva, T. D. J.; Ringold, H. J. *Tetrahedron Lett.* **1967**, 1505.
22. Harris, R. K.; Spragg, R. A. *J. Mol. Spectrosc.* **1967**, *23*, 158.
23. Eliel, E. L.; Allinger, N. L.; Angyal, S. J.; Morrison, G. A. *Conformational Analysis*; Interscience; New York, 1965; pp. 460–466.
24. Bingham, R. C. *J. Am. Chem. Soc.* **1976**, *98*, 535.
25. For a review, see Kirby, A. J. *The Anomeric Effect and Related Stereoelectronic Effects at Oxygen*; Springer-Verlag: New York, 1983; pp. 20–23.
26. Lessard, J.; Phan Viet, M. T.; Martino, R.; Saunders, J. K. *Can. J. Chem.* **1977**, *55*, 1615.
27. Phan Viet, M. T.; Lessard, J.; Saunders, J. K. *Tetrahedron Lett.* **1979**, 317.
28. Lessard, J.; Saunders, J. K.; Phan Viet, M. T. *Tetrahedron Lett.* **1982**, *23*, 2059.
29. Zefirov, N. S.; Barenenkov, I. V. *Tetrahedron* **1983**, *39*, 1769, and references cited therein.
30. Ouédraogo, A.; Phan Viet, M. T.; Saunders, J. K.; Lessard, J. *Can. J. Chem.* **1987**, *65*, 1761.

31. Robins, P. A.; Walker, J. *Chem. Ind. (London)* **1955**, 772.
32. Klyne, W. *Experientia* **1956**, *12*, 119.
33. Gorthey, L. A., Ph.D. Dissertation, Stanford University, 1985.
34. Lambert, J. B.; Clikeman, R. R. *J. Am. Chem. Soc.* **1976**, *98*, 4203.
35. Lambert, J. B.; Taba, K. M. *J. Org. Chem.* **1980**, *45*, 452.
36. Eliel, E. L.; Gilbert, E. C. *J. Am. Chem. Soc.* **1969**, *91*, 5487.
37. Lambert, J. B.; Taba, K. M. *J. Am. Chem. Soc.* **1981**, *103*, 5828.
38. Lambert, J. B.; Marko, D. E. *J. Org. Chem.* **1988**, *53*, 2642.
39. Lambert, J. B.; Xue, L.; Bosch, R. J.; Taba, K. M.; Marko, D. E.; Urano, S.; LeBreton, P. R. *J. Am. Chem. Soc.* **1986**, *108*, 7575.

3

Conformational Analysis of 1,3-Cyclohexadienes and Related Hydroaromatics

Peter W. Rabideau
Andrzej Sygula

3.1 1,3-CYCLOHEXADIENE

The geometry of 1,3-cyclohexadiene (**1**; 1,3-CHD) is expected to be influenced by three factors: (1) angle strain, (2) steric effects, and (3) conjugation of the two double bonds. The first two effects favor a nonplanar structure, whereas the third promotes coplanar π systems. Angle strain in planar **1** arises from the fact that a flat six-membered ring requires 120° bond angles while the preferred angles for the two methylene carbons are expected to be more or less tetrahedral. Similarly, a planar geometry produces eclipsing of the allylic hydrogens leading to an increase in torsional energy.

The first effort to distinguish between planar (**1a**) and puckered (**1b**)

1a

1b

geometries was made in 1965 by Butcher[1] who examined the microwave spectrum of **1**. He determined the ground vibrational state rotational constants and, after assuming the generally accepted bond lengths and valence angles, concluded that the molecule is nonplanar. The "best fit" torsion angle ψ (the angle by which one ethylene group is rotated relative to the other about the C-2—C-3 bond) was determined to be 17.5 ± 2°. This result was subsequently confirmed in a similar study by Luss and Harmony.[2] Later, three different electron diffraction studies on the structure of **1** in the gas phase appeared by Dallinga and Tonneman,[3] Traetteberg,[4] and Oberhammer and Bauer,[5] confirming the previous conclusions about the ring geometry. In all cases, the reported value of the torsion angle ψ was in the range of 17–18° (planarity of the individual ethylene groups was assumed).[3–5] The most important conclusion from the aforementioned studies is

Table 3.1 Proton NMR Results for 1,3-Cyclohexadiene, 1,3-Cyclohexadiene-[1,2,3-D_3], and 1,3-Cyclohexadiene-[1,2,6,6-D_4][a]

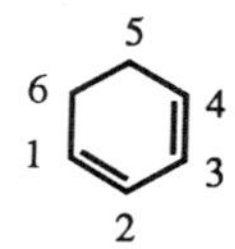

	Chemical Shifts, δ ppm		
	H_1	H_2	H_3
	5.703	5.811	2.086

Coupling Constants, H_z

$J_{1,2}$	$J_{1,3}$	$J_{1,4}$	$J_{1,5}$	$J_{1,6}$	$J_{2,3}$	$J_{2,5}$	$J_{2,6}$	$J_{5,6}$	$J_{5,6'}$
9.61 (9.64)	0.99 (1.02)	1.08 (1.12)	−0.33	4.46	5.13 (5.04)	0.45	−1.97	8.74	10.97

[a] Values from Ref. 8 (Ref. 7).

that a semi-chair with C_2 symmetry represents the most stable conformation of **1** (the C_2 symmetry axis bisects the C-2—C-3 and C-5—C-6 bonds).

In 1973 Carreira et al.[6] reported the results of their investigations of gaseous **1** by means of Raman spectroscopy. In the low-frequency region of the spectrum they localized a series of *Q* branches which were assigned to a twisting mode of the 1,3-cyclohexadiene ring. It was assumed that the twisting could be described by a double potential function, and this led to a barrier for ring inversion of 1099 ± 50 cm^{-1} (i.e., 3.1 ± 0.1 kcal/mol).

The ^{1}H-NMR spectrum of **1** although simplified by fast inversion (on the NMR time scale), is still rather complicated. In 1970 Cooper et al.[7] provided a partial analysis of the proton NMR spectrum of **1** and eight years later a full analysis of the ^{1}H-NMR of **1** along with its partly deuterated analogs, [1,2,3-D_3]-**1** and [1,2,6,6-D_4]-**1**, was reported by Auf der Heyde and Luttke.[8] Both of these groups argued that their NMR results were consistent with a somewhat puckered ring. In one case (Ref. 7) the values of $J_{1,2}$ and $J_{2,3}$ were compared with known planar (cyclopentadiene) and nonplanar (1,3-cyclooctadiene) structures, and in the other (Ref. 8) $J_{1,6}$ and $J_{5,6}$ were subjected to a Karplus-type analysis. These results are presented in Table 3.1.

Conjugation of the π bonds in 1,3-CHD has been considered by a number of investigators. Turner and co-workers[9] suggested that 1,3-cyclohexadiene is

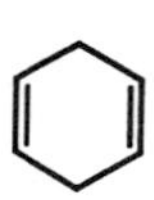

ΔH_{H_2} (kcal/mol)	-53.64 ± 0.29	-53.90 ± 0.33

"devoid of conjugation" in view of the rather small difference in thermodynamic stability between the two isomeric cyclohexadienes. Isomerization studies show a −0.41 kcal/mol difference, favoring **1**, at 110°C. The energies of the π molecular orbitals (MOs) were considered by Bishof and Heilbronner,[10] and by Giordan and co-workers.[11] In such systems, four atomic p_z orbitals give rise to four π MOs, ψ_1, ψ_2, ψ_3^*, and ψ_4^*, two of them being bonding and occupied and two antibonding and unoccupied, in order of increasing energy. [Symmetry assigned above to the

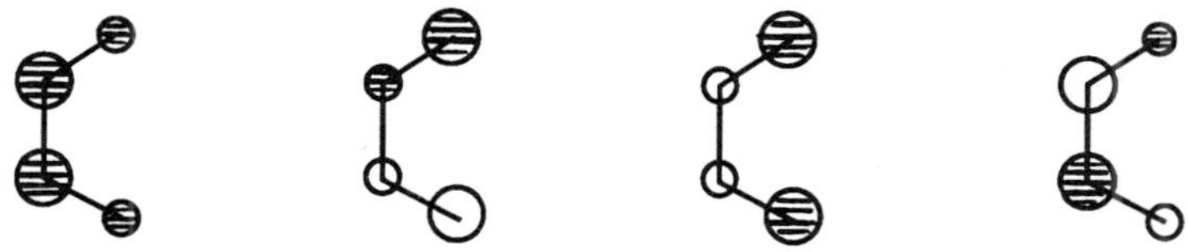

$\psi_1(b_1)$ $\psi_2(a_2)^3$ $\psi_3^*(b_1)$ $\psi_4^*(a_2)$

MOs are valid only for a planar system. Upon distortion (ring puckering), symmetry is reduced to C_2 (from C_{2v}) so the MO symmetries are a and b, instead of a2 and b1, respectively.] The MO splittings, that is, the differences in energies of ψ_1 and ψ_2, on one hand, and ψ_3^* an ψ_4^* on the other, may be a measure of conjugation of the diene system which is, of course, related to a torsion angle ψ. The splitting of both occupied and unoccupied MOs is expected to be a maximum of $\psi = 0°$ and then to decrease with increasing torsion angle. A series of cyclic 1,3-dienes was chosen,[10,11] 1,3-CHD (**1**), 1,3-cycloheptadiene (**2**), and 1,3-cycloocta-diene (**3**), which provided an interesting range of orientation for the double bonds

1 **2** **3**

relative to one another. Photoelectron spectroscopy (PES) gave the order of splitting of the occupied MOs as **1** > **2** > **3**.[9] This result forced Batich, Bishof, and Heilbronner to claim nonplanarity for the diene part of **2**.[10b] However, it is now generally accepted that the diene system in **2** is planar,[12] so the PES results concerning **1** may be in question.

Giordan et al.[11] estimated the splittings of the antibonding MOs ψ_3^* and ψ_4^* for the same series **1–3** using electron transmission spectroscopy. They found the order of the splittings to be **2** > **1** > **3**, in agreement with available structural information (**3** is known to be highly twisted with a torsion angle in the range of 42–60°).[13] MINDO/3 calculations performed on a model 1,3-butadiene showed that the splitting of the MOs is not a simple function of the torsion angle, and the discrepancy between the PES data and the other methods (as far as the structure of **1** is concerned) was at least partly explained.[11]

In summary, the results of experimental investigations are consistent: 1,3-

cyclohexadiene is a slightly puckered half-chair with C_2 symmetry and undergoes rapid ring inversion, presumably through a planar intermediate conformation.

3.2 THEORETICAL STUDIES

As a molecule of importance to the understanding of the concept of conjugation, **1** has been a subject of numerous theoretical studies. Since the force field (or molecular mechanics) methods are reviewed elsewhere in this book, we focus here on recent *ab initio* MO calculations, performed with geometry optimization.[14]

The calculations reported by Birch and co-workers[14a] were performed with a minimal STO-3G basis set. On the other hand, Saebo and Boggs[14b] determined the optimal geometry by the force-relaxation method at the double-zeta level basis set. Despite the different methods, the results obtained by both groups are very close. The optimal conformation for **1** was found to be nonplanar (C_2 symmetry), and the calculated bond lengths and valence angles are in good agreement with the electron diffraction data.[3–5] The only significant (but still small) difference is found in the value of the torsion angle, which is predicted to be 13.9° (Ref. 9) or 13.5°,[14] as opposed to the 17–18° value found by experiment. It is surprising to find that MINDO/3 geometry optimization also predicted a similar value (12°) for the torsional angle in very good agreement with the *ab initio* calculations.

Numerous molecular mechanics calculations predict **1** to be slightly more flattened than found by electron diffraction. Moreover, it should be mentioned that X-ray structure determination of some 1,3-CHD derivatives showed the cyclohexadiene moiety to be less puckered than 17° [e.g., 14° (Ref. 16a) or 14.4° (Ref. 16b). A possible explanation is as follows. Traetteberg[4] pointed out that the experimental data do not rule out the possibility that nonplanarity is due to the combination of slightly nonplanar ethylene groups together with a smaller torsional angle. In fact, calculations do show that there is a small distortion (1–2°) of the ethylene groups,[9,14,15] and this may indicate that 14° is a better estimation of the torsion angle in **1** than 17–18°. Saebo and Boggs[14b] also calculated an energy difference of 1.9 kcal/mol between the optimal C_2 conformer of **1** and the conformer forced to be planar (C_{2v}). This is in reasonable agreement with the value of the inversion barrier (3.1 kcal/mol) measured by Carreira et al.[6]

Finally, the question arises as to *why* 1,3-CHD is a nonplanar molecule since a straightforward consideration of conjugation would lead to a prediction for the planar form to be most stable. It seems that the nonplanarity of **1** is mainly the result of two factors: (1) from the values of the valence angles, it is apparent that the C_2 structure is less strained than the planar form,[14b] and (2) the methylene protons at C-5 and C-6 become eclipsed (Pitzer strain) in the planar form. While these factors oppose planarity, conjugation and a gauche relationship between the vinyl and methylene protons promote planarity. Apparently the first two factors are more important in this system.

3.3 SUBSTITUTED 1,3-CYCLOHEXADIENES

As a consequence of the nonplanar structure of the six-membered ring in **1**, a substituent at C-5 and/or C-6 may be located in two different positions,

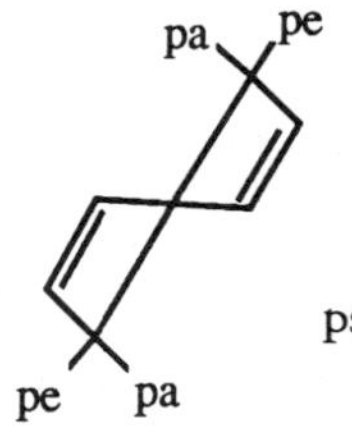

Substituents may be pseudoaxial or pseudoequatorial

pseudoaxial (pa) or pseudoequatorial (pe), and so the question arises as to which position is energetically preferred. In contrast with the many papers devoted to this problem for cyclohexanes, there is a lack of such data for the 1,3-CHD system. Some of the rare studies in the field are associated with diene chirality rules with 2-methyl-5-isopropyl-1,3-cyclohexadiene (α-phellandrene) as the most frequently investigated case.

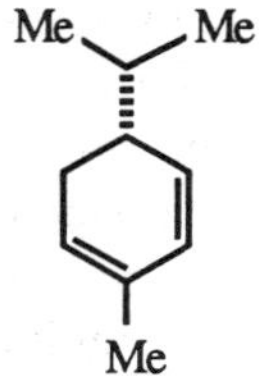

α-Phellandrene

From the Cotton effect, in connection with the diene helicity rules, a qualitative prediction was made for the axial conformer to be more stable.[17] Some time later, however, an opposite conclusion was drawn from the same observation.[18] Finally, an optical rotatory dispersion (ORD) measurement confirmed the pseudoequatorial form to be the most stable.[19]

The first attempt to determine the relative energies between the two conformers was made by Baldwin and Krueger[20] who investigated the photolytic ring opening of α-phellandrene. They suggested that the distribution of the primary products was controlled by the pseudoequatorial/pseudoaxial ratio of the substrate. Using a simplified view of the mechanism, the authors estimated the pseudoequatorial conformer to be more stable by 0.46 kcal/mol.[20] The next estimation came from Snatzke[21] who reported the pseudoequatorial form to be more stable by 0.28 kcal/mol, based on low-temperature circular dichroism (CD) measurements. Recently Lightner et al.[22a] published a reinvestigation of the problem by variable-temperature CD studies, reporting a value of 0.25 kcal/mol, and Rank and Peoples published the results of STO-3G calculations based on MM geometries and similarly found the pseudoequatorial form to be more stable (0.17 kcal/mol).[23]

Similar studies have been performed for 5-methyl and 5-*tert*-butyl-1,3-

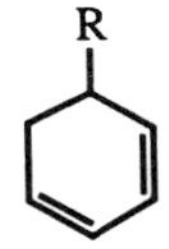

R = Me or *t*-Bu

4

CHD,[22] and in both cases preference for the pe position was found. Spangler and Hennis[22b] used the technique analogous to that applied by Baldwin and Kruegger[20] and estimated $\Delta G°$ to be 0.24 and 2.98 kcal/mol for Me-**4** and *t*-Bu-**4**, respectively, in favor of the pe substituent. On the other hand, the variable-temperature CD measurements suggested that Me-**4** exists as a 1:1 mixture of pe and pa conformers, whereas the *t*-butyl group was found to be slightly more stable in the pe position ($\Delta G° = \sim 0.4$ kcal/mol).[22a]

Despite discrepancies in the qualitative estimation of the relative stabilities of pe versus pa conformers, it is apparent that the conformational preference in 5-R-1,3-CHD systems is analogous to that found in both cyclohexane and cyclohexene systems; the pseudoequatorial substituent is favored over pseudoaxial. Because of the significant flattening of the 1,3-CHD system relative to cyclohexane and cyclohexene, factors that are responsible for pe preference in these latter compounds (e.g., the 1,3-diaxial interactions in cyclohexane) are not so important in CHD and this is why the relative stabilities of pe-5-R-1,3-CHD systems are not as significant as the analogous stabilities in the cyclohexane and cyclohexene systems.

A 1,3-cyclohexadiene of special interest is chorismate,[24] which serves as an intermediate in the biosynthesis of several aromatic amino acids. This process

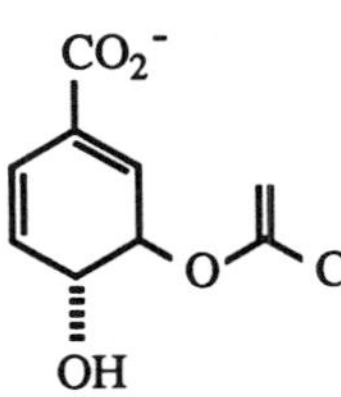

Chorismate (**4**)

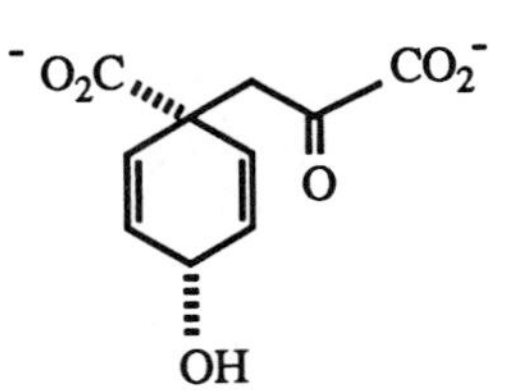

Prephenate (**5**)

involves a Claisen rearrangement, producing prephenate, which presumably must occur through the dipseudoaxial form of chorismate (**4** pa). The value of $J_{5,6}$ = 11.7 Hz would, of course, be most consistent with the dipseudoequatorial

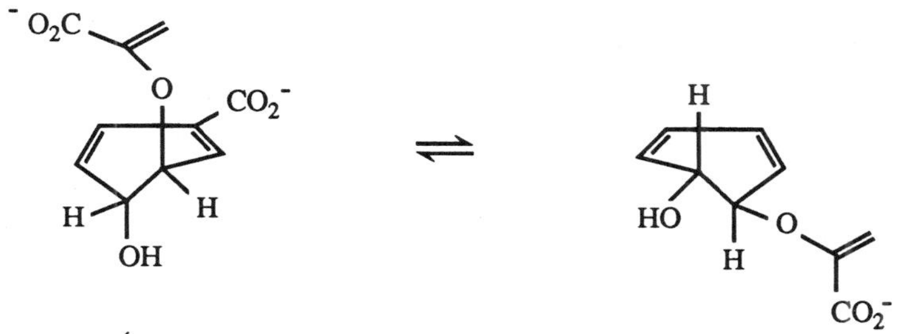

4 pa **4** pe

arrangement **4** pe. However, some solvent dependence (9.6–13 Hz) was later reported for this value suggesting a shift in equilibrium. A more striking example was provided by the related dihydroxy acid **6**, which showed a low value of $J_{5,6}$ = 2.0 Hz in 100% DMSO, and a high of 8.3 Hz in 10% DMSO/90% $CDCl_3$.

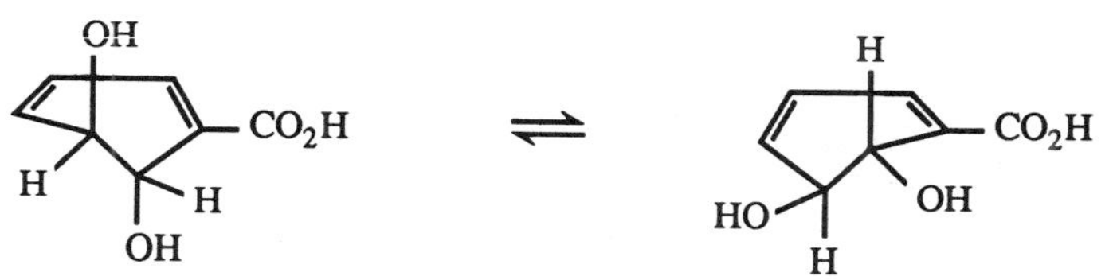

6 pa **6** pe

Presumably, DMSO disrupts intramolecular hydrogen bonding in **6** pe resulting in the predominance of **6** pa. We also note that **6** is quite different from **4** in that the former has a steric interaction between the carboxyl group and the substituent at C-6 which begins to resemble the 9,10-dihydrophenanthrene system (see below) which does show dipseudoaxial preference.

A recent, rigorous reinvestigation by Copley and Knowles[24c] suggests that a significant amount of the pseudodiaxial form of chorismic acid exists in methanol at 25°C (estimated ratio of di-pe to di-pa is 5–10) with a difference in $\Delta G°$ of 0.9–1.4 kcal/mol. Even greater amounts of the dipseudoaxial form are present for the dianionic salt in H_2O although the diequatorial conformation still predominates.

3.4 THE CYCLOHEXADIENYL ANION

The structure of the cyclohexadienyl anion (**7**) has also attracted considerable attention due to the possibility of the homoaromatic species **7b**. The question of **7a** versus **7b** was first considered by Kloonsterziel and van Drunen[25] who studied a

7a

7b

series of cyclic dienyl anions (**7**–**10**) by proton NMR spectroscopy. The most

7

8

9

10

important observation relating to the potential homoaromaticity of **7** was that the chemical shifts for the hydrogen atom attached to the carbon atoms C-2 were almost equal for **7**, **9**, and **10** (5.72, 5.57, and 5.65 ppm, respectively). According to MO theory, C-2 should be uncharged in the dienyl system and so one can conclude that equal ring currents occur in **7**, **9**, and **10**. Since the 1–5 distances are quite different in these systems, however, there must be an approximately zero ring current in all three cases. Anion **7** is thus the cyclohexadienyl anion **7a** and should not be considered to be the homocyclopentadienyl anion **7b**.[25]

Subsequent investigations were made by Bates and co-workers[26] and by Olah et al.[27] who based their conclusions primarily on the analysis of ^{13}C chemical shifts. Again, if the CHD anion is correctly described by the structure **7a**, the negative charge should be localized at the odd-numbered carbon atoms (i.e., C-1, C-3, and C-5), whereas the charge at the even-numbered carbons (C-2 and C-4) should remain almost unchanged when compared, for example, with the analogous CHD cation. On the other hand, in the case of homoaromatic **7b**, the negative charge should be distributed over all sp^2 carbon atoms (C-1–C-5). Actually, the measured ^{13}C chemical shifts of CHD anion showed typical pentadienyl behavior; C-2 and C-4 do not carry negative charge as demonstrated by the similarity of their chemical shifts to those of the corresponding cation (136.9 ppm as opposed to 131.8 ppm for **7**).[27] The total upfield shift for C-1, C-3, and C-5 in **7** is 151.5 ppm, approximately that expected for the location of one additional electron on these atoms.

An additional argument against the homocyclopentadienyl anion was provided by Olah et al.[27] from the analysis of $^1J_{C—H}$ coupling constants. In **7**, $^1J_{C\text{-}6—H}$ was found to be 124 Hz, a typical value for sp^3 hybridized carbon atoms in the absence of angle strain. In a homoaromatic structure like **7b**, however, an angular distortion must be present at C-6 and so $^1J_{C\text{-}6—H}$ is expected to be significantly higher. For example, in the tetramethylhomotropylium ion (**11**), the corresponding J value is 147 Hz,[28] whereas in the planar, unstrained mesitylenium ion (**12**), $^1J_{CH}$ = 122 Hz.[28] The agreement of the latter value with the

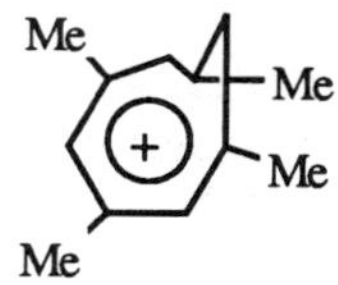

11

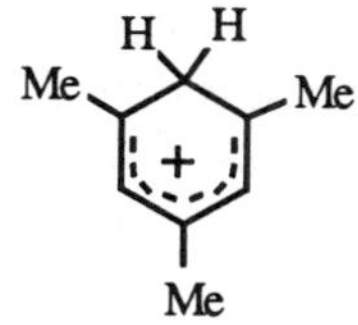

12

corresponding coupling constant in **7** further indicates the absence of 1,5 overlap (homoconjugation).[27]

Recently, Tolbert and Rajca[30] investigated the anion **13** generated from 6,6-dimethyl-2,4-diphenyl-1,3-cyclohexadiene, expecting that the phenyl substituents could stabilize the bicyclic anion **14**. Transformation **13** to **14** should occur through the "negative transition state" **15** of homoaromatic character. However,

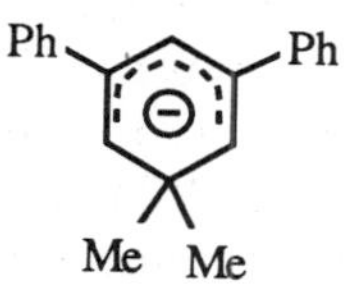

13

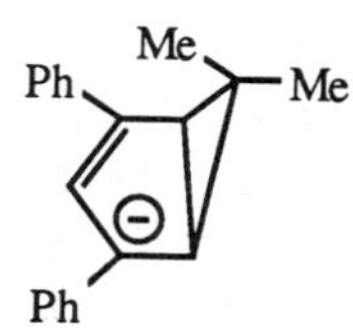

14

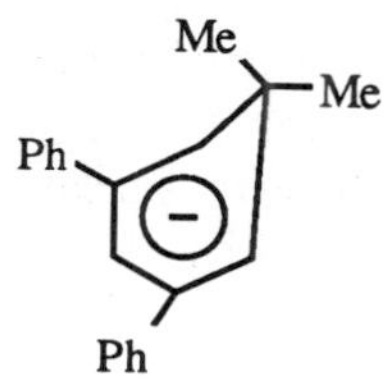

15

the authors did not find any spectroscopic evidence for the existence of a homocyclopentadienyl anion.

On the other hand, Perkins and Ward,[31] studying the base-catalyzed rearrangement and hydrogen isotope exchange reaction of tricyclo[7.1, $0.0^{2,7}$]deca-2(7),3,5-trien (cyclopropindene; **16**), tentatively suggested that the kinetic results, taken together with the low selectivity of deprotonation, may be consistent with participation of homoindenyl intermediate **17** possessing a small degree of homoaromatic character. Compound **16** undergoes a 30-fold rate enhancement of hydrogen exchange compared with indane, and a similar enhancement is observed for diphenylbicyclohexene (**18**) when compared to cyclopentene. According to Perkins and Ward,[31] these enhancements are too large to be explained in terms of cyclopropyl conjugation, and so a small homoaromatic stabilization of the anions (**17** and **19**) was proposed.[31] The authors advised, however, that this proposal be viewed with caution because the measured acidity

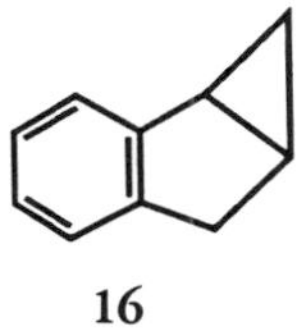

16

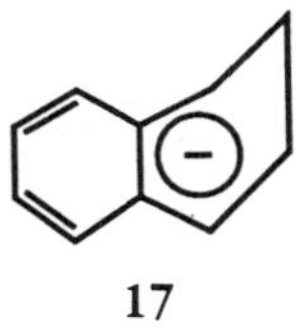

17

18

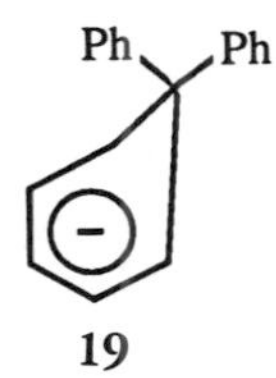

19

enhancements for **16** and **18** are relatively low and may be caused by other factors. Actually, homoaromatic stabilization of the anion **21** causes an increase in the rate of deprotonation of **20**, which is greater by a factor of 10^4 relative to deprotonation of **22** under comparable conditions.[31,32]

20

21

22

The problem concerning the structure of the CHD anion has been the subject of numerous theoretical investigations as well. Earlier studies did not exclude the possibility of homoaromaticity in **7**.[33] Subsequent investigations based on more sophisticated MO methods (mainly MINDO/3 and *ab initio*) have given results in agreement with experiment. First, Olah and co-workers[27] showed that geometry optimization with the MINDO/3 method gave the planar conformation **7a** as a global minimum. This minimum, however, was found to be rather shallow; puckering by 20° (the plane defined by C-1, C-5, and C-6 intersecting with the plane C-1–C-2–C-3–C-4–C-5 at the angle of 20°) causes an energy increase of only 2.2 kcal/mol.[27] Later Haddon[34] reinvestigated the problem confirming **7a** to be a global MINDO/3 minimum. In addition, he was able to find a local minimum with a C-1–C-5 distance equal to 1.586 Å (corresponding to **7b**) lying 36 kcal/mol higher in energy than **7a**. In 1980, an *ab initio* study on the structure of **7** was reported,[35] and the results were consistent with those obtained by MINDO/3. Again STO-3G geometry optimization found the planar conformation **7a** to be the global minimum with a secondary minimum related to the homocyclopentadienyl anion **7b** lying 34 (STO-3G) or 44 kcal/mol (STO-3G//4-31G) higher in energy. In the optimal STO-3G geometry of **7b**, the folding angle (C-6–C-1–C-5–C-4) is 122° and the C-1–C-5 distance is 1.609 Å, both values being close to the MINDO/3 values.

In conclusion, considering the available results of both experimental and theoretical studies, the CHD anion should be considered as structure **7a**. The isolated anion is probably planar, although according to MO calculations, only a small amount of energy is required to pucker the ring. It is therefore not surprising that the cyclohexadiene moiety is not planar in transition metal complexes (**17**). For example, the X-ray structure determination of cyclohexadienyl manganese tricarbonyl (**23**, M = Mn)[36] showed the dihedral angle between the planes C-1–C-

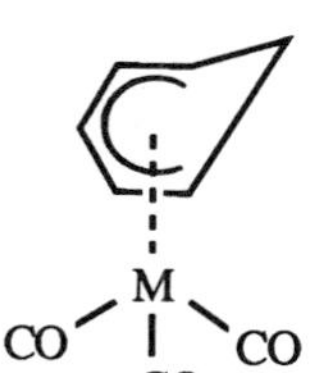

M = Fe^+, Mn, Cr^-

23

5-C-6 and C-1–C-2–C-3–C-4–C-5 to be 43°. Hoffmann and Hofmann[37] suggested on the basis of EHT calculations that the interactions of the metal *d* oribitals with the orbitals of the methylene group are responsible for the puckering and so the folded structure of the CHD moiety does not mean homoaromatic character. The C-1–C-5 distance in **23** (M = Mn) is equal to approximately 2.4 Å, being typical of the dienyl system **7a**, and significantly too large for efficient, 1,5 overlap.

3.5 1,2-DIHYDRONAPHTHALENES

The first proton NMR analysis of 1,2-dihydronapthalene (**24**) was reported by Cook et al. in 1969.[38] (Independently, Cooper et al.[7] published the results of their analysis of the NMR spectrum of **24**, but they did not discuss the structural consequences of the data.) By analogy with both 1,3-CHD and 9,10-dihydrophenanthrene (9,10-DHP), Cook et al. assumed 1,2-DHN to be nonplanar. A consequence of nonplanarity is ring inversion of the nonaromatic ring; however, the NMR spectrum of **24** and 4-phenyl-1,2-DHN did not show any signal

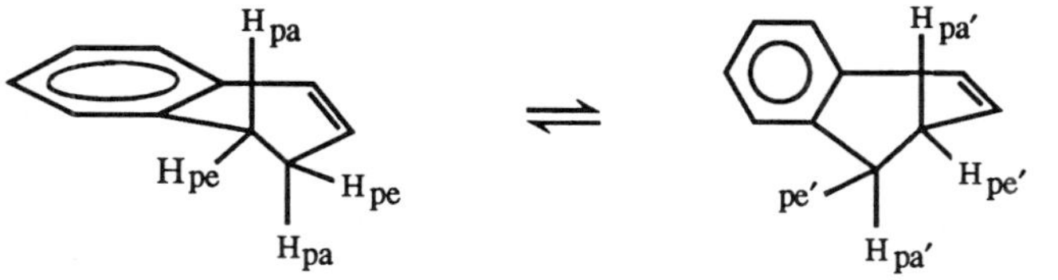

24 **24′**

broadening upon cooling to −55 °C. Again by analogy with 1,3-CHD and 9,10-DHP, it was concluded that ring inversion is rapid on the NMR time scale.[38] Based on the ratio of the coupling constants between the protons located at C-1 and C-2 (i.e., J_{trans}: J_{cis}), Cook and Dassanayake[39] estimated the torsional angle for the CH_2—CH_2 in **24** to be approximately 50°. The same angle in 1,3-CHD is 45°, suggesting that 1,2-DHN may be slightly more puckered than 1,3-CHD.

The rapid inversion of the partially saturated ring in 1,2-DHN and its derivatives causes the coupling constants between the alicyclic protons to be averaged. Thus, two coupling constants, J_{trans} and J_{cis}, may be expressed as

J_{trans} —an average of $J_{pa\text{-}pa}$ and $J_{pe\text{-}pe}$

J_{cis} —an average of $J_{pa\text{-}pe}$ and $J_{pe\text{-}pa}$

(where pa and pe subscripts denote pseudoaxial and pseudoequatorial, respectively). The value of J_{trans} is of special interest since it is expected to be the most sensitive for preferred conformation determination. Thus, for two conformations of the 1,2-DHN system, one with a substituent at C-1 (or C-2) in the pa position and the other in the pe position, J_{trans} will be a weighted average given by the formula

$$J_{\text{trans}} = xJ_{\text{pe-pe}} + (1-x)J_{\text{pa-pa}}$$

where x is a mole fraction of the conformer in which the substituent is pseudoaxial. If one knows the values of $J_{\text{pe-pe}}$ and $J_{\text{pa-pa}}$, then the determination of J_{trans} provides an estimation of the pseudoaxial conformer population. A similar expression may be written for J_{cis},

$$J_{\text{cis}} = xJ_{\text{pe-pa}} + (1-x)J_{\text{pa-pe}}$$

but since $J_{\text{pe-pa}}$ is expected to nearly equal $J_{\text{pa-pe}}$, the value of J_{cis} is more or less independent of x.

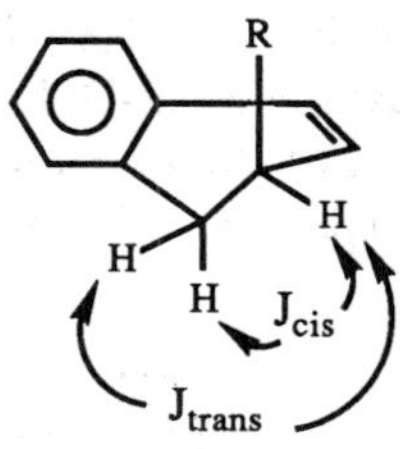

Ring Inversion

J_{cis} remains pa-pe

but J_{trans} becomes pa-pa

To determine the individual J values, the model compounds *cis*- and *trans*-9-phenyl-1,2,3,4,4a,10a-hexahydrophenanthrenes (**25** and **26**) were synthesized by Cook et al.[38] While the cis compound **25** may exist in two conformations, only pa-pe coupling is involved between the protons at positions 4a and 10a. In contrast, the trans compound **26** has a fixed conformation with the 4a and 10a

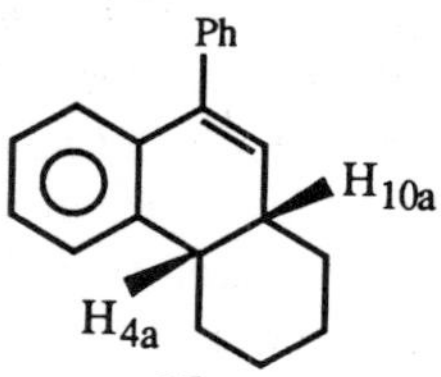

25

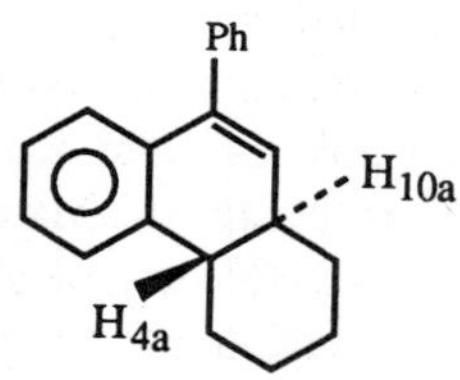

26

protons located in pa positions. The value of $J_{\text{pa-pe}}$ for the cis compound was found to be 6.8 Hz, and this value is very close to one of the J values previously obtained for **24** (7.0 Hz) and its 4-phenyl derivative (6.4 Hz). Unfortunately, the trans compound **26** did not allow direct measurement of $J_{\text{pa-pe}}$ since the chemical shifts of the 4a and 10a protons were too close to those of the fused cyclohexane ring and decoupling was unsuccessful. It was possible, however, to measure the coupling between H_{10a} and the adjacent vinyl proton (2.0 Hz). This measurement, which involves a H_{pa}–H_{vinyl} relationship, then allowed an estimate of H_{pe}–H_{vinyl} coupling to be approximately 6.6–7.4 Hz from values that result from the average of the two relationships. On this basis they concluded **25a** to be favored over **25b** by 0.5 kcal/mol.

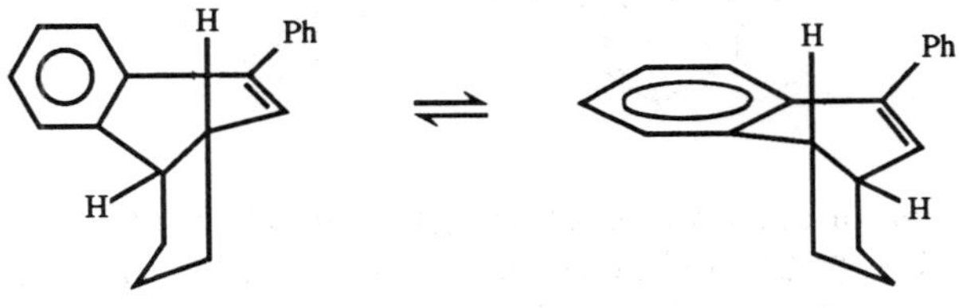

25a **25b**

With compounds **27** and **28**, these same investigators used LAOCOON II computer simulation of the ABCX pattern generated by the nonaromatic protons to generate coupling constants. Values for $J_{1,2}$ and $J_{1,2'}$ were reported to be 7.2 and 3.5 Hz for **27**, and 7.0 and 5.6 Hz for **28** (respectively). Using "best fit"

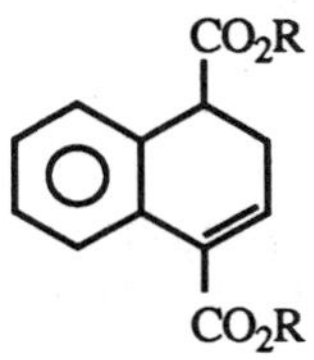

27, R = H
28, R = CH_3

values for $J_{pa\text{-}pa}$ (16.0 Hz) and $J_{pe\text{-}pe}$ (2.0 Hz), they estimated **27** and **28** to be 90 and 75% in conformations with the C-1 carboxyl in the pseudoaxial position (respectively). This was accomplished by the following general procedure:

1. Analysis of the "benzylic" part of the NMR spectrum to provide J_{cis} and J_{trans} for the protons attached to C-1 and C-2.

2. Assignment of one of the above values lying in the range 6.4–7.2 Hz to be J_{cis}.

3. Using the other value, taken as J_{trans}, for calculation of the molar fraction of the conformer with pa substituent.

The preferred pseudoaxial substituent preference in **27** and **28** is in contrast with the equatorial preference in the cyclohexane ring. This discrepancy was explained primarily by lack of 1,3-diaxial interactions in the 1,2-DHN system,[38] the interactions that are responsible for the equatorial preference in cyclohexanes. In the following paper, however, Cook and Dassanayake suggested that destabilization of the pseudoequatorial bond at C-1 by the neighboring aryl proton (*peri* interactions) may be the overriding factor causing pa preference in 1,2-DHN systems.[39] In an attempt to assess the relative importance of these influences, they investigated a series of 2-substituted 1,2-DHA systems (**29–31**).

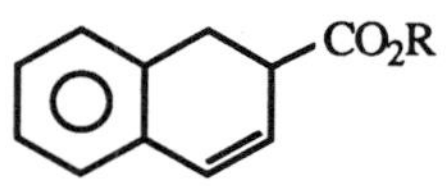

29, R = H
30, R = Me
31, R = Et

Application of the same procedure gave the percentage of the pseudoaxial conformers to be 29, 27, and 26% for **29**, **30**, and **31**, respectively, and so the same substituents that preferentially occupy the pseudoaxial site at C-1 are pseudoequatorial at C-2. This is, of course, in the absence of destabilizing 1,3-diaxial interactions.[39] Evidently the destabilization of the pseudoequatorial substituent by peri hydrogen atoms dominates the conformational equilibria in 1-substituted 1,2-DHN systems. Such an explanation was proposed earlier for pseudoaxial preference in 9-substituted 9,10-dihydroanthracenes.[40]

The pseudoaxial preference of substituents was also found in *trans*-1,2-dichloro-1,2-DHN systems (**32** and **33**).[41] The conformational populations were estimated to be 83 and 84% diaxial for **32** and **33**, respectively.[41] However, there

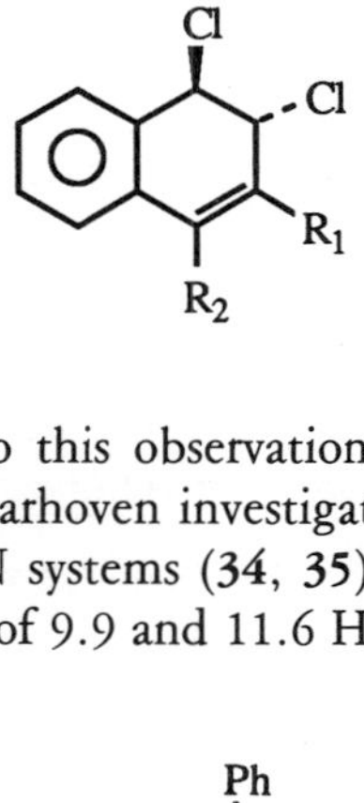

32, $R_1 = Cl$, $R_2 = H$
33, $R_1 = H$, $R_2 = Cl$

are exceptions to this observation of pseudoaxial preference at C-1. Recently, Lamberts and Laarhoven investigated the conformational equilibria of 1- and 2-phenyl-1,2-DHN systems (**34**, **35**)[42] and a full analysis of the ^{1}H-NMR spectra gave J_{trans} values of 9.9 and 11.6 Hz for **34** and **35**, respectively. Using the values

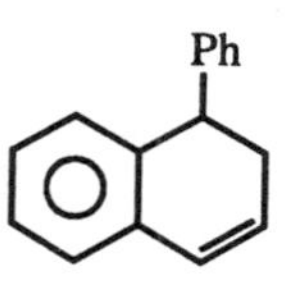

34

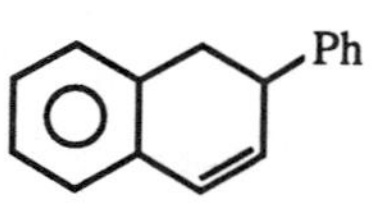

35

for $J_{pe\text{-}pe}$ and $J_{pa\text{-}pa}$ proposed by Cook et al.,[38] they estimated the population of the pseudoaxial conformer of **34** to be 44%, implying a slight conformational preference for the pseudoequatorial conformation. This is in contrast with the aforementioned pseudoaxial preference at C-1. Of course one may question whether or not the values taken from Cook et al. are adequate for these systems. Similarly, there is a possibility that the conformation of the DHN moiety in **34** is changed by the presence of the phenyl substituent. However, this latter possibility can be excluded by comparison of the other coupling constants measured for **34** which would undoubtedly also be affected. In fact, the values of J_{cis} (7.3 Hz) and $J_{2,2'}$ (-17.3 Hz) differ only slightly from those reported by Cook et al.[38] for both unsubstituted (7.0 and -16.8 Hz, respectively) and substituted 1,2-DHN systems (6.9 to 7.2 and -16.8 to -18.2 Hz, respectively), so the slight predominance of pe for the phenyl substituent in **34** does not seem to be an artifact of the method used. Lamberts and Laarhoven concluded that in this case the peri interaction with the aromatic hydrogen atom at C-8 is balanced by "π stacking" between the pseudoequatorial phenyl and the aromatic moiety.[42] It seems, however, from our MM calculations that the pseudoequatorial phenyl does not suffer a serious *peri* interaction because of its perpendicular orientation in respect to the mean plane of the DHN moiety (A. Syguła, unpublished results). This may cause the energies of the pa and pe conformers to be very similar, resulting in an almost equal distribution of the conformers.

The percentage of the pseudoaxial conformer in **35** was found to be 31%, in agreement with the pseudoequatorial preference of substituents at C-2 in the 1,2-DHN system. Hence, in this position a phenyl substituent does not prove to be anomalous.

3.6 9,10-DIHYDROPHENANTHRENES

On the basis of chemical arguments, Beckett and Mulley[43] suggested nonplanarity for the central ring in 9,10-dihydrophenanthrene (9,10-DHP, **36**). At the same time, Braude and Forbes, investigating the UV absorption spectra of a series of biphenyl derivatives, estimated the torsional angle by which the biphenyl moiety in **36** is twisted along the pivotal bond as approximately 18°.[44] This is in

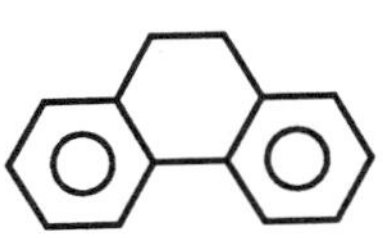

36

good agreement with the value of 20° deduced earlier from molecular model inspection.[45] Various physical methods reported[46] give estimations of the angle in the range 20–24°, whereas refraction measurements predict the angle to be approximately 38°,[47] a value that seems to be too high when compared with other data. Crystal structure determination of 9,10-DHP chromium tricarbonyl showed that both outside rings are planar and the dihedral angle between planes is 15.3°.[48a] This complex can be used as a structural model for **36** because, as it was stated recently,[48b] complexation does not change the torsional angle (and thus the geometry of the ligand) to a significant extent.

The first calculation on this system was reported by Mislow et al.[49] in 1964 and provided a value of 15.3° for the torsion angle in question. Our results with MNDO calculations estimate this angle to be 18.7° (A. Syguła, unpublished results). Thus, the geometry of **36** appears to be well established as nonplanar with a torsion angle of approximately 20°. The nonplanarity implies, as it does also in the case of 1,3-CHD and 1,2-DHN, that the protons attached to C-9 and C-10 should be nonequivalent, that is, pseudoaxial (pa) and pseudoequatorial (pe).

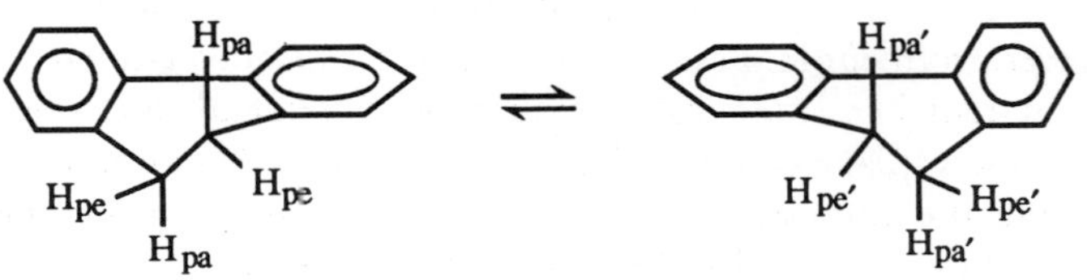

The ^{1}H-NMR spectrum of **36**, first reported by Mislow et al.,[49] shows a narrow singlet for the methylene protons, indicating a rapid inversion process on the NMR time scale similar to 1,3-CHD and 1,2-DHN. Attempts to "freeze out" this process were unsuccessful, and Oki and co-workers[50] observed that **36** retains a singlet for the methylene protons at temperatures as low as −90 °C. This sets the upper limit for the activation barrier at 9 kcal/mol. Recently, Schloegel et al.[51] repeated the experiment cooling down the solution to −130 °C, but again they did not observe any signal broadening. If we assume there is no accidental isochrony, this measurement lowers the upper limit for the inversion barrier of **36** to approximately 7 kcal/mol.[51]

The low barrier to inversion of 9,10-DHP is not surprising, however. An early estimation of the energy difference between the planar and nonplanar conformations predicted the latter to be approximately 4 kcal/mol more stable,[49] and MNDO calculations estimate this difference to be even lower (2.0 kcal/mol) (A. Syguła, unpublished results). Hence, if the planar form is assumed to be the transition state for interconversions, this process should be quite rapid. Although it has not been possible to measure the inversion barrier for **36** itself, it has been achieved for the radical anion. The temperature dependence of the ESR spectrum of the radical union allowed determination of the inversion barrier as 4.5 kcal/mol,[52a] 6.1 kcal/mol,[52b] or 6.3 kcal/mol.[52c] It is still an open question, however, as to what relationship is to be expected between the inversion barrier for the radical anion and the parent neutral hydrocarbon.

3.7 SUBSTITUTED 9,10-DIHYDROPHENANTHRENES

The barrier to inversion in the 9,10-DHP system may be increased if substituents are present which increase the energy of the transition state (presumably the planar or almost planar conformation). Such a case is represented by 4,5-dimethyl-9,10-DHP, **37**. In 1964, it was observed that, in contrast to **36**, the NMR spectrum of **37** exhibits a AA'BB' pattern for the methylene protons. However, the inversion barrier for **37** was determined two years earlier by Mislow

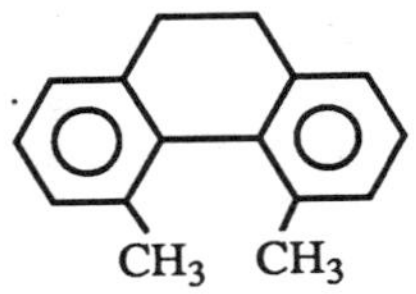

37

and Hoops,[53] who investigated the racemization of the optically active compound in benzene. From the temperature dependence of the rate of racemization (which in this case is simply ring inversion), they found an activation barrier of 23.1 kcal/mol.[53] A similar method applied earlier by Hall and Turner[54] to the related optically active 9,10-dihydro-3:4,5:6-dibenzophenanthrene **38**, gave a value of

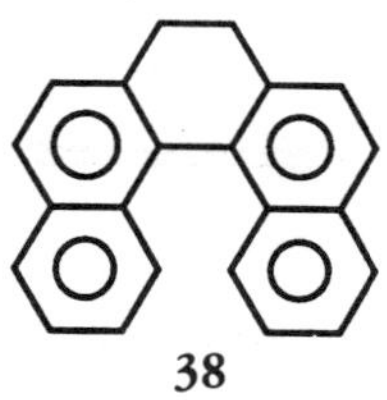

38

approximately 34 kcal/mol. Apparently the steric repulsions in the transition state of **38** are more severe than in **37** as would be expected from the models.

The inversion of **38** was reinvestigated by ^{1}H-NMR, which gave an activation energy of 25.3 ± 1.0 kcal/mol,[55] and more recently by a kinetic method based on the relative intensities of diagnostic bands in the circular dichroism spectra. The latter technique provided free enthalpies of activation at 297 °C of 23.3 ± 0.1 kcal/mol ($CHCl_3$) and 23.5 ± 0.1 kcal/mol (Et_2O).[51]

The inversion barrier in 9,10-DHP systems may also be increased by substitution at C-9 and C-10, again due to overcrowding in the transition state. Rabideau and co-workers[56] investigated a series of cis and trans isomers with like substituents in the 9 and 10 positions. As a consequence of this substitution pattern, both of the interconverting conformers in the cis isomers are identical, and hence of equal energy. If we assume a planar transition state, the steric interaction between R_9 and R_{10} would be expected to increase the inversion barrier. Indeed, with *cis* 39 and 41, temperature-dependent proton NMR spectra were observed and coalescence temperatures of −47, +27, and −70 °C were determined for **39**, **40**, and **41**, respectively. The corresponding free energies of activation ($\Delta G^{\ddagger}$) were calculated to be 10.8, 14.9, and 9.7 kcal/mol, and so the barriers to conformational interconversion are in the unexpected order Ph < Me < Et. The proposed explanation[57] suggests that, in *cis* **41**, the two phenyl rings are in a parallel orientation which greatly reduces the steric interaction in the transition state.

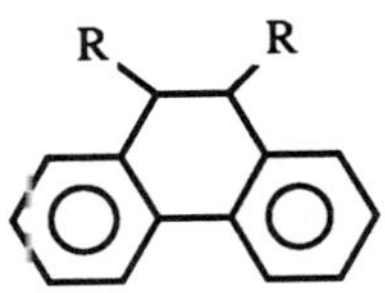

39, R = Me
40, R = Et
41, R = Ph

cis-9,10-Dimethyl-9,10-DHP was subsequently reinvestigated[57] by carbon NMR, and the enthalpy (10.3 ± 0.1 kcal/mol) and entropy [−3.3 ± 0.4 cal/(mol·K)] of activation were determined. Hence, the negative $\Delta S^{\ddagger}$ value supports

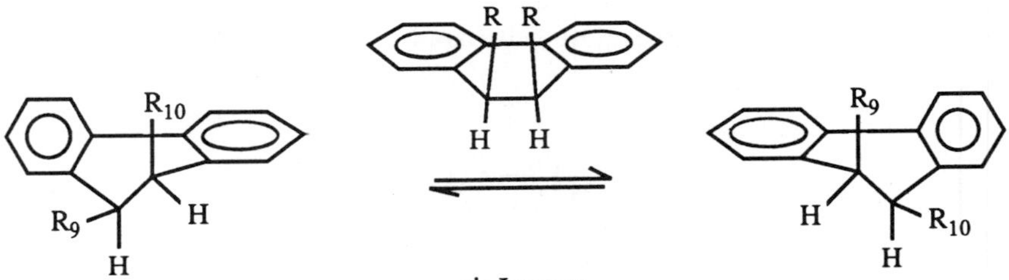

cis Isomers

the restriction in rotational mobility of the methyls as would be expected for the planar transition state discussed above.

In contrast with the cis isomers, the trans compounds do not show temperature-dependent NMR behavior.[56,57] This is not surprising, however, since

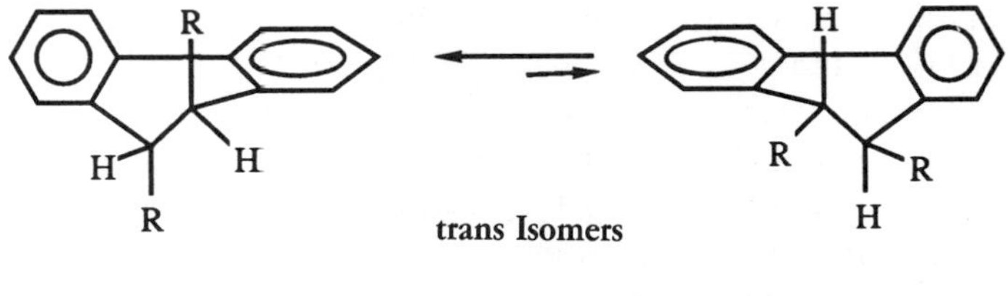

trans Isomers

in this case the two conformers are of unequal energy and the dipseudoaxial geometry is expected to predominate. Of course, this cannot be demonstrated experimentally when R-9 = R-10 since spin coupling between H-9 and H-10 is not observable. However, dipseudoaxial preference has been demonstrated for *trans*-9-acetoxy-10-chloro-9,10-DHP,[41] and for the radical anion of *trans*-9,10-diphenyl-9,10-DHP by ESR measurements.[58]

3.8 MONOSUBSTITUTED 9,10-DIHYDROPHENANTHRENES

In 1967, Cohen et al.[59] reported on their investigation of a series of 9,9′,10,10′-tetrahydro-9,9′-biphenanthryls (**42**) by ^{1}H-NMR and they concluded that

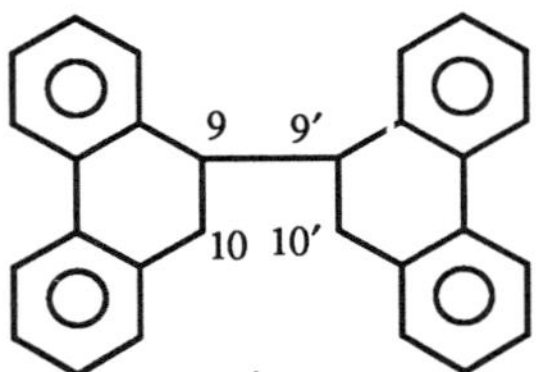

42

the 9,9′ link is formed by pseudoaxial bonds (the protons at C-9 and C-9′ are trans to one another).

Later, Harvey and co-workers[60] investigated a series of eight monosubstituted 9,10-DHP systems by computer analysis of the ABC spectrum originating

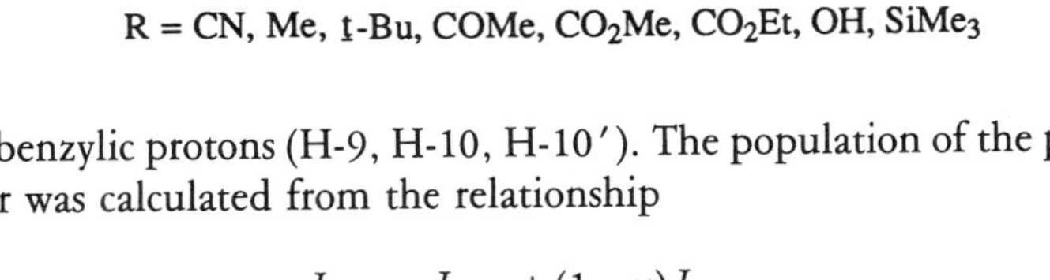

R = CN, Me, t-Bu, COMe, CO_2Me, CO_2Et, OH, $SiMe_3$

from the benzylic protons (H-9, H-10, H-10′). The population of the pseudoaxial conformer was calculated from the relationship

$$J_{9,10} = xJ_{\text{pe-pe}} + (1 - x)J_{\text{pa-pa}}$$

where $J_{\text{pe-pe}}$ was taken to be 2.0 Hz and $J_{\text{pa-pa}}$ as 16.0 Hz as proposed by Cook et al.[38] for the 1,2-DHN system. The use of 1,2-DHN values seemed justified since the $J_{9,10'}$ values (see Table 3.2) are all within 0.9 Hz (usually closer) to the 6.8 Hz value for J_{cis} in 1,2-DHN systems.

The larger groups, *t*-Bu and $SiMe_3$, exhibit a strong pseudoaxial preference

Table 3.2 Coupling Constants and Pseudoaxial Populations for 9-*R*-9,10-Dihydrophenanthrenes[a]

R	Solvent[b]	$J_{9,10}$	$J_{9,10'}$	$J_{10,10'}$	Percentage Pseudoaxial
CN	CCl_4	9.89	6.37	−14.57	44
Me	CS_2	5.24	6.70	−14.73	77
t-BU	CD_3CN	1.50	7.11	−16.57	100
COMe	CCl_4	3.81	6.02	−15.38	86
CO_2Me	CCl_4	5.76	6.64	−15.40	73
CO_2Et	CCl_4	5.38	6.86	−14.57	75
OH	$CDCl_3$	2.9	5.9	−15.2	94
$SiMe_3$	$CDCl_3$	1.95	6.79	−15.5	100

[a] From Ref. 60.
[b] Various solvents were employed to achieve best spectrum for analysis.

as indicated by the small $J_{9,10}$ values (1.5 and 1.95 Hz, respectively). A reduction in substituent size showed diminished pseudoaxial preference as expected (e.g., Me = 77%); however, two groups showed somewhat anomalous behavior. Hydroxy appears to be too large with 94% pseudoaxial preference, but this measurement was made (by necessity) in the presence of $Eu(fod)_3$ shift reagent and so this may have upset the equilibrium balance. On the other hand, cyano appears to be too small and actually prefers the pseudoequatorial position! This is no doubt due to the cylindrical nature of this substituent which greatly reduces the peri interaction with the aryl protons.

Another exception to pseudoaxial preference was reported by op het Veld and Laarhoven[61] for 9-phenyl-9,10-DHP. In this case, they reported a pseudoaxial population of only 39%. This may be rationalized by the lack of serious peri interactions for a phenyl substituent in an orientation perpendicular to the mean plane of the DHP moiety. Interestingly a more recent X-ray structure[62] on 9-methyl-10-phenyl-9,10-DHP shows the phenyl group to be pseudoaxial and the methyl pseudoequatorial. Hence, it appears that an adjacent substituent can have a substantial effect on the steric requirements for phenyl with respect to the DHP system.

ACKNOWLEDGMENTS

Support of this work has been provided by the U.S. Department of Energy, Office of Energy Research.

REFERENCES

1. Butcher, S. S. *J. Chem. Phys.* **1965**, *42*, 1830.
2. Luss, G.; Harmony, M. D. *J. Chem. Phys.* **1965**, *43*, 3768.
3. Dallinga, G.; Tonneman, L. H. *J. Mol. Struct.* **1967**, *1*, 11.

4. Traetteberg, M. *Acta Chem. Scand.* **1968**, *22*, 2305.
5. Oberhammer, H.; Bauer, S. H. *J. Am. Chem. Soc.* **1969**, *91*, 10.
6. Carreira, L. A.; Carter, R. O.; Durig, J. R. *J. Chem. Phys.* **1973**, *59*, 812.
7. Cooper, M. A.; Elleman, D. D.; Pearce, C. D.; Mannat, T. S. *J. Chem. Phys.* **1970**, *53*, 2343.
8. Auf der Heide, W.; Luttke, W. *Chem. Ber.* **1978**, *111*, 2384.
9. Turner, R. B.; Mallon, B. J.; Tichy, M.; Doering, W. von E.; Roth, W. P.; Schröder, G. *J. Am. Chem. Soc.* **1973**, *95*, 8605.
10. (a) Bishof, P.; Heilbronner, E. *Helv. Chim. Acta* **1970**, *53*, 1677. (b) Batich, C.; Bishof, P.; Heilbronner, E. *J. Electron Spectrosc. Relat. Phenom.* **1972-73**, *1*, 333.
11. Giordan, J. C.; McMillan, M. R.; Moore, J. H.; Staley, S. W. *J. Am. Chem. Soc.* **1980**, *102*, 4871.
12. Schrader, B.; Ansmann, A. *Angew. Chem. Int. Ed. Engl.* **1975**, *14*, 364 and references therein.
13. Traetteberg, M. *Acta Chem. Scand.* **1970**, *24*, 2285.
14. (a) Birch, A.J.; Hinde, A. L.; Radom, L. *J. Am. Chem. Soc.* **1981**, *103*, 284. (b) Saebo, S.; Boggs, J. E. *J. Mol. Struct.* **1981**, *73*, 137.
15. Komornicki, A.; McIver, J. W., Jr. *J. Am. Chem. Soc.* **1974**, *96*, 5798.
16. (a) Beecham, A. F.; Fridrichson, J.; Matieson, A. *Tetrahedron Lett.* **1969**, *27*, 3131. (b) Paaren, H.; Moriarty, R. M.; Flippen, J. *J. Chem. Soc., Chem. Commun.* **1976**, 114.
17. Burgstahler, A. W.; Ziffer, H.; Weiss, U. *J. Am. Chem. Soc.* **1961**, *83*, 4660.
18. Ziffer, H.; Charney, E.; Weiss, U. *J. Am. Chem. Soc.* **1962**, *84*, 2961.
19. Horsman, G.; Emeis, C. A. *Tetrahedron* **1966**, *22*, 167.
20. Baldwin, J. E.; Krueger, S. M. *J. Am. Chem. Soc.* **1969**, *91*, 6444.
21. Snatzke, G. *Angew. Chem. Int. Ed. Engl.* **1968**, *7*, 14.
22. (a) Lightner, D. A.; Bouman, T. D.; Gawronski, J. K.; Gawronska, K.; Chappuis, J. L.; Crist, B. V. *J. Am. Chem. Soc.* **1981**, *103*, 5314. (b) Spangler, C. W.; Hennis, R. P. *J. Chem. Soc., Chem. Commun.* **1972**, 24.
23. Rank, A.; Peoples, H. A. *J. Comp. Chem.* **1980**, *1*, 240.
24. (a) Edwards, J. M.; Jackman, L. M. *Aust. J. Chem.* **1965**, *18*, 1227. (b) Batterham, T. J.; Young, I. G. *Tetrahedron Lett.* **1969**, 945. (c) Copley, S. D.; Knowles, J. R. *J. Am. Chem. Soc.* **1987**, *109*, 5008 (and references therein).
25. Kloonsterziel, H.; van Drunen, J. A. A. *Recl. Trav. Chim. Pays-Bas* **1970**, *89*, 368.
26. (a) Bates, R. B.; Brenner, S.; Cole, C. M.; Davidson, E. W.; Forsythe, G. D.; McCombs, D. A.; Roth, A. S. *J. Am. Chem. Soc.* **1973**, *95*, 926. (b) Bates, R. B.; Gosselink, D. W., Kaczynski, J. A. *Tetrahedron Lett.* **1967**, 205.
27. Olah, G. A.; Asensio, G.; Mayr. H.; Schleyer, P. v. R. *J. Am. Chem. Soc.* **1978**, *100*, 4347.
28. Olah, G. A.; Stural, J. S.; Liang, G.; Paquette, L. A.; Melega, P. W.; Carmody, M. J. *J. Am. Chem. Soc.* **1977**, *99*, 3349.
29. Olah, G. A.; Spear, R. J.; Messina, G.; Westerman, P. W. *J. Am. Chem. Soc.* **1975**, *97*, 4051.
30. Tolbert, L. M.; Rajca, A. *J. Org. Chem.* **1985**, *50*, 4805.
31. Perkins, M. J.; Ward, P. *J. Chem. Soc., Perkin Trans. 1* **1974**, 667.
32. Radlick, P.; Rosen, W. *J. Am. Chem. Soc.* **1967**, *89*, 5308.
33. (a) Goldstein, M. J.; Hoffman, R. *J. Am. Chem. Soc.* **1971**, *93*, 6193. (b) Haddon, R. C. *Tetrahedron Lett.* **1974**, 2797. (c) Haddon, R. C. *J. Am. Chem. Soc.* **1975**, *97*, 3608.
34. Haddon, R. C. *J. Org. Chem.* **1979**, *44*, 3608.
35. Birch, A. J.; Hinde, A. L.; Radom, L. *J. Am. Chem. Soc.* **1980**, *102*, 6430.
36. Churchil, M. R.; Scholer, F. R. *Inorganic Chem.* **1969**, *8*, 1950.
37. Hoffmann, R., Hofmann, P. *J. Am. Chem. Soc.* **1976**, *98*, 598.
38. Cook, M. J.; Katritzky, A. R.; Pennington, F. C.; Semple, B. M. *J. Chem. Soc. B* **1969**, 523.

39. Cook, M. J. Dassanayake, N. L. *J. Chem. Soc., Perkin Trans. 2* **1972**, 1901.
40. Brinkmann, A. W.; Gorden, M.; Harvey, R. G.; Rabideau, P. W.; Stothers, J. B.; Ternay, A. L., Jr. *J. Am. Chem. Soc.* **1970**, *92*, 5912.
41. Burton, G. W.; Carr, M. D.; de la Mare, P. B. D.; Rosser, M. J. *J. Chem. Soc., Perkin Trans. 2* **1972**, 710.
42. Lamberts, J. J. M.; Laarhoven, W. H. *Recl. Trav. Chim. Pays-Bas* **1983**, *102*, 181.
43. Beckett, A. H.; Mulley, B. A. *Chem. Ind. (London)* **1955**, 146.
44. Braude, E. A.; Forbes, W. F. *J. Chem. Soc.* **1955**, 3776.
45. Beaven, G. H.; Hall, D. M.; Lessie, M. S.; Turner, E. E. *J. Chem. Soc.* **1952**, 854.
46. (a) Suzuki, H. *Electronic Absorption Spectra and Geometry of Organic Molecules*; Academic: New York, 1967. (b) Unanue, A.; Bothorel, P. *Bull. Chem. Soc. Chim. Fr.* **1966**, 1640.
47. Zaitsev, B. A.; Hramova, G. I. *Izv. Acad. Nauk SSSR, Ser. Khim.* **1974**, 2722.
48. (a) Muir, K. W.; Ferguson, G. *J. Chem. Soc. B* **1968**, 476. (b) Kalchhauser, H.; Schloegl, K.; Weissensteiner, W.; Werner, A. *J. Chem. Soc., Perkin Trans. 1* **1983**, 1723.
49. Mislow, K.; Glass, M. A. W.; Hoops, H. B.; Simon, E.; Wahl, G. W. *J. Am. Chem. Soc.* **1964**, *86*, 1710.
50. Oki, M.; Iwamura, H.; Hayakawa, N. *Bull. Chem. Soc. Jpn.* **1963**, *36*, 1542.
51. Schloegl, K.; Werner, A.; Widhalm, M. *J. Chem. Soc., Perkin Trans. 1* **1983**, 1731.
52. (a) Adam, F. C. *Can. J. Chem.* **1971**, *49*, 3524. (b) van der Kooij, J.; Gooijer, C.; Velthorst, N. H.; Maclean, C. *Recl. Trav. Chim. Pays-Bas* **1971**, *90*, 732. (c) Iwaizumi, H.; Matsuzaki, T.; Isobe, T. *Bull. Chem. Soc. Jpn.* **1972**, *45*, 1030.
53. Mislow, K.; Hoops, H. B. *J. Am. Chem. Soc.* **1962**, *84*, 3018.
54. Hall, D. M.; Turner, E. E. *J. Chem. Soc.* **1955**, 1242.
55. Yamamoto, O.; Nakanishi, H. *Tetrahedron* **1973**, *29*, 781.
56. Rabideau, P. W.; Harvery, R. G.; Stothers, J. B. *J. Chem. Soc., Chem. Commun.* **1969**, 1005.
57. Morin, F. G.; Horton, W. J.; Grant, D. M.; Pugmire, R. J. *J. Org. Chem.* **1985**, *50*, 3380.
58. Bauld, N. L.; Yound, J. D. *Tetrahedron Lett.* **1974**, *36*, 3143.
59. Cohen, D.; Millar, I. T.; Heaney, H.; Constantine, P. R.; Katritzky, A. R.; Semple, B. M.; Sewell, M. J. *J. Chem. Soc. B* **1967**, 1248.
60. Harvey, R. G.; Fu, P. P.; Rabideau, P. W. *J. Org. Chem.* **1976**, *41*, 3722.
61. op het Veld, P. H. G.; Laarhoven, W. H. *J. Chem. Soc., Perkin Trans. 2* **1978**, 915.
62. Lapouyade, R.; Koussini, R.; Nourmamode, A.; Courseille, C. *J. Chem. Soc. Chem. Commun.* **1980**, 740.

4

Conformational Analysis of 1,4-Cyclohexadienes and Related Hydroaromatics

Peter W. Rabideau

The preferred geometries of 1,4-cyclohexadiene (**1**), 1,4-dihydronaphthalene (**2**), 9,10-dihydroanthracene (**3**), and their derivatives have been the source of

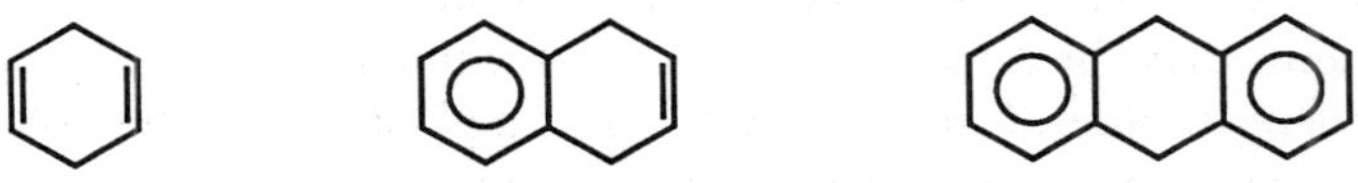

1 **2** **3**

considerable controversy. At issue are (1) planar versus boat-shaped conformations for the cyclohexadiene ring, (2) dynamic processes like ring inversion, and (3) preferred locations of substituents. In addition, the long-range (homoallylic), proton–proton coupling constants across the six-membered ring (^5J) have also been a matter of concern in that these values serve as an especially useful method for the determination of geometries in these systems.

Drieding mechanical models indicate a similar degree of puckering in the 1,4-cyclohexadiene ring throughout the series **1** to **3** with an angle between the two planes containing the π bonds of $\alpha = 145°$. Moreover, 9,10-dihydroanthracene is boat shaped in the solid state (X-ray analysis) with a deviation from

planarity more or less identical to that predicted by mechanical models. It is perhaps for these reasons that there seemed to be a reluctance to accept early experimental evidence indicating 1,4-cyclohexadiene itself to be planar (i.e., $\alpha = 180°$).

As we shall see, there is certainly a trend toward nonplanar structures in the direction **1** → **3**. However, the main characteristic of 1,4-dihydroaromatic rings

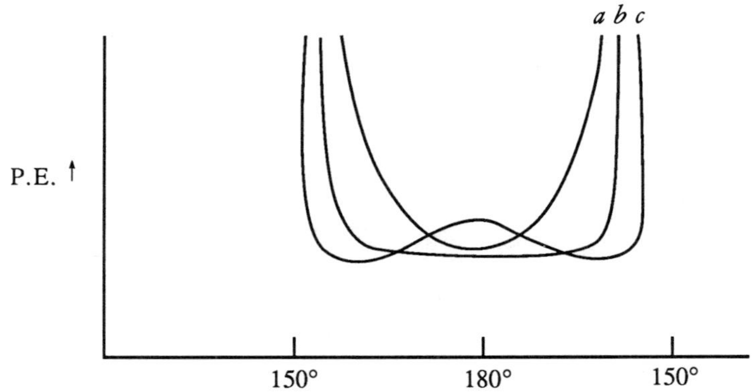

Figure 4.1 Changes in potential energy profiles resulting from relative energy differences between planar and folded structures (planar = 180°).

appears to be a very low barrier toward perturbation so that a wide variety of ring conformations may be expected. Figure 4.1 represents several hypothetical potential energy profiles. Curve *a* corresponds to a planar ring system while curve *c* represents two nonplanar, interconverting rings with some energy barrier separating them. However, there is a "grey" area (curve *b*) where the well can be quite broad and the planar state may be either slightly higher or slightly lower in energy across this region. Many of the structures to be discussed here fit into this latter category.*

4.1 1,4-CYCLOHEXADIENE

The conformation of 1,4-cyclohexadiene has been the subject of numerous investigations. Vibrational,[1,2] rotational Raman,[3] and an early NMR study by Garbisch and Griffith[4] all concluded D_{2h} symmetry (planarity). However, two different electron diffraction studies appeared with the first[5] favoring the planar

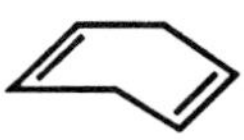
Electron diffraction results

D_{2h} (Ref. 5) C_{2h} (Ref. 6)

form as suggested by previous workers, but the second,[6] in contrast with all previous studies, indicated C_{2h} symmetry with a folding angle of $\alpha = 159.3°$. A far-infrared study by Laane and Lord[7] concluded that the molecule favors the planar form *or* that the barrier for the interconversion of boat forms is very high.

---

* It is important to note several terms which have been used throughout the literature, and, as a consequence, are used herein. A "fully puckered" 1,4-cyclohexadiene (or dihydroaromatic) refers to a value of $\alpha = 145°$ as described above. "Flattened boat" or "slightly puckered" geometries refer to α values somewhere between 145° and 180° (planar), although probably closer to the latter.

They did favor the planar form, however, and we also note that a high barrier to inversion would lead to a nonequivalence of allylic protons in the NMR spectrum which is not observed.[4] More recent studies by NMR[8] and Raman[9,10] also support a planar conformation as do *ab initio*,[11,12] molecular orbital,[13,14] and force-field calculations[13,15] (discussed elsewhere in detail in this book). Interestingly, the most recent Raman investigation[10] also examined low-temperature liquid and solid states of 1,4-cyclohexadiene and they concluded that the planar form was favored under all conditions including the solid state. Apparently conclusions supporting the boat geometry failed to take into account that average interatomic distances will be shorter due to vibrational distortion from planarity.[8] Hence, 1,4-cyclohexadiene is planar!

Laane and Lord[7] have provided a "qualitative" description of the ring-puckering potential of 1,4-cyclohexadiene and this result is reproduced in Figure 4.2, where 2Z represents the displacement of each aliphatic carbon atom out of

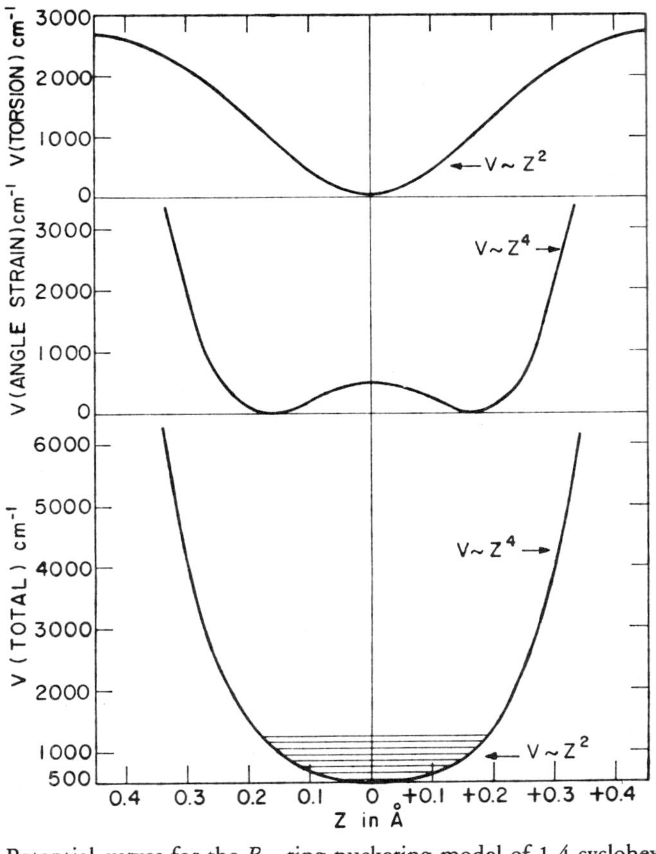

Figure 4.2 Potential curves for the B_{2u} ring-puckering model of 1,4-cyclohexadiene: (top) potential due to torsion about the C—C single bonds; (center) potential due to C—CH_2—C angle strain; (bottom) sum of these two. Reproduced with permission from *J. Mol. Spectrosc.* **1971**, *39*, 340. Copyright © 1971 Academic Press, Inc.

the plane of the other four carbons (in Å). Thus, the molecule is planar for $Z = 0$ and puckered into a boat conformation for nonzero values of Z. The upper curve shows the torsional potential resulting from the barrier to internal rotation around the C—C single bonds. This potential is at a minimum in the planar form when the olefinic CH is bisecting the allylic CH_2's and is maximized as the ring puckers and the CH and (one of the) CH_2 hydrogens become eclipsed. They estimate this torsional barrier to be about 8.0 kcal/mol.

The center curve in Figure 4.2 represents the angle strain considering the average angle of 120° in a planar six-membered ring and the presumed desirable tetrahedral angle (109.5°). However, this angle strain was also addressed in a later

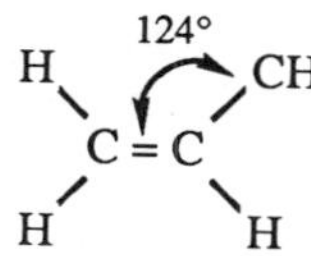

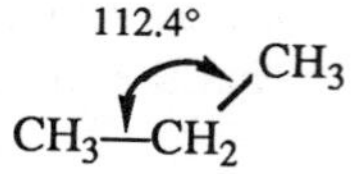

report[13] where propene and propane were used as models. The sum of the interior angles in a planar ring is, of course, $6 \times 120° = 720°$, and since four "propene" type angles ($4 \times 124° = 496°$) plus two "propane" angles ($2 \times 112.4° = 224.8°$) adds up to 720.8°, this study concluded that planar 1,4-cyclohexadiene should have no angle deformation. Of course, angle strain becomes very important for large puckering angles.

The combination of these potentials leads to the bottom curve, which represents the overall potential. Several curves similar in appearance to this one are to be found in the literature as a result of molecular mechanics[13,15] or molecular orbital calculations.[13,14] This description of 1,4-cyclohexadiene as an essentially planar compound with vibrational motion involving boat conformations has also received experimental support from the observed temperature dependence of the trans homoallylic NMR coupling constant[8] (see next section).

4.2 NMR HOMOALLYLIC COUPLING CONSTANTS

An important method for the conformational analysis of **1–3**, and especially their derivatives, has involved the use of the proton–proton homoallylic coupling constants across the six-membered ring. This ^5J coupling, which can be quite large (12 Hz), is transmitted though the p orbitals and consequently decreases as the orbital overlap decreases so the values are largest for a dipseudoaxial (di-pa) relationship in a "fully puckered" boat geometry, and smallest in the dipseudoequatorial (di-pe) arrangement. Flattening of the ring will cause di-pa coupling to decrease and di-pe coupling to increase until they become identical in

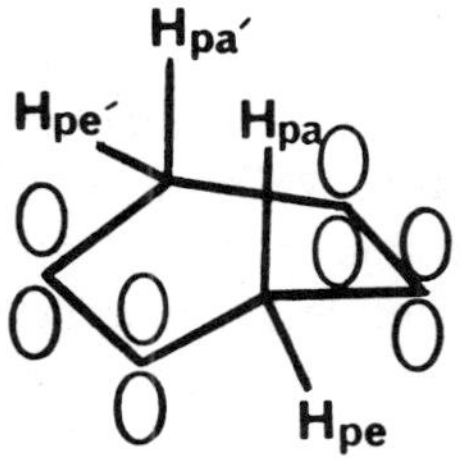

(1) $J_{pa/pa'} > J_{pa/pe'} > J_{pe/pe'}$

(2) 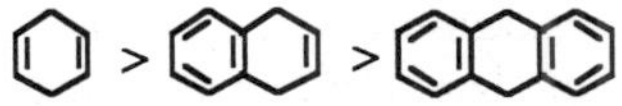

the planar ring. As would be expected, pa/pe interactions are intermediate in value and are not as sensitive to geometry changes. However, in contrast to the cis coupling, the trans values are at a maximum with a planar conformation. The magnitude of these couplings for any given geometric relationship also decreases in the order DHB > DHN > DHA (**1** > **2** > **3**). This is presumably a consequence of the decreasing bond order throughout this series.

Like many aspects of the conformational analysis of **1–3**, however, this has been a controversial topic.[16] These long-range couplings were first observed by Durham et al.[17] for the isomeric 1,4-dihydrobenzenes, **4c** and **4t**. The coupling was largest for the cis compound **4c** ($^5J_{1,4}$ = 11 Hz) and somewhat smaller for the trans **4t** $^5J_{1,4'}$ = 7.5 Hz). Nonetheless, these are *large* values for long-range

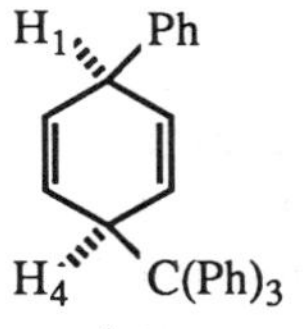

4c

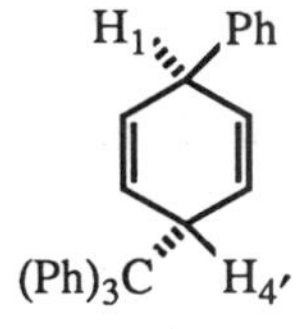

4t

coupling! It was the work of Garbisch and Griffith,[4] however, that really brought attention to the use of these values for the determination of preferred conformations in 1,4-cyclohexadienes. They were successful in determining the cis and trans homoallylic coupling constants for 1,4-cyclohexadiene-d_6 in a rather clever manner. Birch reduction of benzene-d_6 provided a mixture of **5c** and **5t**,

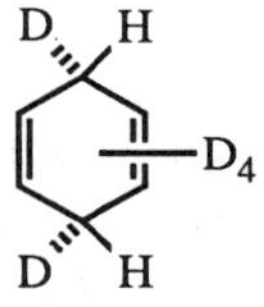

5c

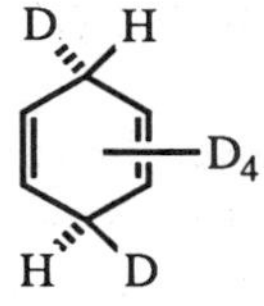

5t

and so the ^{1}H-NMR spectrum showed a single resonance for the allylic hydrogens. However, observation of the ^{13}C—H satellite spectrum in the ^{1}H-NMR allowed the determination of J_{cis} (9.63 Hz) and J_{trans} (8.04 Hz). Using a Karplus-type approach, they provided an equation relating the ratio of J_{cis}/J_{trans} to the extent of ring puckering (α), and on this basis they concluded **1** to be planar or nearly so, and this work has been subsequently cited as the first NMR evidence[7,12,15] for planar **1**. Atkinson and Perkins[18] questioned the equation of Garbisch and Griffith, who later published a correction.[19] In fact, their recalculation produced a value of $\alpha = 165°$, and so they felt their results were now supportive of the Oberhammer and Bauer electron diffraction study suggesting a boat-shaped conformation.[19]

Based on the approach of Barfield and Sternhall[20] used to determine the homoallylic coupling in *cis*-2-butene, Atkinson and Perkins,[18] and later Grossel and Perkins,[21] calculated values for $^5J_{cis}$ and $^5J_{trans}$ in 1,4-cyclohexadiene and concluded that $^5J_{trans}$ should be somewhat larger than $^5J_{cis}$ for a planar ring. Marshall et al.,[22] on the other hand, took issue with these results and their SCF-INDO-FPT[23] calculated results are reproduced in Table 4.1 (note that α has a different meaning here). As we will see later, these results are too large for the ^5J couplings as recognized by these investigators who attributed it to an overestimation of σ-π contributions as noted previously.[24] In any event, the issue at question was the $^5J_{cis}/^5J_{trans}$ ratio, and their calculated ratio for a planar ring (1.29) was in excellent agreement with the experimental results of Garbisch and Griffith[4] (1.20). Later calculations using the Dallinga and Toneman[5] geometry as well as a "standard" geometry produced similar results with ratios of 1.22 and 1.28, respectively.[25]

It was later pointed out by Grossel[26] that both theoretical methods give similar, qualitative descriptions of the changes in $^5J_{cis}/^5J_{trans}$ ratios with changing ring geometry. He conceded, however, that the available experimental evidence seemed to support a $^5J_{cis}/^5J_{trans}$ ratio of slightly greater than unity. In an interesting experiment, Grossel and Perkins[8] also demonstrated a temperature dependence in $^5J_{trans}$ ($\sim 5 \pm 2\%$) but not $^5J_{cis}$. This was attributed to a reduction of the boat-to-boat vibrational motion which would be expected to reduce trans coupling since the protons in question become less axial.

4.3 DERIVATIVES OF 1,4-CYCLOHEXADIENE (1,4-DIHYDROBENZENE)

As discussed above, the energy required to pucker the otherwise planar 1,4-cyclohexadiene is relatively small and it is quite conceivable that substituents can lead to a boat-shaped ring with groups at the 1 or 4 position (i.e., 1,4-dihydrobenzene),* occupying either a pseudoaxial or a pseudoequatorial position.

* Most workers in this field number the allylic positions 1 and 4. This is, of course, correct for 1,4-dihydrobenzene, but these positions are 3 and 6 if the ring is considered to be 1,4-cyclohexadiene. We use the former system since it is most consistent with the literature on this subject.

Table 4.1 Calculated[a] Proton–Proton Coupling Constants of 1,4-Cyclohexadiene at Various Puckering Angles

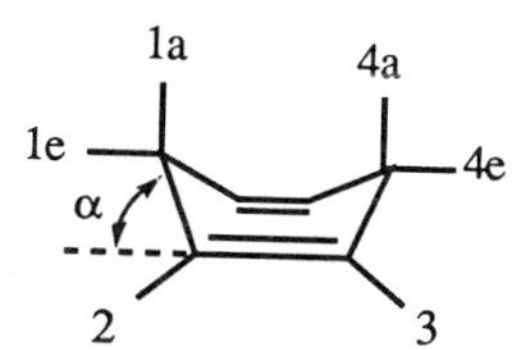

Puckering Angle, α, deg	Calculated J, Hz								Energy, kcal/mol
	1a, 4a (cis)	1e, 4e (cis)	1a, 4e (trans)	1a, 2	1e, 2	1a, 3	1e, 3	1a, 1e	
0	16.06	16.06	12.45	4.10	4.10	−2.30	−2.30	−13.85	0.0
10	21.96	9.73	11.22	3.20	5.28	−3.26	−1.10	−13.01	0.67
20	25.44	4.55	8.05	2.86	6.41	−3.85	0.18	−10.85	3.08
30	25.13	1.23	4.29	3.16	7.14	−4.02	1.37	− 8.24	8.08
35	23.41	0.31	2.65	3.53	7.26	−3.95	1.89	− 7.04	11.89
35[b]	22.83		2.79	3.58		−4.01		− 6.52	
2-Butene, O°[c]	4.78	4.78	4.59	3.85	3.85	−2.18	−2.18	− 9.44	

[a] Calculated by SCF—INDO—FPT; geometry taken from an electron diffraction study. All bond lengths and independent bond angles are not changed, except for the puckering angle α, defined as the dihedral angle of the plane of C(2)—C(3)—C(5)—C(6) and the plane of C(2)—C(1)—C(6).
[b] Done as for the previous 35° case except that all overlap integrals between C(1) and its hydrogens, and C(4) and its hydrogens, have been reduced to zero. [c] Geometry exactly the same as for 1,4-cyclohexadiene at $\alpha = 0°$.

Reproduced with permission from *J. Am. Chem. Soc.*, *1971*, 99, 321. Copyright (1977) ACS. Geometry used was from electron diffraction study of Oberhammer and Bauer (reference 6); α as used here should be subtracted from 180° to compare values used elsewhere in the text.

versus

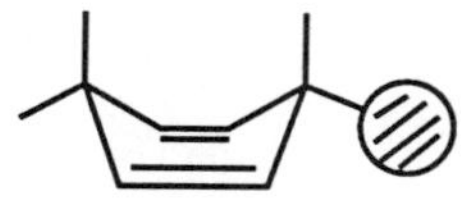

pseudoaxial pseudoequatorial

On the basis of space-filling models, Atkinson and Perkins[18] argued that a single bulky substituent should be preferentially located in the pseudoequatorial position (however, see below).

Among the earliest derivatives studied were the isomeric *cis*- and *trans*-1,4-dihydro-4-tritylbiphenyls (**4c** and **4t**).[17,18] It was concluded from the large

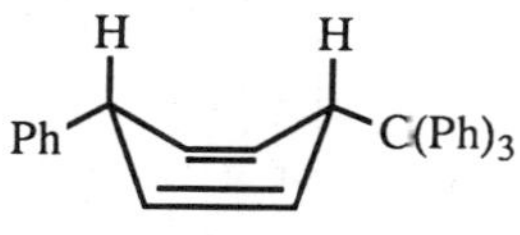

4c

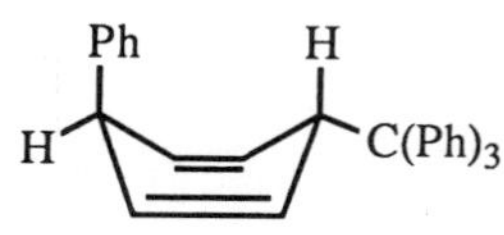

4t

homoallylic coupling constant in the cis isomer (11 Hz) that the cyclohexadiene ring is highly puckered with the substituents located in pseudoequatorial positions. Further evidence for the trityl group "locking" the conformation in these systems was provided by a "preliminary" X-ray crystallographic study of *trans*-4′-bromo-4, which concluded a puckered ring (α = 165°) and the trityl group in the pseudoequatorial position. However, the value of $^5J_{trans}$ in **4t** was recognized later[27] as being too large[16] for a highly puckered ring, and this prompted Grossel et al.[27] to determine its crystal structure. This brought to light two important features in contrast to the previous studies: (1) the 1,4-cyclohexadiene ring was found to be flattened (α = 171.8°), and (2) the trityl group was found to be pseudoaxial! Subsequently, a crystal structure of the cis isomer appeared, and it was found to be even *more planar* (α = 175°) than the trans isomer (although somewhat distorted)[28] with both substituents slightly pseudoequatorial.

Marshall et al.[29] investigated the behavior of 1,4-dihydrobenzoic acid (**5**), which represents a 1,4-cyclohexadiene with a single, less bulky substituent. Based on the current thinking at that time about **4c** and **4t**, they felt that **5** should also

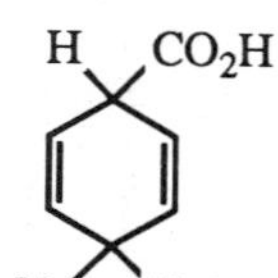

5

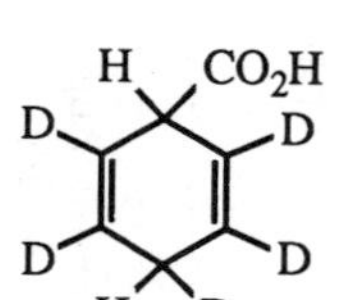

5 -d_5 (cis and trans)

be puckered but to a lesser extent. Unfortunately, the NMR analysis of **5** was complicated by the fortuitous chemical shift equivalence of H_4 and $H_{4'}$, resulting in a deceptively simple ABX system (although an A_2X system where $^5J_{1,4} = {^5J_{1,4'}}$ could not be ruled out). In the hope of overcoming this problem, they prepared *cis*- and *trans*-**5**-d_5 by the Birch reduction of perdeuterobenzoic acid. Unfortunately, the NMR spectrum of **5**-d_5 showed only two sharp doublets and so any differences in ^5J between the cis and trans isomers could not be resolved. In any event, it was clear that J_{cis}/J_{trans} was close to unity and according to the previous work of Atkinson and Perkins,[18] $J_{cis} = J_{trans}$ only for a somewhat puckered ring. Thus, they concluded not only that **5** must be puckered,* but the parent compound as well. They also added that a single moderately sized substituent can be fitted comfortably into the pseudoequatorial position.

A remarkable number of "accidental" chemical shift equivalences in a variety of derivatives led to serious problems with the NMR conformational analysis of these systems. The NMR spectra (albeit at 60 or 100 MHz) of the

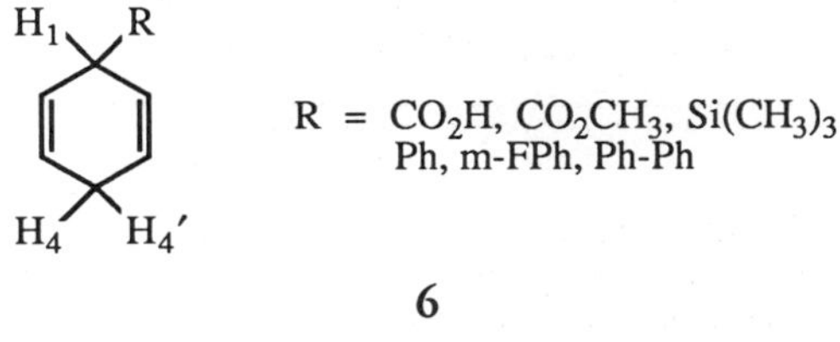

6

assorted derivatives illustrated as **6** all look more or less the same[16] with H_1 as a broad triplet, H_4 and $H_{4'}$ together as a broad doublet, and the vinyls (often) as a broad singlet and so the vicinal and allylic coupling constants (which are also geometrically dependent) could not be measured.

In an attempt to overcome this difficulty, Paschal and Rabideau[31,32] examined the dihydrobenzyl alcohols **7** and **8** using $Eu(fod)_3$ shift reagent, which

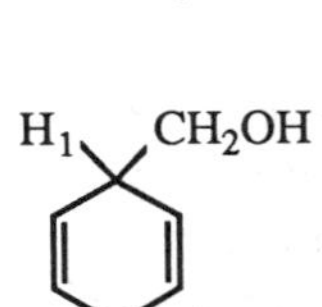

7

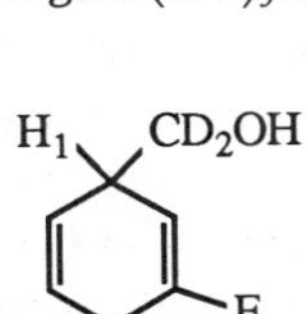

8

allowed the complete ^{1}H-NMR analysis of these 1,4-cyclohexadiene derivatives. It was, of course, necessary that at least one conformationally dependent coupling constant be measured both in the presence and absence of shift reagent so as to

---

* Later in a paper focusing in 1,4-dihydronaphthoic acid, Marshall and Folsom refer to **5** as "nearly planar."[30]

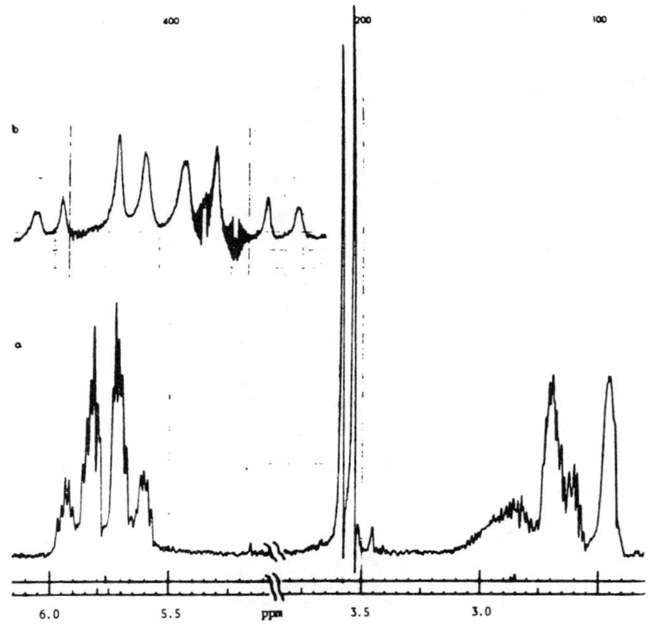

Figure 4.3 The 100 MHz NMR spectrum of **7**. Upper trace is H_4 and $H_{4'}$ with added Eu(fod)$_3$ and triple irradiation of the vinyls (i.e., each vinyl decoupled separately and simultaneously). Reproduced from *J. Am. Chem. Soc.* **1975**, 97, 5700 with permission. Copyright © 1977 American Chemical Society, Washington, DC.

rule out the possibility of conformational change induced by the Eu(fod)$_3$. The spectrum of **7** (Figure 4.3) beautifully illustrates these long-range couplings (upper trace); the values are $^5J_{cis} = 8.6$ and $^5J_{trans} = 7.4$ Hz for **7**, and $^5J_{cis} = 8.3$ and $^5J_{trans} = 7.5$ Hz for **8**. Thus, the $^5J_{cis}/^5J_{trans}$ ratios (1.16 and 1.11, respectively) were in close agreement with the INDO calculated result[20,33] for a completely planar ring (1.29). Moreover, the vicinal coupling constants between both H_4 and $H_{4'}$ with the adjacent vinyl(s) were equal, indicating a planar or time-averaged planar ring. The latter question was the reason for the fluorine in **8**. It was argued that a single puckered conformation or an equilibrium between (substantially) puckered conformations (**8a** ⇌ **8e**) would lead to a large vicinal $J_{H,F}$ coupling (the

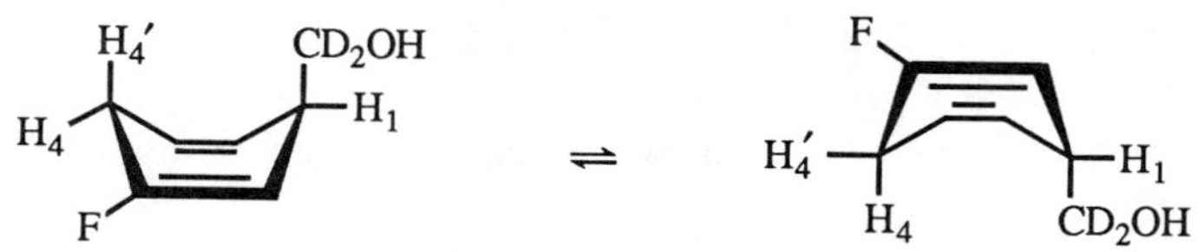

8a **8c**

"averaged" coupling in 2-fluoropropene is[36] J_{F,CH_3} = 16.0 Hz). In fact, this coupling was unobservable and could be estimated at less than 2 Hz, which is consistent with the fluorine bisecting[34] H_4 and $H_{4'}$ in a planar ring with only shallow vibrations.

Subsequently, Marshall et al.[22] reinvestigated **5**-d_5 wherein the carboxyl

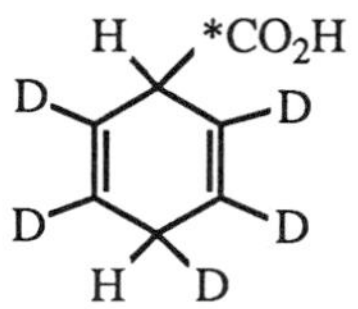

5 -d_5 (cis and trans) with carbon-13 label

carbon carried carbon-13 enrichment. In contrast to the earlier study,[29] they were now able to measure accurately the proton–proton homoallylic coupling constants due to enhanced resolution gained by deuterium decoupling. The values were $^5J_{cis}$ = 9.19 and $^5J_{trans}$ = 7.56 Hz, providing a $^5J_{cis}/^5J_{trans}$ ratio of 1.22, which is consistent with both calculation and the previous results for **7** and **8** indicating a planar ring. Interestingly, they also measured cis $^5J_{CH}$ and trans $^5J_{CH}$ coupling constants (finding values of 5.75 amd 4.65 Hz, respectively) and suggested that the ratio of these values (also about 1.2) may also be used as a conformational probe in 1,4-cyclohexadienes.

Attempts to demonstrate the variation of the homoallylic coupling constants with ring puckering led to the investigation of rigid models.[32] The dihydrotriptycene derivative **9c** shows one of the largest $^5J_{cis}$ values ever observed (^5J = 12.0 Hz). This is a big number for long-range coupling! The ^5J coupling in **9t**

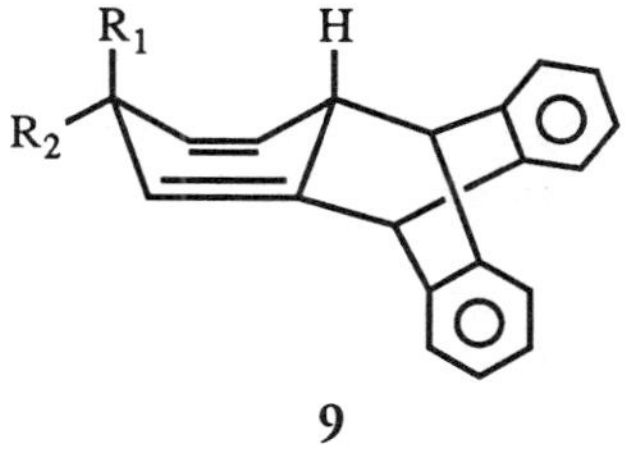

9

9c $R_1 = H$, $R_2 = CO_2CH_3$
9t $R_1 = CO_2CH_3$, $R_2 = H$

represents a pseudoaxial/pseudoequatorial arrangement and so the smaller value of 4.7 Hz is not surprising. As expected, the $^5J_{cis}/^5J_{trans}$ ratio (2.6) for **9c**/**9t** is quite different from the usual value found for planar rings (1.2). The vicinal and allylic coupling constants are also geometrically dependent, of course, and should also provide a conformational tool, although as mentioned previously, they often cannot be measured. The use of Eu(fod)$_3$, however, facilitated such measurements once again with **9c** and **9t**, and a "composite" set of values is shown in Table 4.2. A comparison of the values in Table 4.2 with the calculated values of Table 4.1 for

Table 4.2 Experimental Proton–Proton Coupling Constants in 1,4-Cyclohexadienes from Data for Compounds *9c* and *9t*.[a]

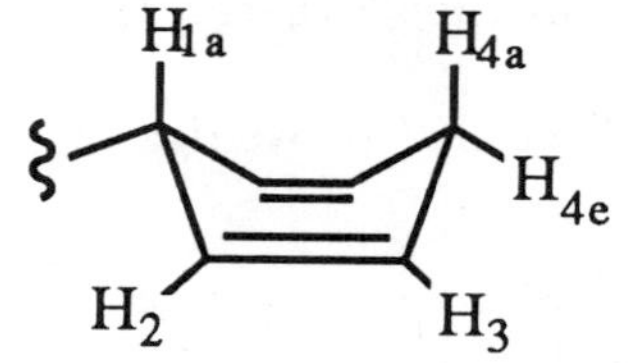

Coupling Type	1a, 4a	1a, 4e	J Values (Hz) 1a, 2	3, 4e	1a, 3	2, 4e
Homoallylic	12.0	4.7				
Vicinal			2.5	5.7		
Allylic					3.0	<1.0

[a] Based on puckered ring; $\alpha = 145°$

a 35° puckering angle is quite interesting (note that Table 4.1 uses a different definition for α- their $\alpha = 35°$ corresponds to $\alpha = 145°$ as defined in this chapter). The INDO calculated results are considered to be too large[22] but their relative values are presumed to have merit. In fact, the actual values for the vicinal and allylic couplings (Table 4.2) would compare favorably with theory (Table 4.1) if the calculated results were uniformly reduced by 25%. However, agreement with the experimental homoallylic coupling constants is poor no matter how it is approached: the calculated result for $^5J_{cis}$ is too high by almost 100%, and the value for $^5J_{trans}$ is too low.

In view of these results, it might be supposed that the calculated $^5J_{cis}/^5J_{trans}$ ratio for a planar ring that seems to agree so well with experiment could be fortuitous. A substituent effect is possible, of course, but the $^5J_{cis}/^5J_{trans}$ ratios for 1,4-dihydrobenzyl alcohol and 1,4-dihydrobenzoic acid agree rather well (1.16 and 1.22, respectively). The appearance of additional models such as **10** and **11**

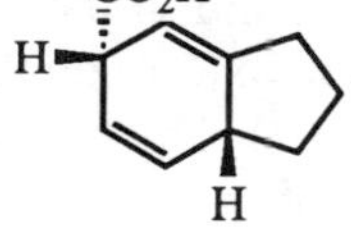

10

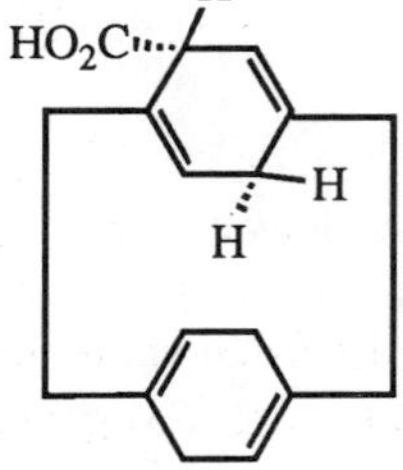

11

have not helped much since they also have a carboxyl substituent. $^5J_{cis}$ was measured as 12 Hz for **10**, but unfortunately only one isomer was available so the corresponding $^5J_{trans}$ remains unknown.[21] Moreover, the value for **11** (8.75 Hz) could only be approximated since it was measured from the methine proton doublet, which the authors recognized as an ABX spectrum.[35] Grossel and Perkins[21] observed a considerable decrease in $^5J_{cis}$ values for the 2,6-disubstituted-1,4-dihydrobenzoic acids **12** and **13** (6.3 and 6.0 Hz, respectively), and mainly on

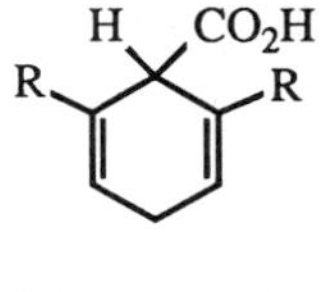

12, R=OCH$_3$
13, R=CD$_3$

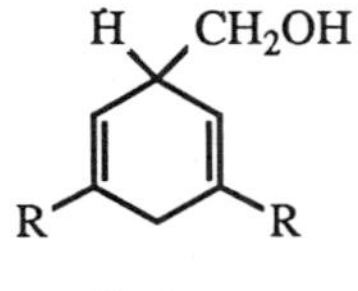

7, R=H
14, R=CH$_3$

the basis of the vicinal coupling between H$_3$ and H$_4$ (H$_{4'}$) (**12**: $J_{3,4}$ = 4.0, $J_{3,4'}$ = 3.3 Hz; **13**: $J_{3,4}$ = 4.0, $J_{3,4'}$ = 3.25 Hz), favored an "approximately planar time-averaged geometry." The significant difference in $^5J_{cis}/^5J_{trans}$ ratios (0.93 for **12** and 0.96 for **13**) as compared with compounds previously determined to be planar was attributed to be due to a substituent effect rather than ring puckering. This prompted Rabideau et al.[25] to examine the effect of methyl substituents on dihydrobenzyl alcohol. Moreover, the substituents were placed at the 3 and 5 positions so as not to create any steric perturbation. In fact, the $^5J_{cis}/^5J_{trans}$ ratios for **7** and **14** are essentially identical (1.16 and 1.14, respectively) and these workers suggested that **12** and **13** do in fact prefer a shallow boat conformation with the substituent pseudoaxial. A later X-ray diffraction study by Grossel et al.[36] found that, in the solid state, **13** (but CH$_3$ not CD$_3$) is indeed slightly puckered, with α = 171.6°, and 3,5-dimethyl-1,4-dihydrobenzoic acid is planar.

The fact that substituents would prefer the pseudoaxial position when the adjacent vinyl positions were substituted was not surprising, of course, but this did not resolve the issue of the location of a single substituent at C-1 (provided it was bulky enough to cause ring puckering). As discussed above, the pseudoequatorial position was favored by some workers[18] on the basis of space-filling models (and an X-ray study) and an early investigation of *trans*-4-trityl-1,4-dihydrobiphenyl indicating that the larger trityl group was located pseudoequatorially. However, a subsequent contrasting X-ray study[27] suggested a pseudoaxial trityl group. In addition, pseudoaxial preference for the *t*-butyl-1,4-dihydrobenzene was predicted by force-field calculations,[37] and this prompted an NMR investigation of 1-(2-hydroxy-2-propyl)-1,4-dihydrobenzene (**15**).[38] The measurements were done

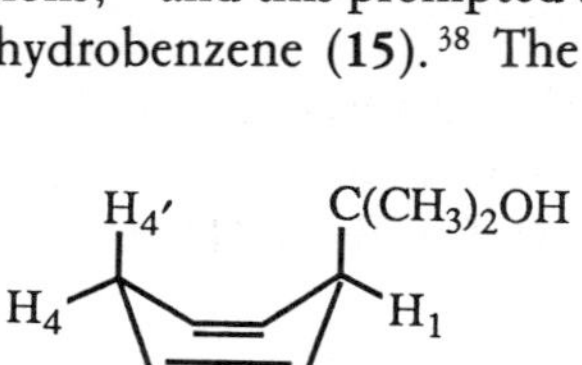

15

Table 4.3 Proton–Proton Coupling Constants for Various Structural Types of 1,4-Dihydrobenzenes.[a]

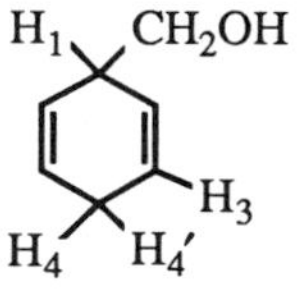

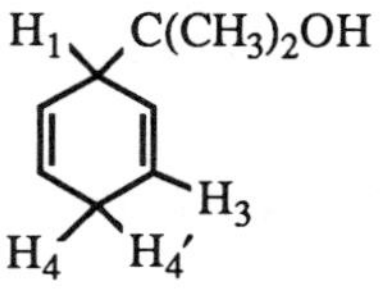

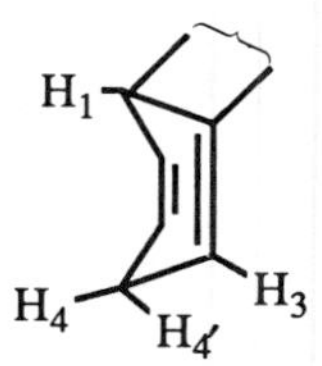

Coupling Constant	Planar	Shallow Boat	Rigid Full-Boat
$J_{1,4}$ (cis)	8.3	7.4	12.0
$J_{1,4'}$ (trans)	7.5	7.7	4.7
$J_{3,4}$	3.1	4.0	2.5
$J_{3,4'}$	3.1	2.9	5.5
${}^5J_{cis} / {}^5J_{trans}$	1.11	0.96	2.55

[a] Data from reference 40 and references therein.

in the presence of $Eu(fod)_3$, but it was possible to measure $J_{1,2}$ (3.4 Hz) both before and after the addition of shift reagent and so any effect of $Eu(fod)_3$ on the geometry of **15** could be precluded. The observation that $J_{3,4}$ (4.0 Hz) $\neq J_{3,4'}$ (2.9 Hz) indicated a nonplanar ring, with the smaller value assigned to $H_{4'}$ and the larger to H_4 based on the usual Karplus relationship. The location of the substituent was established as pseudoaxial since (1) the ${}^5J$ coupling to $H_{4'}$ (established as pseudoaxial) is too small (7.4 Hz) for a dipseudoaxial coupling, and (2) $H_{4'}$ moved downfield faster than H_4 upon addition of $Eu(fod)_3$, suggesting that it is in closer proximity to the site of complexation (the OH group).*

Although the value of ${}^5J_{cis}$ for a dipseudoequatorial relationship is unknown, these authors estimated of value of 2.6 Hz or less. On this basis it may be concluded from ${}^5J_{cis}$ = 7.4 Hz for **15** that the molecule is in a rather shallow boat conformation. This allows a summary of coupling constant values for a variety of 1,4-cyclohexadiene structural types, and such data are presented in Table 4.3. These figures may now serve as a guide for the conformational analysis of new systems.

4.4 1,4-DIHYDRONAPHTHALENES

We suggested earlier that angle strain for planar conformations is expected to increase in the series **1** < **2** < **3**, and so 1,4-dihydronaphthalene (**2**) should be intermediate between 1,4-cyclohexadiene (**1**) and 9,10-dihydroanthracene (**3**) on

* The molecule undergoes vibration, of course, but we prefer something like boat ⇌ planar over boat ⇌ boat! In addition, $J_{1,2}$ in **15** (3.4 Hz) is slightly smaller than $J_{3,4}$ (4.0 Hz) and this may be suggesting an unsymmetrical distortion of the six-member ring (i.e., flatter on the substituted side).

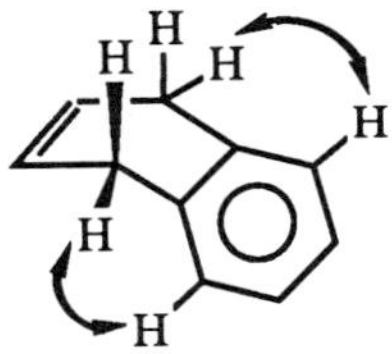

peri interactions

2

this basis (alone). This means that **2** should be more likely to assume boat-shaped conformations, but a puckered geometry for **2** also introduces a peri interaction between H_{pe} and the ortho aromatic hydrogens, an effect that is minimized with a completely planar structure. It would appear that both of these effects are important since calculations suggest that **2** also has a planar minimum but that less energy should be required for displacement to puckered geometries.

Marshall and Folsom[30] examined the NMR behavior of 1,4-dihydronaphthoic acid (**16**) in an attempt to understand better the data associated with 1,4-dihydrobenzoic acid. They initially suggested a boat-shaped geometry with the substituent pseudoaxial, on the basis of vicinal and allylic, proton–proton coupling constants. However, in a subsequent investigation,[33] a nearly planar structure was favored on the basis of the (previously undetermined) homoallylic coupling constants. Later, **17** and **18** were studied[39] to learn the effect of

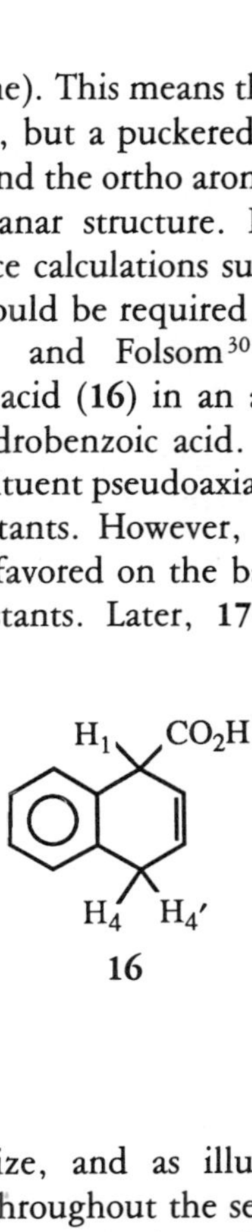

16

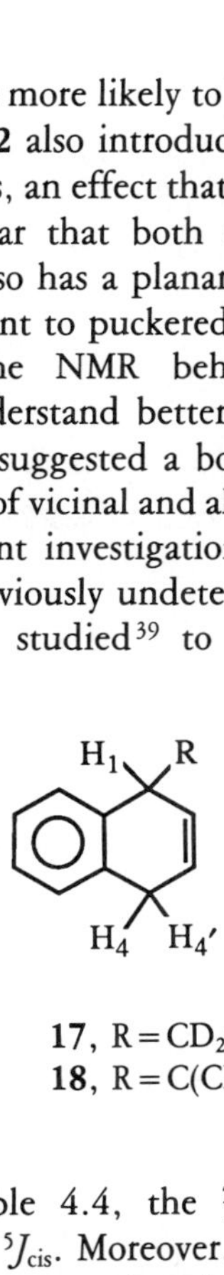

17, R = CD_2OH
18, R = $C(CH_3)_2OH$

substituent size, and as illustrated in Table 4.4, the ^5J couplings change significantly throughout the series, especially $^5J_{cis}$. Moreover, the $^5J_{cis}/^5J_{trans}$ ratios for **16** (0.86), **17** (0.95), and **18** (0.60) suggest puckered geometries for all if indeed $^5J_{cis}/^5J_{trans} \approx 1.2$ for a planar ring. This is also supported by the vicinal couplings and the reduction (actually less negative) in the geminal value. The rather low value for $^5J_{cis}$ in **18** suggests that it is considerably more puckered than **16** or **17**, and this suggests that ring geometry in 1,4-dihydronaphthalenes can be variable depending on the size of the substituent.

To investigate further the extent of puckering in **18**, rigid models were prepared that would provide a "fully puckered" boat conformation (i.e., $\alpha \approx 145°$).[39] A pair of dihydroacenaphthenes were prepared with one (**19**) having a cis geometry and the other (**20**) trans, but unfortunately synthetic problems did not allow for identical substituents. In any event, as illustrated in brackets, these

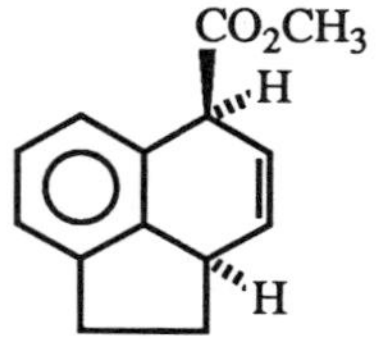
19

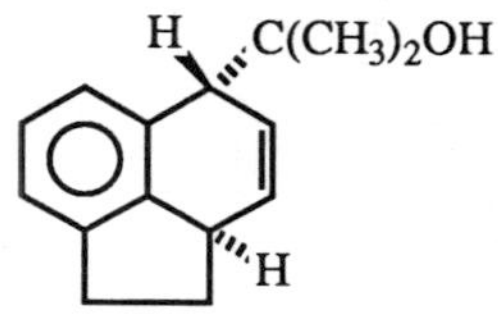
20

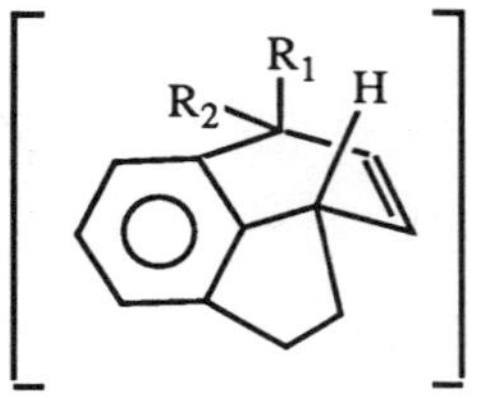

models are locked into boat conformations. Compound **19** provides what is expected to represent a maximum value for $^5J_{cis}$ (8.5 Hz) in 1,4-dihydronaphthalenes (Table 4.4); a similar value (8.4 Hz) was obtained for the corresponding carboxylic acid.[40] The favorable comparison of $J_{3,4}$ in **18** (5.0 Hz) with $J_{1,2}$ in **20** (5.0 Hz), and $J_{1,4'}$ in **18** (3.0 Hz) with $J_{1,4'}$ in **20** (3.2 Hz), suggests that **18** is indeed "fully puckered." The more modest deviation from planarity for **16** was subsequently confirmed for the solid state by X-ray diffraction, which provided a value of $\alpha = 169.2°$.[41]

The steric effect of bulky groups was also investigated by Holy et al.[42] who examined the *t*-butyl derivatives **21** and **22**. They were not able to determine ^5J

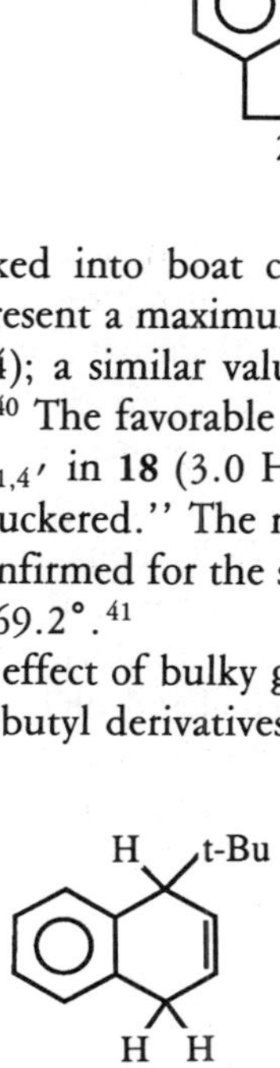
21

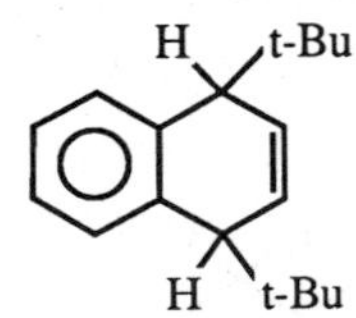
22

values for **21**, however, and suggested that these values must be less than one. This would be highly unusual, especially in view of the normal behavior of **18**. They suggested that twisting of the six-membered ring may be responsible for the peculiar behavior. A homoallylic coupling cannot be measured for **22**, of course, due to the equivalence of H_1 and H_4. Unfortunately, they were not able to make a

Table 4.4 Proton–Proton Coupling Constants for 1-Substituted 1,4-Dihydronaphthalenes[a]

Compound	J Values (Hz) 1,4 (cis)	1,4′ (trans)	3,4	3,4′	4,4′
16	3.8	4.4	4.6	2.4	21.9
17	3.5	3.7	3.7	2.1	21.0
18	1.8	3.0	5.0	2.0	20.5
19	8.5		1.4		
20		3.2		1.2[b]	

[a] Data from reference 42. [b]The vicinal coupling on the other side of the ring ($J_{1,2}$) = 5.0.

cis/trans assignment, but the $J_{3,4}$ (3.7 Hz) value prompted them to suggest a flattened cyclohexadiene ring.

More recently, Lamberts et al.[43] examined the 500 MHz spectrum of 1-phenyl-1,4-dihydronaphthalene (**23**), and relatively small differences in the

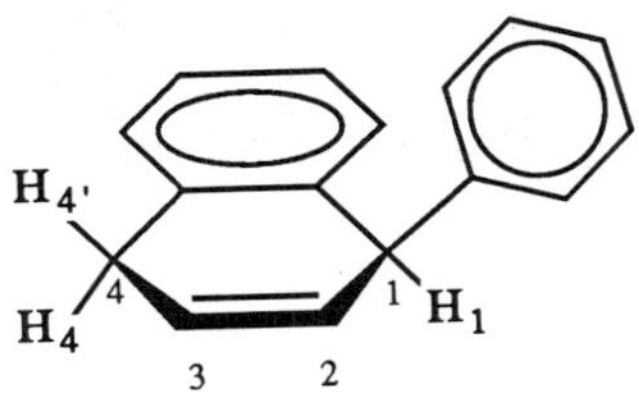

23

homoallylic ($^5J_{cis}$ = 4.6 Hz; $^5J_{trans}$ = 4.9 Hz), allylic ($J_{2,4}$ = 1.6 Hz; $J_{2,4'}$ = 2.0 Hz), and vicinal ($J_{3,4}$ = 3.75 Hz; $J_{3,4'}$ = 2.95 Hz) coupling constants indicated a rather flattened cyclohexadiene ring. They noted that this serves as an example of relatively small steric demand by a phenyl substituent. Previous examples of this type of behavior have been observed for other structures considered in this book such as 9-phenyl-9,10-dihydroanthracene[44] and 9,10-diphenyl-9,10-dihydrophenanthrene.[45] A recent molecular mechanics calculation[46] also predicts a rather flat structure for **23** with the phenyl substituent in a plane perpendicular to the dihydronaphthalene system.

Although rigid models can be helpful in making a conformational assignments for cyclohexadienes, they generally suggest "fully puckered" rings, and many of the dihydronaphthalenes studied appear to be "flattened boats." This made the synthesis[46] of **24** quite interesting since Dreiding mechanical models suggested that one of the three possible conformations (**24a**) would possess a flattened dihydronaphthalene ring ($\alpha \approx 170°$). In fact, molecular mechanics calculations predicted **24a** to be the most stable, and, moreover, **24b** and **24c** to be sufficiently higher in energy (2.6 and 4.4 kcal/mol, respectively) so as not to contribute at all.

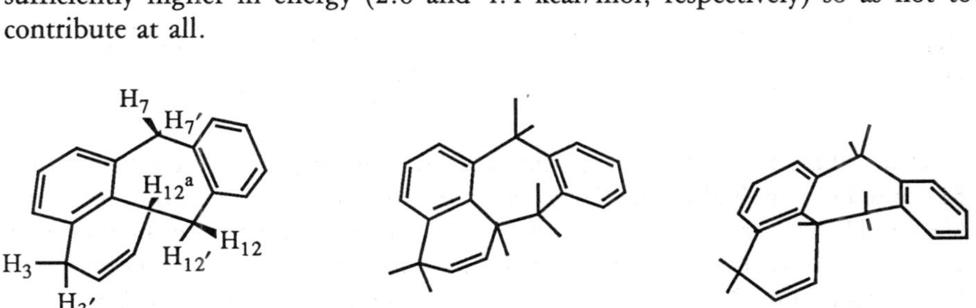

24a **24b** **24c**

The two-dimensional COSY spectrum of **24** is shown in Figure 4.4. The AB pattern for $H_7H_{7'}$ was assigned on a chemical shift basis. Similarly, a single hydrogen just past 4 ppm could only be H_{12a}, and the cross peaks in the COSY

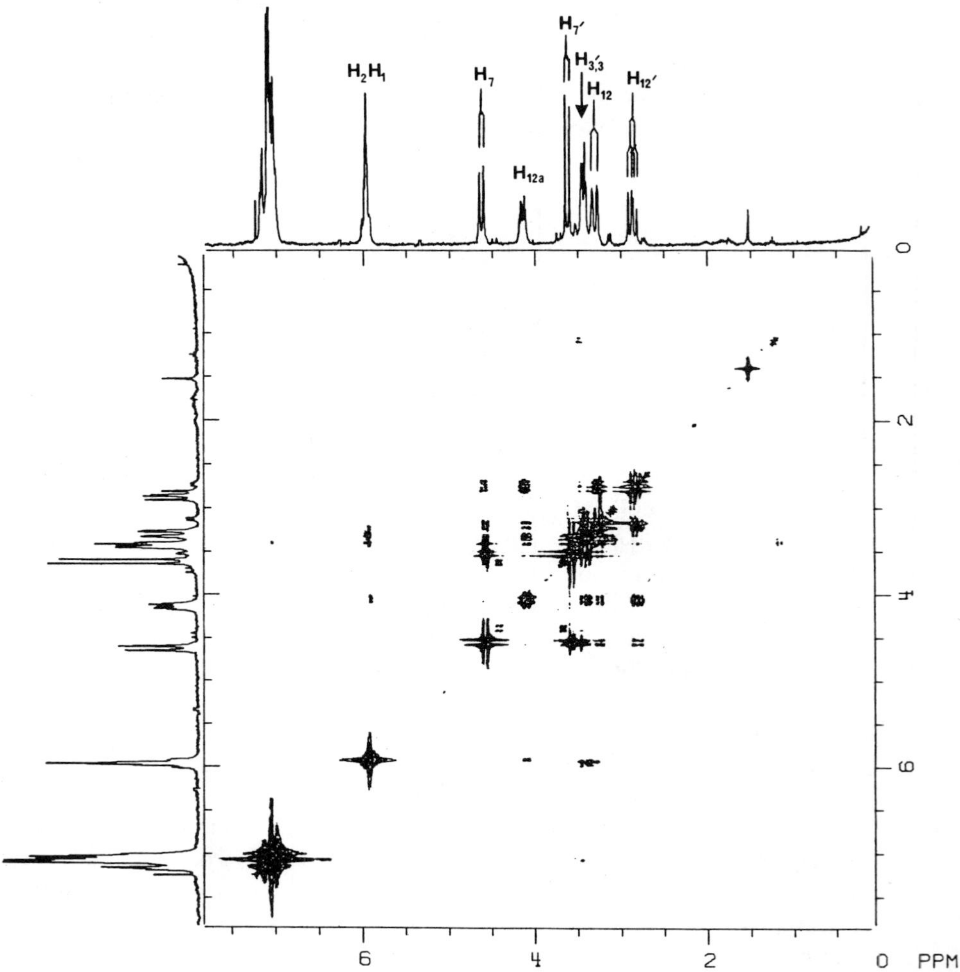

Figure 4.4 Two-dimensional correlation spectrum (COSY) of 3,12a,7,12-tetrahydropleiadene. Reproduced from Ref. 46 with permission.

spectrum at higher field must represent H_3, $H_{3'}$, and $H_{12}H_{12'}$. The assignment of H_{12} and $H_{12'}$ to higher field was based on the fact they are singly benzylic, whereas H_3, $H_{3'}$ are benzylic plus allylic. Once again we note little chemical shift separation between H_3 and $H_{3'}$, which is characteristic for these systems. The COSY spectrum also indicates a long-range coupling between H_7 and $H_{12'}$, representing a single-path homoallylic interaction.

A large coupling was observed for $J_{12a,12'}$ (or $J_{12a,12}$) and this value (12.9 Hz) was only consistent with a ~180° or 0° relationship. Thus, **24c** could be ruled out since in this case H_{12a} more or less bisects $H_{12}H_{12'}$. Distinction between the remaining conformers was accomplished by the observation of an ~10% nuclear

Overhauser enhancement at H_{12} when H_7 was irradiated (H_7 was distinguished from $H_{7'}$ due to broadening from weak aryl hydrogen coupling which is, of course, maximized at a 90° relationship—see also Ref. 44).

The homoallylic coupling constants in **24a** ($^5J_{cis}$ = 4.1 Hz, $^5J_{trans}$ = 4.4 Hz) are remarkably consistent with several 1,4-dihydronaphthalenes considered to be "flattened boats." The $^5J_{cis}/^5J_{trans}$ ratio for **24c** is 0.93 and the ratios for carboxy, hydroxymethyl, and phenyl are 0.86, 0.95, and 0.94, respectively. Therefore, we may conclude that these values will serve as a useful conformational probe and ratios of ~0.9 will correspond to an α value of around 170°.

Attempts to study 1,4-disubstituted-1,4-dihydronaphthalenes have not been common. This may be due, in part, to a problem with a separation of cis/trans isomers which appears to be common for this system.[47] However, a recent study examined a series of cis/trans mixtures involving carbonyl substitution at C_1 (**25–29**). In each case, $Eu(fod)_3$ was added until the signals for H_1 were separate for

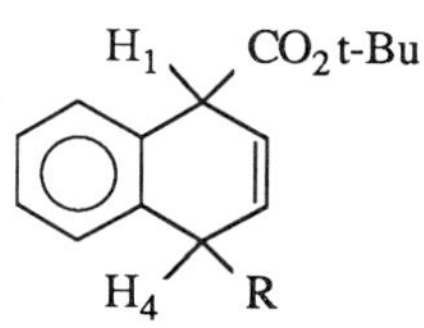

cis and trans

25, $R = CH_3$
26, $R = CH_2CH_3$

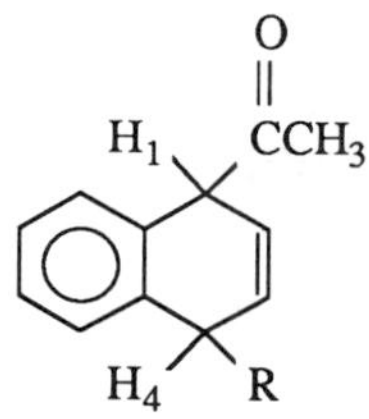

cis and trans

27, $R = CH_3$
28, $R = CH_2CH_3$
29, $R = CH(CH_3)_2$

each of the isomer pairs and then the vinyls were decoupled, resulting in a pair of doublets corresponding to $^5J_{cis}$ and $^5J_{trans}$. Assignment of the trans isomer could be accomplished by determining which H_4 ($H_{4'}$) signal was moving fastest with $Eu(fod)_3$ addition, since in the trans isomer $H_{4'}$ moves fastest (being on the same side as the coordination site). The results are presented in Table 4.5.

Table 4.5 Homoallylic Coupling Constants in *cis*- and *trans*-1,4-Disubstituted-1,4-dihydronaphthalenes[a]

Compound	$^5J_{cis}$ (Hz)	$^5J_{trans}$ (Hz)
25	3.0	3.8
26	3.1	3.1
27	4.0	3.9
28	3.6	4.2
29	3.0	3.8

[a] Data from reference 50.

In several cases, ${}^5J_{cis}$ and ${}^5J_{trans}$ are equal or nearly equal and so ${}^5J$ couplings may not always serve reliably as a method to distinguish between cis and trans isomers. However, where there is a significant difference, say 0.5 Hz or greater, the smaller coupling is expected to be ${}^5J_{cis}$. We saw previously that most of the "flattened" 1,4-dihydronaphthalenes had ${}^5J_{cis}$ values in the 3.5–4.5 Hz range. This suggests that the *t*-butyl esters, **25** and **26**, and the isopropyl substituted ketone, **29**, may be slightly more puckered. We would expect the trans compounds to be flatter than their cis isomers especially when the steric requirements of the two groups are similar. The ${}^5J_{trans}$ values in Table 4.5 are all consistent with relatively flat structures with the possible exception of **26**.

One of the more reliable methods to determine cis/trans assignments in both 1,4-dihydronaphthalenes and 9,10-dihydroanthracenes appears to be carbon chemical shifts. This has been demonstrated for methyl and ethyl substituted derivatives.[47,48] As illustrated for **26** and **28**, the trans substituents show an upfield

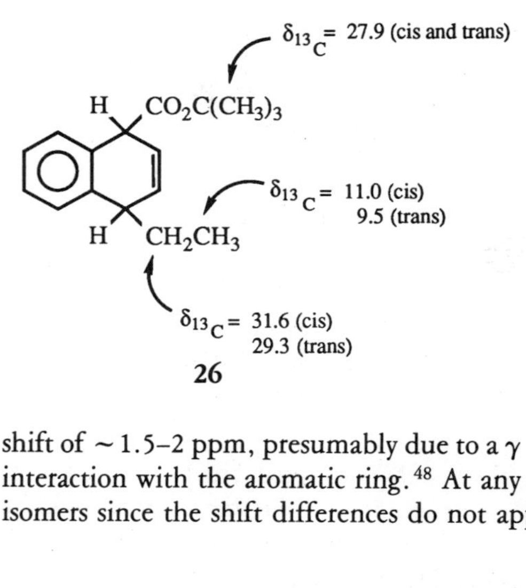

26

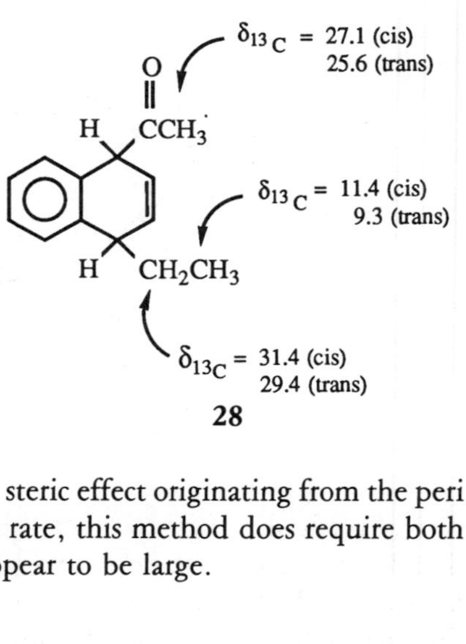

28

shift of ~ 1.5–2 ppm, presumably due to a γ steric effect originating from the peri interaction with the aromatic ring.[48] At any rate, this method does require both isomers since the shift differences do not appear to be large.

4.5 9,10-DIHYDROANTHRACENES

9,10-dihydroanthracene (**3**) has been shown by X-ray diffraction to be nonplanar in the solid state ($\alpha = 145°$).[49,50] It has been assumed that the boat conformation also predominates in solution, and the failure to observe separate

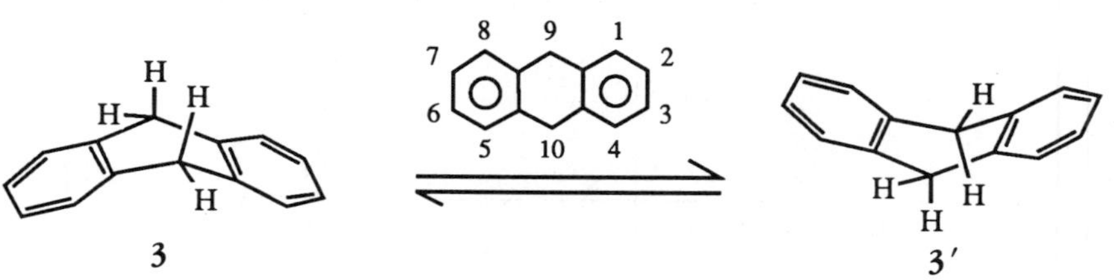

3 **3′**

^{1}H-NMR resonances for the pseudoaxial and pseudoequatorial protons at C_9 and C_{10}, even at low temperatures, has been attributed to rapid boat-to-boat inversion.[51] As we pointed out above, **3** should have the greatest tendency for a boat conformation in the series **1** → **3**. However, molecular mechanics calculations,[13] as well as MNDO and *ab initio* molecular orbital calculations,[14] indicate only small differences between boat and planar conformations and so variation of ring conformation with substituent pattern may also be expected with **3**.

The derivatives of **3** are not unlike those of **1** and **2** in that certain aspects of their conformational studies have been controversial. Perhaps the most unusual claim[52] was the suggestion that introduction of methoxy substituents on the aromatic rings (**30**, **31**, **32**) inhibits ring inversion (relative to the NMR time scale)

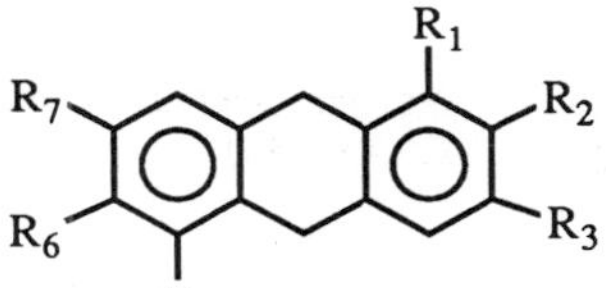

30, $R_2 = R_7 = OCH_3$
$R_1, R_3, R_5, R_6 = H$

31, $R_2, R_3, R_6, R_7 = OCH_3$
$R_1, R_5 = H$

32, R_1, R_2, R_3, R_5
$R_6, R_9 = OCH_3$

so that separate pseudoaxial and pseudoequatorial hydrogen resonances became observable. However, this prompted[53] the synthesis of 2,3,6,7-tetramethoxy-9,10-dihydroanthracene (**31**) by a new route, resulting in a product that was different from Ref. 52. This latter product did not show separate resonances at C_9/C_{10}, even at low temperatures. Of course, one would expect that substituents far removed from the central ring could not have much impact on ring inversion.

It was recognized that a single substituent at C_9 could shift the proposed equilibrium of boat conformations based on preferred location. Initial chemical studies[54] led to the suggestion of pseudoaxial preference (i.e., **33a**), but later

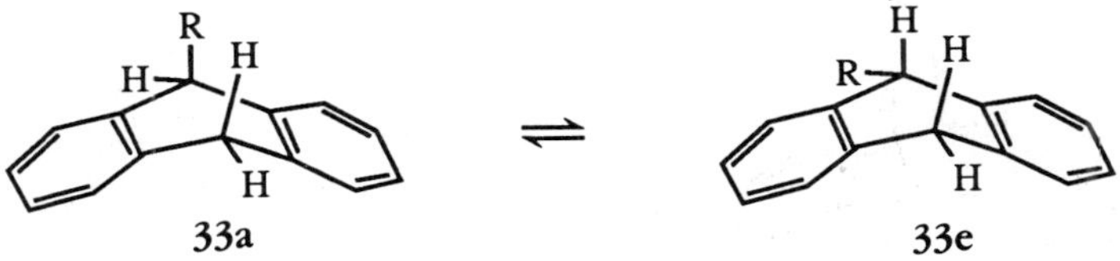

33a **33e**

workers favored the pseudoequatorial position based on NMR analysis.[55–57] Their reasoning was based on the observation that axial protons absorb at higher field than equatorial protons in cyclohexane, and the expectation that this should also be true for **33**. In fact, the behavior in 9,10-dihydroanthracene is just the opposite.

Two reports[44,58] appeared in 1970 indicating that a single substituent at C_9 occupies the pseudoaxial position. This was demonstrated by the homoallylic coupling, and even more convincingly by nuclear Overhauser enhancements (as illustrated). It was assumed that large groups like isopropyl and *t*-butyl are 100% pseudoaxial, whereas methyl was estimated at 75% pseudoaxial from the $^5J_{9,10}$ and $^5J_{9,10'}$ values (0.7 and 1.0 Hz, respectively), and phenyl appeared to exhibit little preference ($^5J_{9,10} = {}^5J_{9,10'} = 1.2$ Hz). Thus, phenyl may be less sterically demanding then methyl, especially if low-energy rotameric geometries are available (i.e., a phenyl group can turn "sideways" to avoid steric congestion).

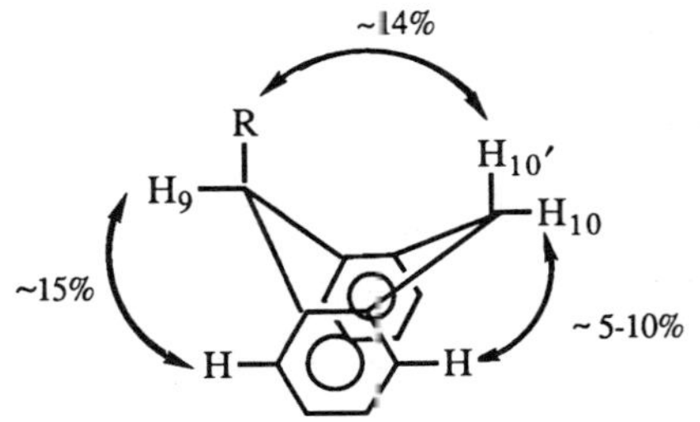

Nuclear Overhauser enhancements
(R = t - Bu)

A summary of ^5J values and nuclear Overhauser enhancement (NOE) results appears in Table 4.6. It is clear that the NOE method provides the most reliable approach to conformer and isomer assignments, although we cannot expect to differentiate degrees of ring folding in this way (i.e., $\alpha = 145°$ versus $\alpha = 160°$; see below). Brinkman et al.[44] suggested values of 2.5, 1.5, and 0.5 Hz for $^5J_{cis}$ dipseudoaxial, $^5J_{trans}$, and $^5J_{cis}$ dipseudoequatorial, respectively, and conformational assignments/equilibrium positions have been made on this basis. However, as illustrated in Table 4.6, later work[47,60] does not fit into this pattern. For example, many cis disubstituted compounds have ^5J values that equal and even exceed their trans counterparts.[59] These values are clearly *too high* for a dipseudoequatorial relationship in a "normal" boat-shaped structure (i.e., $\alpha = 145°$). This raises questions about the presumed equilibrium between two boat conformations (**34a** ⇌ **34e**) which in this case would heavily favor **34a**. In fact,

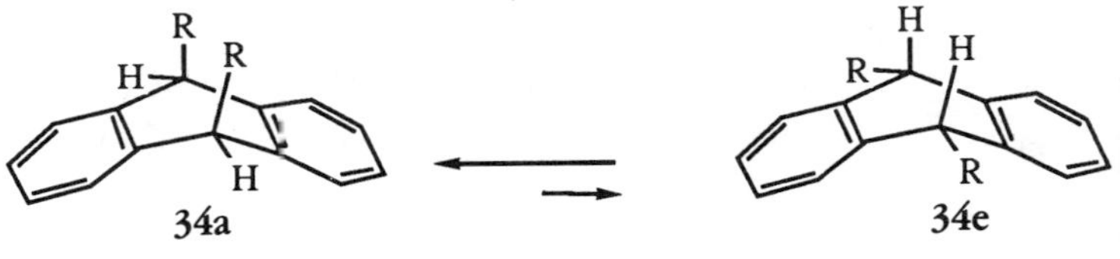

34a **34e**

molecular mechanics calculations (described in detail elsewhere in this book) do not predict "normal" boat conformations ($\alpha = 145°$) but rather suggest a flatter structure with $\alpha = 155°$ for *cis*-dimethyl, and $\alpha = 153°$ for *cis*-di-*t*-butyl.

Table 4.6 Proton NMR Data for Substituted 9,10-Dihydroanthracenes

	R_9	R_{10} or $R_{10'}$	$^5J_{homo}$	Signal Irradiated	NOE Results Signal Observed	% Enhancement
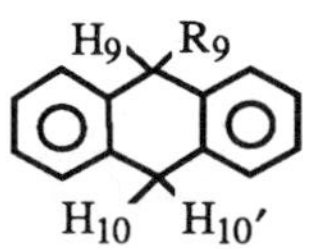	Me[a]	H	0.7 (cis) 1.0 (trans)	aromatics	H_9	15
	iPr[a]	H	0.4 (cis) 1.1 (trans)	aromatics	H_9	13
	tBu[a]	H	0.4 (cis) 1.3 (trans)	aromatics	H_9	16
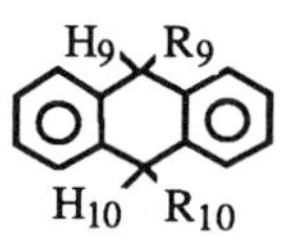	iPr[b]	Et	<0.3	aromatics	$H_9(H_{10})$	18(18)
	tBu[b]	Me	1.3	aromatics	$H_9(H_{10})$	12(10)
	tBu[b]	Et	1.3	aromatics	$H_9 + H_{10}$	15
	tBu[b]	iPr	d	aromatics	$H_9(H_{10})$	11(13)
cis isomer	CO_2tBu[c]	Me	0.9	aromatics	$H_9(H_{10})$	13(8)
	CO_2tBu[c]	Et	0.6	aromatics	$H_9(H_{10})$	23(7)
	COMe[c]	Me	1.1	aromatics	$H_9(H_{10})$	20(15)
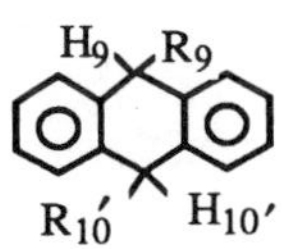	iPr[b]	Et	1.1	aromatics	$H_9(H_{10'})$	16(4)
	tBu[b]	Me	1.2	aromatics	H_9	17
				tBu	$H_{10'}$	18
	tBu[b]	Et	1.2	aromatics	H_9	14
				tBu	$H_{10'}$	20
trans isomer	tBu[b]	iPr	1.2	aromatics	H_9	8
				tBu	$H_{10'}$	21

[a] Data from reference 47. [b] Data from reference 62. [c] Data from reference 50. [d] not resolvable.

Table 4.7 X-Ray Structures of Substituted 9,10-Dihydroanthracenes

	R_9	$R_{9'}$	R_{10}	$R_{10'}$	Central Ring (α)	Reference
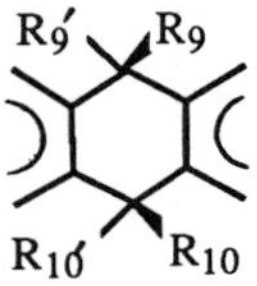	t-Bu	H	H	H	Boat (146.6°)	64
	Me	H	Et	H	Boat (152°)	65
	Me	H	H	iPr	Boat (129°)	66
	Bz	H	H	NO_2	Boat (148.5°)	69
	CH_2OH	H	CH_2Ar^a	H	Boat (151°)	74
	$SiMe_3$	H	$SiMe_3$	H	Boat (150°)^{ab}	68
	$SiMe_3$	H	H	$SiMe_3$	Chair/Planar^c	68
	Cl	Cl	Cl	Cl	Planar	63
	OH	Ph	Ph	OH	Planar	70
	Pr	OH	Pr	OH	Boat (149°)	71
	Pr	OH	OH	Pr	Planar	71
	Et	H	Et	H	Boat (147°)	71

^{a} Ar = anthracene. ^{b} Distorted. ^{c} A more or less planar conformation with a "chair-like" distortion.

Moreover, the calculations are not supportive of an equilibrium either since the dipseudoequatorial conformation (**34e**) is too high in energy to contribute at all. Thus, the best (calculated) structure is represented by a parabolic potential well reminiscent of 1,4-cyclohexadiene but displaced from the 180° position. These calculations also indicate considerable distortion in the trans isomers, and, in some cases, even "chair" conformations!

The solid-state geometry of numerous dihydroanthracene derivatives has been investigated by X-ray crystallography.[50,60–68] Herbstein et al.[53] point out that "in broad terms there are two preferred conformations for the DHA molecular skeleton, either folded with the dihedral angle distributed around 146.2° (sample standard deviation 9.2°), or planar (or very nearly so)." Data for a number of derivatives are shown in Table 4.7 and it does appear that, in general, mono substituted and cis disubstituted derivatives may be expected to be boat shaped, whereas trans derivatives (where $R_9 = R_{10'}$) are planar. Moreover, in the latter case, distortions into a "chairlike" geometry are possible if the substituents are large.

A number of methylated derivatives were investigated by Dalling et al.[48] using solid-state ^{13}C-NMR. Boat conformations were unambiguously assigned to **35** and **36** since they showed distinct signals for pseudoaxial and pseudoequatorial

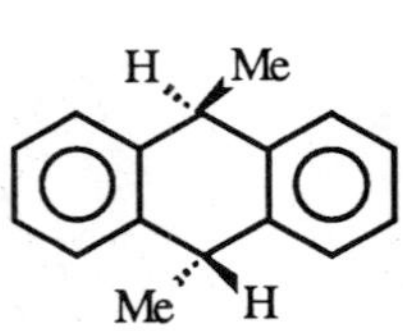

35

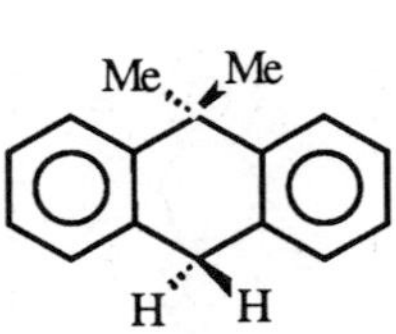

36

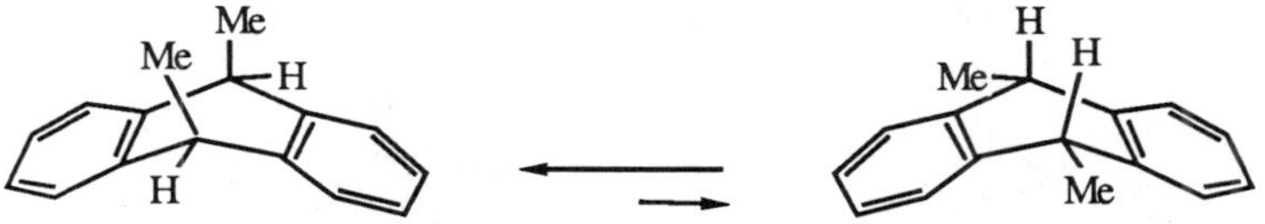

37a **37e**

methyl groups. However, in solution only average methyl resonances were observed, even at −70 °C, and this was attributed to rapid, boat-to-boat ring inversion. A similar process was also suggested for **37** in solution, and the observed temperature dependence (~1.5 ppm) in the methyl carbon resonances (+40 to −70 °C) was interpreted to mean a shift in the equilibrium between **37a** and **37e**. They did, however, favor a planar or nearly planar geometry for 9,9,10,10-tetramethyl-DHA since clearly distinguishable pseudoaxial or pseudoequatorial carbons could not be found.

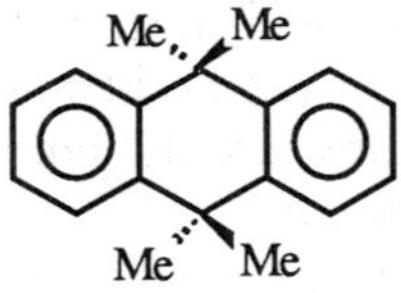

planar

Subsequently, the concept of equilibria such as **37a** ⇌ **37e** was questioned based on the results of molecular mechanics calculations.[69] Theory suggests the now familiar parabolic well with $\alpha \approx 150°$ for **37**, and the methyls dipseudoaxial: a second minima for the dipseudoequatorial form does exist on the potential surface, but it is more than 5 kcal/mol higher in energy than the other conformer. The question, of course, is whether or not the solid-state conformations are the same as in solution since the former are affected by crystal packing forces.

In an attempt to answer this question, models of known, rigid geometry were examined.[69] For example, **38** can be used as a model for **39** since the latter

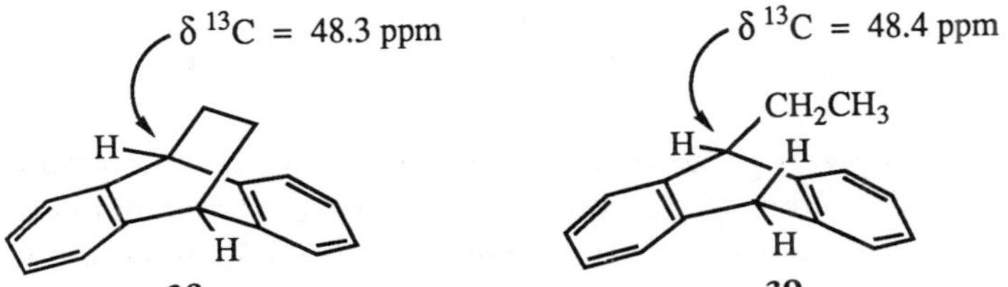

38 **39**

monosubstituted derivative is expected to be a boat conformation with an axial two-carbon substituent. In fact, the C_9 carbon resonances are almost identical. The sensitivity of the bridge carbon, C_9, to geometry was demonstrated by the

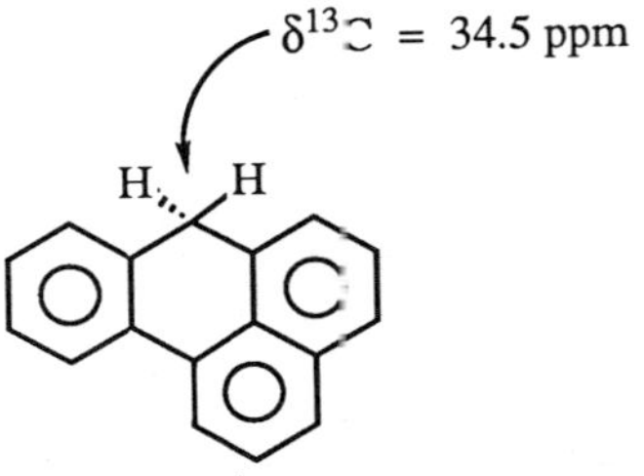

40

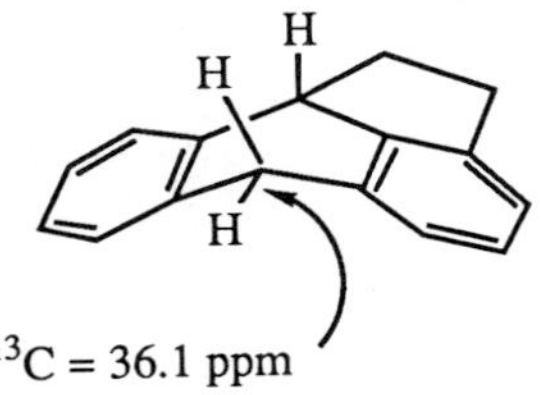

41

comparison of **40** and **41**. They appear to be reasonable models without significant perturbation from the "add-on" units needed to fix geometry. For example, the methyl carbons of toluene and 2-methylnaphthalene are within 0.1 ppm of each other so the naphthalene unit of **40** should not be a problem. In the case of **41**, the extra ethyl bridge is equivalent to a pseudoequatorial ethyl group and this is known not to affect the carbon resonance across the ring.[69] Thus, it appears that ring folding causes a ~ 1.6 ppm downfield shift at the bridge carbon. The C_9 chemical shift of 9,10-dihydroanthracene itself is identical to that of the puckered model **41**, suggesting that boat conformations do indeed contribute significantly to the geometry of the parent hydrocarbon.

It would appear that a *t*-Bu substituent will "lock" the central ring into a boat conformation, and so the examination of **42** and **43** allowed the study[69] of a

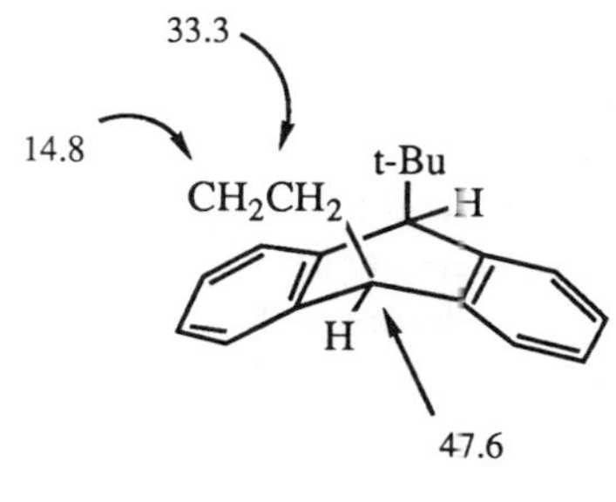

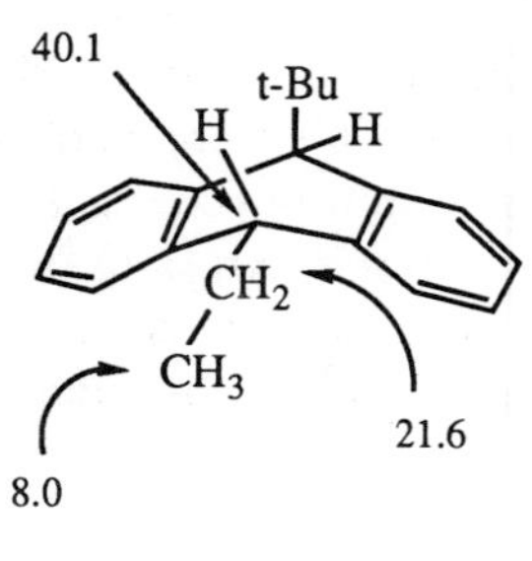

[$\delta^{13}C$(ppm)]

42 **43**

fixed central ring with pseudoaxial and pseudoequatorial ethyl groups (respectively). The bridge carbon (C_9) absorbs at higher field (-7.5 ppm) in **43** reflecting the effect of a pseudoequatorial substituent (see Table 4.8), and this is exactly the same value found by Dalling et al.[48] for the difference between C_9's with pseudoaxial versus pseudoequatorial methyls in the solid state. As seen in Table 4.8, this would appear to be a method for the distinction between cis and trans isomers. Therefore, in trans isomers where $R_9 \neq R_{10}$, the bridge carbon bearing the smallest group (i.e., pseudoequatorial) will be at higher field as compared

Table 4.8 Carbon Chemical Shifts for 9,10-Dihydroanthracene and Derivatives

Compound	C-1, C-4	C-2, C-3	C-4a, C-8a	C-9	C-10	CH_2CH_3	CH_2CH_3	CH_3	Other	Reference
DHA[a]	127.2	126.0	136.0	36.1						72
9-EtDHA[a]	127.8, 127.6[b]	125.9, 125.8[b]	135.5, 139.9[b]	48.8	31.3	35.3	12.1			72
cis-9,10-diEtDHA[a]	128.6	125.7	139.3	48.3		35.5	13.4			72
trans-9,10-diEtDHA[a]	126.7	125.8	138.6	43.8		27.0	10.0			72
cis-9-Et-10-*t*-BuDHA[a]	129.0, 130.9[b]	124.6, 125.8[b]	136.5, 140.1[b]	47.6	55.8	33.3	14.8		3.51[c], 30.0[d]	72
trans-9-Et-10-*t*-BuDHA[a]	125.9, 130.5[b]	124.6, 125.4[b]	138.1, 138.3[b]	40.1	57.2	21.6	8.0		38.9[c], 28.5[d]	72
9-*t*-Bu-DHA[a]	130.7, 127.9[b]	126.1, 125.1[b]	137.8, 138.2[b]	57.5	37.2				39.1[c], 28.1[d]	72
9-MeDHA[e]	127.1, 127.9	126.6, 126.2[b]	136.0, 142.0	41.1	35.1			23.6		51
9,9-diMeDHA[e]	124.4, 128.0	126.7, 126.0[b]	135.8, 145.2	39.1	35.8			28.8		51
cis-diMeDHA[e]	128.0	126.4	140.6	40.0				28.6		51
trans-diMeDHA[e]	126.0	126.4	141.5	38.5				18.8		51
9,9,10-triMeDHA[e]	126.2, 128.1	126.6, 126.4[b]	139.5, 143.7	38.3	38.9			36.3 (9a)		51
								30.4 (9e)		51
								29.0 (10)		51
9,9,10,10-tetraMeDHA[e]	127.0	126.4	142.1	37.2				35.1		51

[a] In CS_2 with internal ME_4Si. [b] Signals could be reversed. [c] *t*-Butyl quaternary. [d] *t*-Butylmethyl. [e] In $CDCl_3$ with internal Me_4Si.

with the cis isomer. When $R_9 = R_{10}$, the overall geometry of the central ring is more or less flat. However, the substituents in this position are "more" pseudoequatorial than in the cis isomer so an upfield shift is expected, although of a smaller magnitude (1.5 ppm in diMe and 4.5 ppm in diEt).

The geometry of the central ring should also have an influence on the one-bond $^{13}C_9$—H coupling constants. It has been shown by the Mislow group[70] that coupling constants in hydrocarbons of the type RCH_2R are given by the equation

$$^1J = \frac{250(1 + \cos\theta_0)}{1 - \cos\theta_0}$$

where the interorbital angle θ_0 is related to the C—C—C internuclear bond angle (θ_n) by the expression

$$\theta_0 = 84.82 + 0.3853\,\theta_n - 0.001473\,\theta_n^2$$

Using MM2 calculated values for the C—C_9—C bond angle in both puckered (θ = 111.0°) and planar (θ = 116.9°) 9,10-dihydroanthracene, Rabideau et al.[72] calculated coupling constants of 125.2 and 123.8 Hz, respectively. The experimental values for a number of systems are shown in Table 4.9, and the difference between the planar (**40**) and puckered (**41**) model compounds is 1.2 Hz with the puckered system larger, and this agrees rather well with the 1.4 Hz value predicted by calculation. Similarly, a reduction of 1.6 Hz in the *trans*-9,10-diEt derivative as compared to the cis is consistent with a relatively planar geometry for the former. The values for 9-Et-10-*t*-Bu are lower than expected, but this may be due to distortions as have been suggested earlier when large groups are involved.[59]

Comparison of 9,10-dihydroanthracene itself with the fixed boat model is quite interesting. Both the bridge carbon chemical shift and the ^{13}C—H coupling

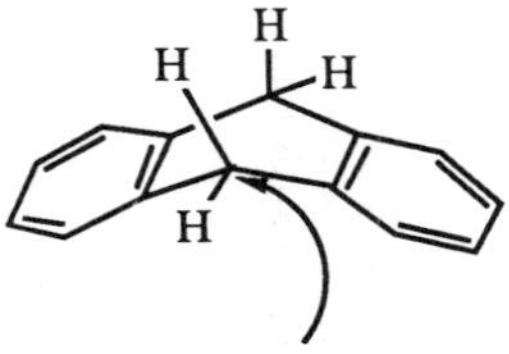

$\delta^{13}C = 36.1$ ppm
$J_{C\text{-}H} = 127.6$ Hz

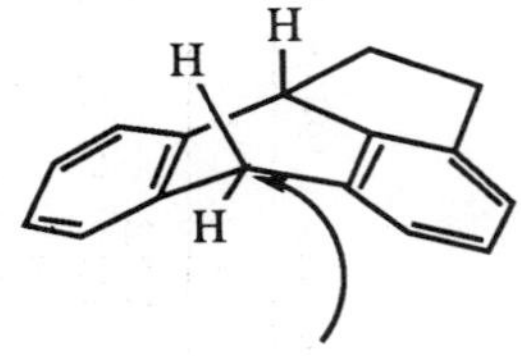

$\delta^{13}C = 36.1$ ppm
$J_{C\text{-}H} = 127.9$ Hz

are essentially identical. This strongly suggests a significant contribution from boat conformations for the parent compound in solution. However, this does not rule out the calculated description involving the planar conformation as plus or minus a few tenths of a kilocalorie relative to the boat geometry.

So far we have seen that many 9,10-dihydroanthracene derivatives do indeed fall into the category of either boat shaped, with $\alpha \approx 145$–155°, or more or less

Table 4.9 One-Bond ^{13}C—H Coupling Constants for DHA and Related Compounds in Hz[a]

Compound	$J_{^{13}C-H}$	Compound	$J_{^{13}C-H}$
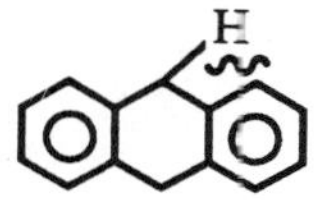	127.6 (128 ± 2)[b]	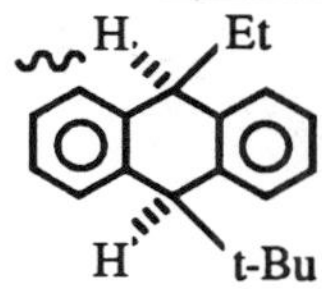	125.9
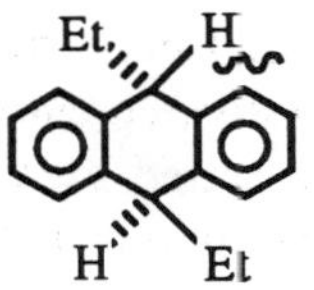	126.9	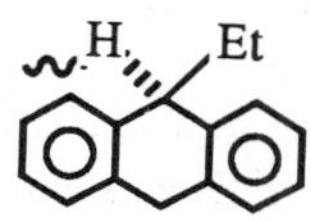	128.0
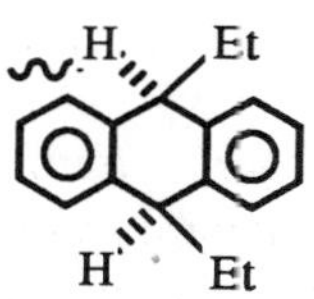	128.5	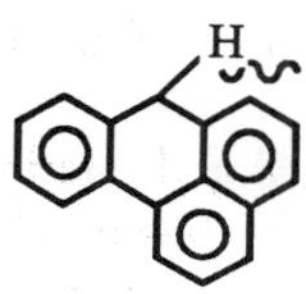	126.7
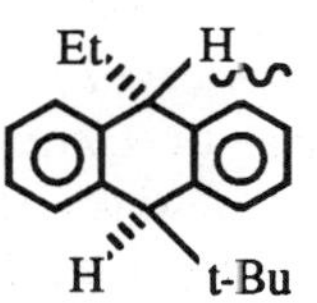	122.6	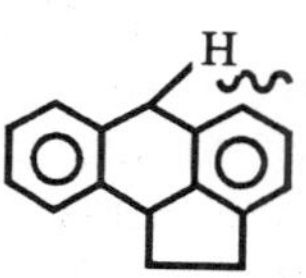	127.9[c]
		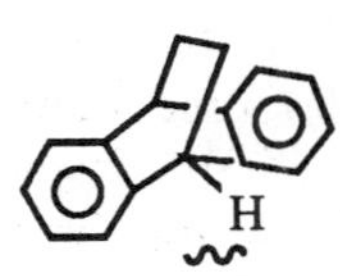	138.7[d]

[a] Run as ~0.5 M solutions in CS_2 with internal Me_4Si as reference and coaxial D_2O as lock [b] Values from: Smith, W. B.; Shoulders, B. A. *J. Phys. Chem. 1965*, 69, 2022. [c] Measured from the triplet although it is, strictly, an ABX. [d] Run in $CDCl_3$. Reproduced from *J. Am. Chem. Soc., 1986*, 108, 8130 with permission. Copyright (1986) ACS.

planar. One report that does not fit is the X-ray study of *trans*-9-isopropyl-10-methyl-9,10-dihydroanthracene, suggesting the unusual value of $\alpha = 129°$.* In fact, the unusual behavior of isopropyl groups in DHAs was noted in 1969 by Zieger et al.,[71] who observed an unusually large coupling constant (9.5 Hz) between the isopropyl methine hydrogen(s) and H_9 (H_{10}) in **44**. This was attributed to a preferred rotational conformation of the isopropyl groups, due to a

* See Ref. 63. This X-ray study was performed as a method to distinguish between the cis and trans isomers.

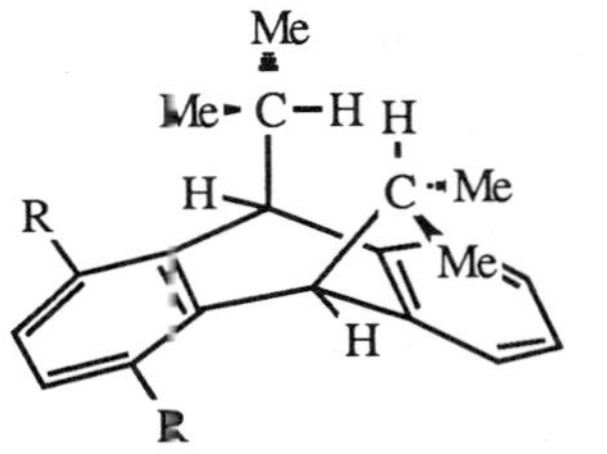

44, R = H
45, R = Me
46, R = OMe

transannular steric effect, leading to a ~180° dihedral angle. Later, the 1,4-dimethyl (**45**) and 1,4-dimethoxy (**46**) derivatives were investigated and also suggested preferred geometries, but with the isopropyl groups rotated away from the ring substituents. This was evident from the drop in the coupling constant (J = 2.0 Hz), and the appearance of one of the diastereotopic doublets at −0.13 ppm. Evidently two of the four methyls are rotated out over the unsubstituted aromatic ring.

The vicinal isopropyl methine/H_9 coupling constants for a number of isopropyl derivatives are shown in Table 4.10. As expected, the largest J values are observed for the cis isomers with large substituents at C_{10}. In the trans isomers, large C_{10} substituents cause the C_9 isopropyl to become pseudoequatorial and so it must rotate "sideways" to avoid peri interactions resulting in decreased J values. As the C_{10} substituent becomes smaller (e.g., ethyl), the isopropyl becomes axial and the J value returns to normal (7.8 Hz). There is an anomaly, however, in that the same J value for the *trans*-9-isopropyl-10-methyl (**47**) derivative is 9.6 Hz! A closer examination of this compound and comparison with the 9-isopropyl derivative **48** provides some interesting results[72] in addition to the difference in J values. Irradiation of the isopropyl methine in **48** produces only a small enhancement (NOE) across the ring at $H_{10'}$ (~2%). With **47**, however, this figure jumps to 13%. This, of course, suggests a closer proximity of $H_{10'}$ to the isopropyl group, and, if true, would result in a greater immobilization of the

Table 4.10 Isopropyl Methine/H_9 Vicinal Coupling Constants for 9-Isopropyl-10-R-9,10-dihydroanthracene[a]

	Geometry	R	J isopropyl methine/H_9 (Hz)
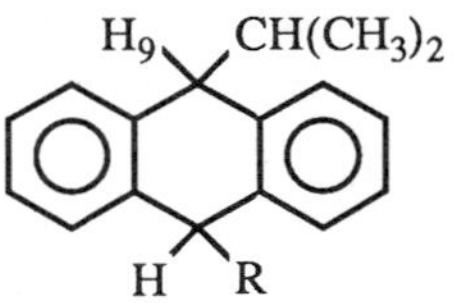	cis	t-Bu	9.9
	cis	i-Pr	9.5
	cis	Et	9.2
	cis	Me	8.8
	—	H	7.0
	trans	t-Bu	2.6
	trans	i-Pr	5.0
	trans	Et	7.8
	trans	Me	9.6

[a] See reference 62 and references therein.

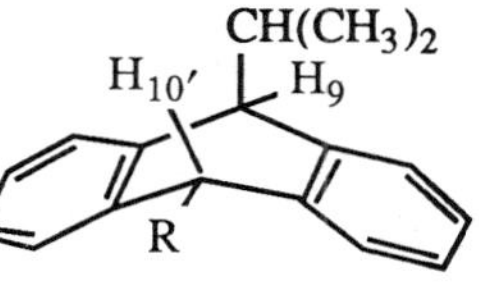

47, R = CH_3
48, R = H

	R = H_{10}	R = CH_3
J_{H_9}, Methine H	7.0	9.6
NOE (CH to $H_{10'}$)	2%	13%
Relaxation Time (T_1) for CH	1.05	2.73

latter substituent. This is substantiated by the almost threefold increase in the carbon relaxation time. All this can be explained by a geometry wherein the pseudoequatorial methyl group is "lifted up" above the peri interactions with the ortho hydrogens, forcing $H_{10'}$ toward the isopropyl across the ring. Thus, this "superfolded" geometry, which is also predicted by molecular mechanics,[72] agrees well with the aforementioned X-ray study.[63]

4.6 EXOCYCLIC DOUBLE BONDS

The conformational analysis of a number of dihydroanthracenes containing an exocyclic double bond has been investigated. On the basis of bond angles alone, **49** should have a greater tendency toward planarity since the additional sp^2

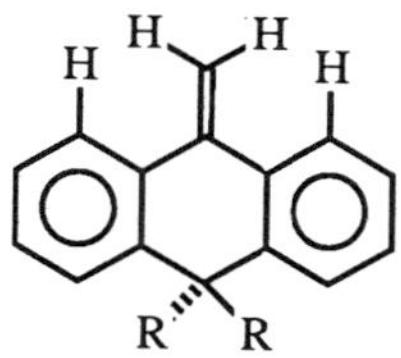

49, R = H
50, R = Me

carbon ion in the central ring incorporates the optimum value of ~120°. However, one would expect this effect to be more than offset by the steric repulsion between the vinyl hydrogens and the ortho aryl hydrogens. Curtin et al.[73] investigated the low-temperature ^{1}H-NMR behavior of **50** and concluded a rapid interconversion of boat conformations (−27 °C). The isopropylidene

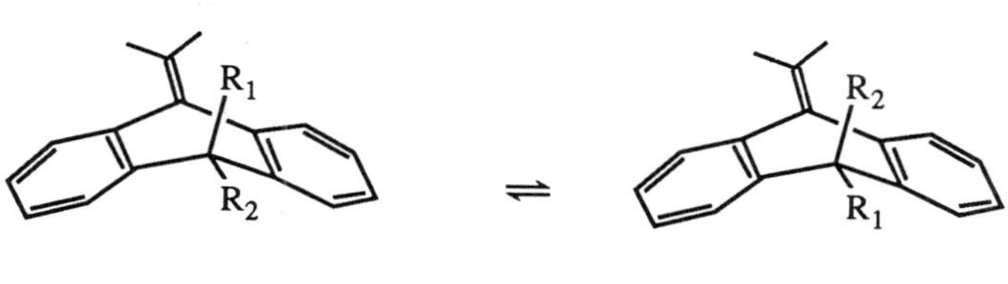

51, $R_1 = R_2 = H$
52, R_1 = Me, R_2 = H
53, R_1 = Et, R_2 = H
54, $R_1 = R_2$ = Me

derivative **51**, on the other hand, has been shown[74] unambiguously to exist as equilibrating boat conformations since at low temperatures H_{10} and $H_{10'}$ appear as an AB quartet. A coalescence temperature of 31.5 °C was observed and ΔG^* was calculated to be 15.1 kcal/mol for the interconversion. A single methyl at C_{10} (i.e., **52**) resulted in an equilibrium distribution of 85% pseudoaxial and 15% pseudoequatorial methyl populations (c.f. 9-methyl-9,10-dihydroanthracene). The 10-ethyl derivative (**53**), however, showed temperature independence down to − 100 °C and so the slight increase in steric demand is sufficient to totally favor the pseudoaxial conformer.

Interestingly, the incorporation of two methyls at C_{10} (**54**) has little effect on the inversion barriers (P. W. Rabideau, unpublished results) ($\Delta G^* = 15.1$ for **51** and 15.9 kcal/mol for **54**). This is due to the fact that the ortho aryl hydrogens would really like to bisect the C_{10} attachments and so larger groups do not raise inversion barriers. In fact, even larger groups might lower the barrier (i.e., steric acceleration).

4.7 COMPLEXES AND ORGANOMETALS

As one would expect, complexation can have a dramatic effect on the geometry of 9,10-dihydroanthracene. Herbstein et al.[50] determined the structure of 9,10-dihydroanthracene-bis(1,3,5-trinitrobenzene) and found it to be planar. On the other hand, bis(9,10-dihydroanthracene)chromium (**55**) involves boat conformations even in solution as evidenced by the presence of H_9H_{10} as an AB quartet.[75]

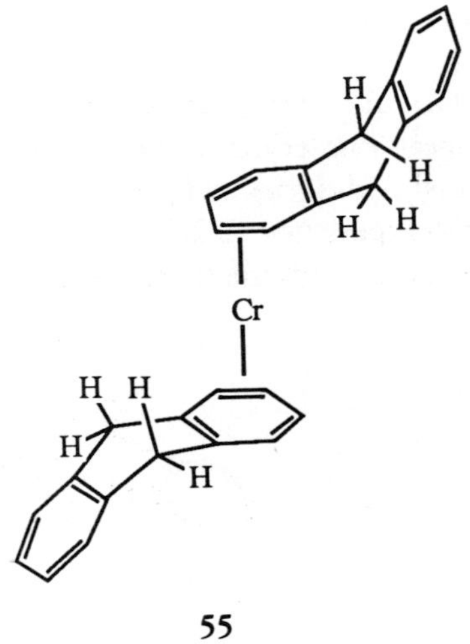

55

The carbon-NMR spectra of lithium derivatives (**56**, R = H and R = *t*-Bu) have been investigated and the upfield shifts for the para carbons (17.7 for R = H and 19.1 ppm for R = *t*-Bu) were taken to mean considerable flattening of the central ring (i.e., delocalization).[76] MNDO calculations on the 10-litho-9-methyl derivative (i.e., **56**, R = Me) predict a minimum energy structure with α =

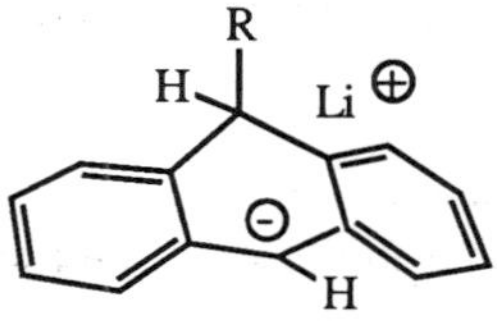

56

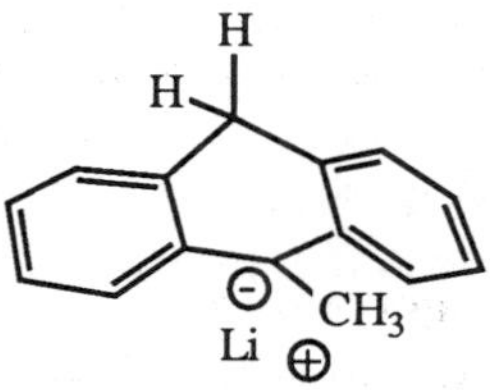

57

149.8° and both methyl and lithium in pseudoaxial positions. The global minimum for the 9-lithio isomer (**57**), on the other hand, is predicted to be almost planar (176.5°) with the methyl slightly pseudoaxial and the lithium "underneath."[77]

4.8 CONCLUSION

Both experimental and theoretical evidence support the tendency toward planarity in the 1,4-cyclohexadiene, 1,4-dihydronaphthalene, and 9,10-dihydroanthracene systems. Of the parent compounds, 1,4-cyclohexadiene is certainly planar as is, in all likelihood, 1,4-dihydronaphthalene. The 9,10-dihydroanthracene case is not quite as clear, but the *ab initio* calculations which suggest an inversion barrier of 2 kcal/mol or less is probably not far off the mark.[14b] Symmetrical trans or gem disubstitution, of course, greatly increases the probability of planar geometries. Nonplanar structures are favored by monosubstitution and (especially) by cis disubstitution with substituent groups preferentially in pseudoaxial positions. Moreover, experimental evidence does not require arguments for significant amounts of pseudoequatorial conformers in equilibrium for these latter cases.

It also appears that many geometries may be possible which lie between the planar, α = 180°, and normally puckered, α = 145°, conformations. These structures are normally in the range of 160–170° and have been called both "flattened boats" and "somewhat puckered" ring systems. The reason for this apparent inconsistency in the literature is due to the particular author's frame of reference, either the planar form or the boat form. In any event, dihydroaromatics must be treated as a system wherein the ring geometry may not be assumed to be a constant.

ACKNOWLEDGMENTS

Support of this work has been provided by the U.S. Department of Energy, Office of Basic Energy Research.

REFERENCES

1. Gerding, H.; Haak, F. A. *Recl. Trav. Chim. Pays-Bas* **1949**, *68*, 293.
2. Stidham, H. D. *Spectrochim. Acta* **1965**, *21*, 23.
3. Monostori, B. J.; Weber, A. *J. Mol. Spectrosc.* **1964**, *12*, 129.
4. Garbisch, E. W., Jr.; Griffith, M. G. *J. Am. Chem. Soc.* **1968**, *90*, 117. However, see discussion below for reconsideration of these results.
5. Dallinga, G.; Toneman, L. H. *J. Mol Struct.* **1967**, *1*, 117.
6. Oberhammer, H.; Bauer, S. H. *J. Am. Chem. Soc.* **1969**, *91*, 10.
7. Laane, J.; Lord, R. C. *J. Mol. Spectrosc.* **1971**, *39*, 340. See also Strube, M. M.; Laane, J. *J. Mol. Spectrosc.* **1988**, *129*, 126.
8. Grossel, M. C.; Perkins, M. J. *Nouv. J. Chim.* **1979**, *3*, 285.
9. Carreira, L. A.; Carter, R. O.; Durig, J. R. *J. Chem. Phys.* **1973**, *59*, 812.
10. Hagemann, H.; Bill, H.; Joly, D.; Müller, P.; Pautex, N. *Spectrochim. Acta* **1985**, *41A*, 751.
11. Ahlgren, G.; Akermark, B. Backvall, J. E. *Tetrahedron Lett.* **1975**, 3501.
12. Saeboe, S.; Boggs, J. E. *J. Mol. Struct.* **1981**, *73*, 137.
13. Lipkowitz, K. B.; Rabideau P. W.; Raber, D. J.; Hardee, L. E.; Schleyer, P. v. R.; Koss, A J.; Kahn, R. A. *J. Org. Chem.* **1982**, *47*, 1002.
14. (a) Syguła, A.; Holak, T. A *Tetrahedron Lett.* **1983**, *24*, 2893. (b) Schaefer, T.; Sebastian, R. *J. Mol. Struct., Theochem.* **1987**, *153*, 55.
15. Allinger, N. L.; Sprague, J. T. *J. Am. Chem. Soc.* **1972**, *94*, 5734.
16. For a previous review, see Rabideau, P. W. *Acc. Chem. Res.* **1978**, *11*, 141.
17. Durham, L. J.; Studebaker, J.; Perkins, M. J. *Chem. Commun.* **1965**, 456.
18. Atkinson, D. J.; Perkins, M. J. *Tetrahedron Lett.* **1969**, 2335.
19. (a) Garbisch, E. W., Jr.; Griffith, M. G. *J. Am. Chem. Soc.* **1970**, *92*, 1107; (b) For a later redetermination of J_{cis} and J_{trans}, see Heyde, W. A. D.; Lüttke, W. *Chem. Ber.* **1978**, *111*, 2384.
20. Barfield, M.; Sternhell, S. *J. Am. Chem. Soc.* **1972**, *94*, 1905.
21. Grossel, M. C.; Perkins, M. J. *J. Chem. Soc. Perkin Trans. 2* **1975**, 1544.
22. Marshall, J. L.; Faehl, J. G.; McDaniel, C. R.; Ledford, N. D. *J. Am. Chem. Soc.* **1977**, *99*, 321.
23. Self-consistent-field intermediate neglect of differential overlap finite perturbation: Pople, J. A.; McIver, J. W.; Ostlund, N. S. *J. Chem. Phys.* **1968**, *49*, 2960, 2965; computer program No. 224 (author Ostlund, N. S.), Quantum Chemistry Exchange program, Indiana University, Bloomington, IN.
24. Marshall, J. L.; Faehl, L. G ; Ihrig, A. M.; Barfield, M. *J. Am. Chem. Soc.* **1976**, *98*, 3406.
25. Rabideau, P. W.; Paschal, J. W.; Marshall, J. L. *J. Chem. Soc., Perkin Trans.* **1977**, *2*, 842.
26. Grossel, M. C. *Tetrahedron Lett.* **1980**, *21*, 1075.
27. Grossel, M. C.; Cheetham, A. K.; Newsam, J. M. *Tetrahedron Lett.* **1978**, 5229.
28. Grossel, M. C.; Cheetham, A. K.; Hope, D. A. O.; Lam, K. P.; Perkins, M. J. *Tetrahedron Lett.* **1979**, 1351.
29. Marshall, J. L.; Erickson, K C.; Folsom, T. K. *J. Org. Chem.* **1970**, *35*, 2038.
30. Marshall, J. L.; Folsom, T. K. *J. Org. Chem.* **1971**, *36*, 2011.
31. Paschal, J. W.; Rabideau, P. W. *J. Am. Chem. Soc.* **1974**, *96*, 272.
32. Rabideau, P. W.; Paschal, J. W.; Patterson, L. E. *J. Am. Chem. Soc.* **1975**, *97*, 5700.
33. Marshall, J. L.; Ihrig, A. M.; Jenkins, P. N. *J. Org. Chem.* **1972**, *37*, 1863.
34. DeWolf, M. Y.; Baldeschwieler, J. D. *J. Mol. Spectrosc.* **1964**, *13*, 344.
35. Marshall, J. L.; Song, B.-H *J. Org. Chem.* **1975**, *40*, 1942.

36. Grossel, M. C.; Cheetham, A. K.; James, D.; Newsam, J. W. *J. Chem. Soc., Perkin Trans. 2* **1979**, 545.

37. Raber, D. J.; Hardee, L. E.; Rabideau, P. W.; Lipkowitz, K. B. *J. Am. Chem. Soc.* **1982**, *104*, 2843.

38. Rabideau, P. W.; Wetzel, D. M.; Paschal, J. W. *J. Org. Chem.* **1982**, *47*, 3993.

39. Rabideau, P. W.; Burkholder, E. G.; Yates, M. J.; Paschal, J. W. *J. Am. Chem. Soc.* **1977**, *99*, 3596.

40. Grossel, M. C.; Hayward, R. C. *J. Chem. Soc., Perkin Trans. 2* **1976**, 851.

41. Cheetham, A. K.; Grossel, M. C.; James, D. *J. Org. Chem.* **1982**, *47*, 566.

42. Holy, N. L.; Vail, H. P.; Nejad, A. H.; Huang, S. J.; Marshall, J. L.; Saracoglu, O.; McDaniel, C. R. *J. Org. Chem.* **1980**, *45*, 4271.

43. Lamberts, J. J. M.; Haasnoot, C. A. G.; Laarhoven, W. H. *J. Org. Chem.* **1984**, *49*, 2490.

44. Brinkman, A. W.; Gordon, M.; Harvey, R. G.; Rabideau, P. W.; Stothers, J. W.; Ternay, A. L. *J. Am. Chem. Soc.* **1970**, *22*, 5912.

45. Rabideau, P. W.; Harvey, R. G.; Stothers, J. B. *J. Chem. Soc., Chem. Commun.* **1969**, 1005.

46. Rabideau, P. W.; Mooney, J. L.; Hardin, J. N. *J. Org. Chem.* **1985**, *50*, 5737.

47. Rabideau, P. W.; Day, L. M.; Husted, C. A.; Mooney, J. L.; Wetzel, D. M. *J. Org. Chem.* **1986**, *51*, 1681.

48. Dalling, D. K.; Zilm, K. W.; Grant, D. M.; Heeschen, W. A.; Horton, W. J.; Pugmire, R. J. *J. Am. Chem. Soc.* **1981**, *103*, 4817.

49. Ferrier, W. G.; Iball, J. *Chem. Ind. (London)* **1954**, 1296.

50. Herbstein, F. H.; Kapon, M.; Reisner, G. M. *Acta Crystallogr., Sect. B* **1986**, *42*, 181.

51. Smith, W. B.; Shoulders, B. A. *J. Phys. Chem.* **1965**, *69*, 2022.

52. Jimenez, F. G.; Perezamador, M. C.; Alcayde, J. R. *Can. J. Chem.* **1969**, *47*, 4489.

53. Rabideau, P. W. *J. Org. Chem.* **1971**, *36*, 2723.

54. Beckett, A. M.; Mulley, B. A. *J. Chem. Soc.* **1955**, 4159.

55. Carruthers, W.; Hall, G. E. *J. Chem. Soc. B* **1966**, 861.

56. Nicholls, D.; Szwarc, M. *J. Am. Chem. Soc.* **1966**, *88*, 5757.

57. Nicholls, D.; Szwarc, M. *Proc. R. Soc. London, Ser. A* **1967**, *301*, 231.

58. Lapouyade, R.; Labandibar, P. *Tetrahedron Lett.* **1970**, 1589. See also, Redford, D. A., Ph.D. Thesis, The University of Saskatchewan, 1967; *Diss. Abstr. B* **1968**, *28*, 4074.

59. Fu, P. P.; Harvey, R. G.; Paschal, J. W.; Rabideau, P. W. *J. Am. Chem. Soc.* **1975**, *97*, 1145.

60. Yannoni, N. F.; Silvermann, J. *Acta, Crystallogr.* **1966**, *21*, 390.

61. Brennan, T.; Putkey, E. F.; Sundaralingam, M. *Chem. Commun.* **1971**, 1490.

62. Bordner, J.; Stanford, R. H., Jr.; Ziegler, H. E. *Acta Crystallogr., Sect. B* **1973**, *29*, 313.

63. Stanford, R. H., Jr. *Acta Crystallogr., Sect. B* **1973**, *29*, 2849.

64. Chu, S. S. C.; Chung, B. *Acta Crystallogr. Sect. B* **1976**, *32*, 836.

65. Leory, F.; Courseille, C.; Daney, M.; Bouas-Laurent, H. *Acta Crystallogr. Sect. B* **1976**, *32*, 2792.

66. Bartoli, G.; Bosco, M.; Pozzo, R. D.; Sgarabotto, P. *J. Chem. Soc., Perkin Trans. 2* **1982**, 929.

67. Toda, F.; Tanaka, K.; Mak, T. C. W. *Tetrahedron Lett.* **1984**, *25*, 1359; see also *J. Inclusion Phenom*, **1985**, *3*, 225.

68. Ahmad, N.; Goddard, R. J.; Hatton, I. K.; Howard, J. A. K.; Lewis, N. J.; MacMillan, J. *J. Chem. Soc., Perkin Trans 1* **1985**, 1859. A planar geometry for the trans compound is also suggested from NMR. See Ahmad, N.; Cloke, C.; Hatton, I. K.; Lewis, N. J.; MacMillan, J. *J. Chem. Soc., Perkin Trans 1* **1985**, 1849.

69. Rabideau, P. W.; Mooney, J. L.; Lipkowitz, K. B. *J. Am. Chem. Soc.* **1986**, *108*, 8130.

70. Baum, M. W.; Guenzi, A.; Johnson, C. A.; Mislow, K. *Tetrahedron Lett.* **1982**, *23*, 31.

71. Zieger, H. E.; Schaeffer, D. J.; Padronaggio, R. M. *Tetrahedron Lett.* **1969**, *57*, 5027.
72. Rabideau, P. W.; Smith, W. K.; Ray, B. D. *Mag. Res. Chem.*, in press.
73. Curtin, D. Y.; Carlson, C. G.; McCarthy, C. G. *Can. J. Chem.* **1964**, *42*, 565.
74. Cho, H.; Harvey, R. G.; Rabideau, P. W. *J. Am. Chem. Soc.* **1975**, *97*, 1140.
75. Elschenbroich, C.; Möckel, R.; Bilger, E. *Z. Naturforsch. Teil B* **1984**, *39*, 375.
76. Rabideau, P. W.; Wetzel, D. M.; Lawrence, J. R.; Husted, C. A. *Tetrahedron Lett.* **1984**, *25* 31.
77. Sygula, A.; Rabideau, P. W. *J. Org. Chem.* **1987**, *52*, 3521.

NOTE ADDED IN PROOF

Very recently, the X-ray crystal structure of 1,4-cyclohexadiene (mp 223 K) has been determined at 153 K by direct crystallization of the X-ray diffractometer. The carbon atom ring was determined to be planar within 0.002 Å with no evidence of any disorder involving boat or chair conformation.*

* Jeffrey, G. A. *J. Am. Chem. Soc.* 1988, *110*, 7218.

5

Use of Carbon-13 Nuclear Magnetic Resonance in the Conformational Analysis of Hydroaromatic Compounds

Frederick G. Morin
David M. Grant

Since the original observation[1,2] of nuclear magnetic resonance (NMR), the technique of solution-state NMR has become a premier method of structural and conformational analysis of organic molecules. With the introduction of Fourier transform methods, ^{13}C-NMR has exceeded ^{1}H-NMR as a tool for conformational analysis and a number of books have become standard reference texts in this field.[3–5]

The chemical shift experienced by a ^{13}C nucleus is determined by several electronic factors such as hybridization and inductive effects of substituents. Relatively smaller but highly informative changes in chemical shift result from even subtle changes in conformation, for example, the difference in shift of an equatorial and an axial methyl group in methylcyclohexane.[6] The utility of conformationally dependent ^{13}C shifts in hydroaromatic compounds is the focus of this chapter.

Hydroaromatic compounds such as tetralin (1,2,3,4-tetrahydronaphthalene), while of significant importance in the field of fuel science as reactive components in the liquefaction and gasification of coals, have not received their share of NMR attention over the years compared with the well-studied and understood saturated ring systems such as cyclohexane. These mixed cyclic systems (both aromatic and aliphatic) possess interesting differences compared with either the pure aliphatic or aromatic ring systems. Therefore, a good understanding of the conformational properties and the energetics involved in these compounds would be valuable. An organized and systematic study of the conformationally dependent ^{13}C chemical shifts of these molecules is aided by the use of a multiparameter regression analysis. To introduce this technique and to provide a comparison of the conformational properties of the cyclohexanes with hydroaromatics, we begin with a brief historical review of the use of ^{13}C chemical shifts in the conformational analysis of alkanes and cyclohexanes.

5.1 ALKANES

The first use of substituent parameters in ^{13}C chemical shifts came in the study of the simple *n*-alkanes[7] from methane to decane where the following linear

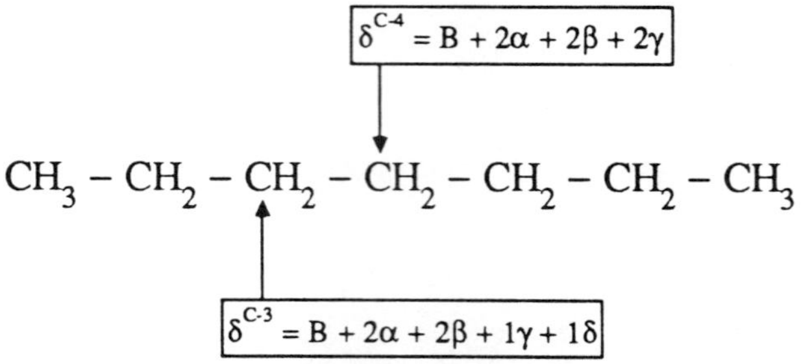

Figure 5.1 Examples of the application of substituent parameters to *n*-alkanes. The C-3 and C-4 carbons of heptane are shown.

relationship was found to correlate the shifts:

$$\delta^i = B + \sum n_{ij} A_j \qquad [5.1]$$

Here δ^i is the chemical shift of the ith carbon, B is the chemical shift of methane, in this case scaled to the reference compound TMS, A_j is the additive substituent parameter of the jth carbon, and n_{ij} is the number of carbons in the jth position relative to the ith atom being observed. In other words, the chemical shift of a particular nucleus would be determined by the number of carbons α, β, γ, and so forth relative to the carbon of interest. For instance, as shown in Figure 5.1, the chemical shift of C-3 of *n*-heptane is estimated by

$$\delta^{\mathrm{C-3}} = B + 2\alpha + 2\beta + 1\gamma + 1\delta$$

and the shift of C-4 by

$$\delta^{\mathrm{C-4}} = B + 2\alpha + 2\beta + 2\gamma$$

where α, β, γ, and so on are the substituent parameters that add to (or subtract from) B, the scaling parameter. More examples are shown in the complete list of the linear alkanes given in Table 5.1. Such tables are required as input for a multiparameter linear regression analysis.[8,9]

While in the case of the normal alkanes approximate values for the parameters may be determined with a simple graph or by inspection, their "best" numerical values are obtained with this computer analysis of the entire data set.

5.1.1 Linear Regression Analysis

Normally, in simple least-squares analysis one possesses a set of data that consists of an observation (the dependent variable) that is a function of an independent variable. For example, in the case of chemical kinetics, the kinetic

Table 5.1 Example Input to a Multiple Linear Regression Analysis of the ^{13}C Chemical Shifts of *n*-Alkanes in Terms of the Substituent Parameters α, β, γ, δ, and ϵ

Compound	Carbon	Shift	α	β	γ	δ	ϵ
Methane		−2.30	0	0	0	0	0
Ethane		5.70	1	0	0	0	0
Propane	C-1	15.40	1	1	0	0	0
	C-2	15.88	2	0	0	0	0
Butane	C-1	13.02	1	1	1	0	0
	C-2	24.82	2	1	0	0	0
Pentane	C-1	13.53	1	1	1	1	0
	C-2	22.37	2	1	1	0	0
	C-3	34.33	2	2	0	0	0
Hexane	C-1	13.68	1	1	1	1	1
	C-2	22.72	2	1	1	1	0
	C-3	31.84	2	2	1	0	0
Heptane	C-1	13.75	1	1	1	1	1
	C-2	22.80	2	1	1	1	1
	C-3	32.20	2	2	1	1	0
	C-4	29.27	2	2	2	1	0
Octane	C-1	13.75	1	1	1	1	1
	C-2	22.83	2	1	1	1	1
	C-3	32.18	2	2	1	1	1
	C-4	29.48	2	2	2	1	0
Nonane	C-1	13.83	1	1	1	1	1
	C-2	22.83	2	1	1	1	1
	C-3	32.25	2	2	1	1	1
	C-4	29.65	2	2	2	1	1
	C-5	29.92	2	2	2	2	1
Decane	C-1	13.95	1	1	1	1	1
	C-2	22.78	2	1	1	1	1
	C-3	32.24	2	2	1	1	1
	C-4	29.72	2	2	2	1	1
	C-5	30.07	2	2	2	2	1

rate of the reaction depends on temperature. The independent variable, the temperature in this example, is controlled by the investigator and a computer program then may be used to obtain an activation energy for the reaction from a least-squares fit of $\ln(k_r)$ versus $1/T$.

The additivity relationship shown in Equation (5.1) differs in that there is more than one independent variable (hence the term "multiple") which affects the shift at a particular site. The carbon atom population variables take only the discrete values of 0, 1, or 2, indicating the absence of or the presence of one or two carbons, respectively. When the independent variables are of this nature, they are sometimes called "indicator variables" because they indicate only whether a certain molecular group is present or not. The calculation of the best-fit parameters is carried out in the usual manner, by varying them until the total sum of squares of the deviations of the calculated shifts from the observed shifts is a minimum.

The final computer printout from the calculation of the data contained in Table 5.1 is given in Table 5.2, including the best-fit values for the parameters,

Table 5.2 Computer Printout of the Final Step of the Linear Regression Analysis of the Data of Table 5.1

STEP NUMBER = 5
VARIABLE ENTERED = 6

MULTIPLE R = 0.9998 STANDARD ERROR OF THE ESTIMATE = 0.2188

ANALYSIS OF VARIANCE

	DF	SUM OF SQUARES	MEAN SQUARE	F RATIO
REGRESSION	5	2457.1069	491.4214	10267.8545
RESIDUAL	24	1.1486	0.0479	

BASE VALUE = -2.75255

VARIABLE		COEFFICIENT	STD. ERROR	F TO REMOVE
ALPHA	2	9.12165	0.09139	9962.0195
BETA	3	9.38400	0.09568	9618.1816
GAMMA	4	-2.58646	0.11379	516.6536
DELTA	5	0.36799	0.13836	7.0737
EPSILON	6	0.20371	0.11243	3.2830

COMPUTED VALUES AND RESIDUALS

CASE	EXPERIMENT	CALCULATED	RESIDUAL
1	-2.3000	-2.7525	0.4525
2	5.7000	6.3691	-0.6691
3	15.4000	15.7531	-0.3531
4	15.8800	15.4908	0.3892
5	13.0200	13.1666	-0.1466
6	24.8200	24.8748	-0.0548
7	13.5300	13.5346	-0.0046
8	22.3700	22.2883	0.0817
9	34.3300	34.2588	0.0712
10	13.6800	13.7384	-0.0584
11	22.7200	22.6563	0.0637
12	31.8400	31.6723	0.1677
13	13.7500	13.7384	0.0116
14	22.8000	22.8600	-0.0600
15	32.2000	32.0403	0.1597
16	29.2700	29.4538	-0.1838
17	13.7500	13.7384	0.0116
18	22.8300	22.8600	-0.0300
19	32.1800	32.2440	-0.0640
20	29.4800	29.4538	0.0262
21	13.8300	13.7384	0.0916
22	22.8300	22.8600	-0.0300
23	32.2500	32.2440	0.0060
24	29.6500	29.6575	-0.0075
25	29.9200	30.0255	-0.1055
26	13.9500	13.7384	0.2116
27	22.7800	22.8600	-0.0800
28	32.2400	32.2440	-0.0040
29	29.7200	29.6575	0.0625
30	30.0700	30.0255	0.0445

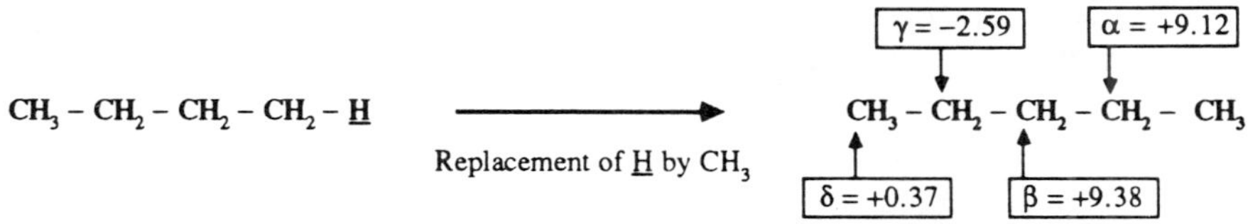

Figure 5.2 Illustration of the results of the regression analysis of the ^{13}C chemical shifts of the *n*-alkanes. The effects of the replacement of a terminal hydrogen with a methyl group are shown.

their standard errors, the calculated shifts, and statistical information (F ratio) for testing the significance of the parameters. The substituent effects are illustrated in Figure 5.2 for the replacement of a terminal hydrogen in butane with a methyl group.

The base value obtained is within the predictive error of the shift of methane. The substitution of a hydrogen by a methyl group generally causes a shift to lower field except in the case of the γ carbon where the methyl substituent produces an upfield shift. It has been established that the magnitude of the γ effect in the alkanes is due to the relative population of gauche and trans conformers. The gauche conformation in contrast to the trans structure brings the γ carbon into close spatial proximity to the observed carbon, and the resulting upfield γ shift has been attributed to this steric perturbation arising from the chain coiling back upon itself. The larger positive shifts found in the α and β parameters are in part explained by inductive and electronic effects. The longer range effects, δ and ϵ, are relatively small in keeping with the attenuation of inductive effects as the chain is lengthened between the observed and perturbing atoms. For example, the calculated shift of C-3 of heptane (case 15 of Table 5.2) is found to be

$$\begin{aligned}\delta^{C-3} &= B + 2\alpha + 2\beta + 1\gamma + 1\delta \\ &= -2.75 + 2(+9.12) + 2(+9.38) + 1(-2.59) + 1(+0.37) \\ &= 32.04\ \text{(calc)} \quad \text{or} \quad 32.20\ \text{(obs)}\end{aligned}$$

In order to extend the analysis to include the branched isomers of alkanes, new parameters were added to account for those structural features, such as tertiary and quaternary carbons and carbons attached to these centers, as linear relationships in the number of carbon atoms at positions α, β, γ, and so forth are no longer preserved.

5.2 CYCLOHEXANES

Subsequently, the same method was used in the analysis of the ^{13}C shifts of the conformationally important methylcyclohexanes.[6,10] These molecules differ from the continuous chain alkanes in two aspects: (1) the substitution of a methyl for a hydrogen may be done in two ways on a given ring carbon, namely, at the equatorial or at the axial position, and (2) rigid ring structures for many of the methylated cyclohexanes preclude the rotational isomerism of the *n*-alkanes. One consequence of this extension is a moderate increase in the total number of parameters required to characterize the methylcyclohexanes. Fortunately, this is counterbalanced in the analysis by an increase in the number of chemical shift values which are dependent on the conformational features and free from motional averaging. The α parameter for the alkanes thus became α_e and α_a, β became β_e and β_a, and so on.

While the chair structure of the cyclohexane ring prevents rotational isomerism about C—C bonds, rapid conformational inversion of the ring in some methylcyclohexanes still occurs, with exchange of the equatorial and axial positions and vice versa, even at room temperature. The inversion process is relatively rapid in all methyl substituted derivatives but depending on the nature of substitution the methylcyclohexanes may be separated into three groups:

1. Those that possess a single highly favored conformer due to unfavorable steric interactions in the alternative chair structure (e.g., *cis*-1,3-dimethylcyclohexane).

2. Those that interconvert between identical forms of equal energy (e.g., *cis*-1,2- and *cis*-1,4-dimethylcyclohexane).

3. Those that have a significant but unequal contribution to the equilibrium from both chair conformations (e.g., 1,1,2-trimethylcyclohexane).

The axial form of methylcyclohexane is destabilized by the addition of two unfavorable interactions similar to that found in *gauche*-butane while the equatorial form has strain energy that is only slightly more than cyclohexane itself because of the absence of unfavorable gauche interactions. The strain energy involved in one gauche interaction is about 0.9 kcal/mol and therefore the axial form of methylcyclohexane is about 1.8 kcal/mol higher in energy and contributes $<5\%$ to the equilibrium. The increase in strain energy brought about by 1,3-diaxial methyl groups is about 3.7 kcal/mol and the equilibrium of *cis*-1,3-dimethylcyclohexane consists of $> 99\%$ diequatorial form.

More than 60 ^{13}C chemical shifts for the ring carbons of methylcyclohexanes of groups (1) and (2) were measured and analyzed[6,10] in the same manner as the *n*-alkanes but with a parameter set that reflected the specific structural features present. Figure 5.3 defines some of the parameters used in the analysis of the methylcyclohexanes, demonstrating the importance of different conformational parameters for axial or equatorial substitution. Some results of the regression are given in Figure 5.4. The scaling parameter B in this situation is the shift of cyclohexane.

Both the α_e (+5.96 ppm) and α_e (+1.40) parameters for the cyclohexanes are smaller than the α parameter determined for the alkanes. More importantly, α_e and α_a are significantly different from each other, reflecting the different

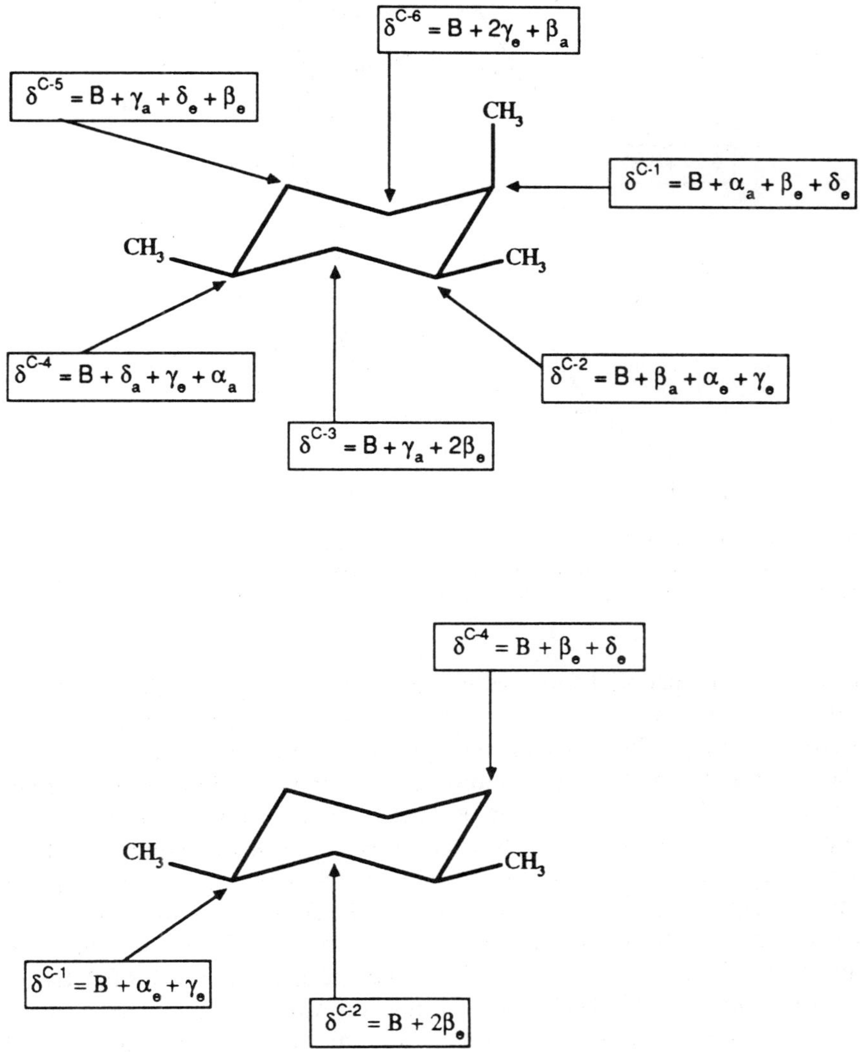

Figure 5.3 Definitions of some of the parameters necessary to analyze the ^{13}C shifts of methylcyclohexanes. Note the different conformational parameters for axial (a) and equatorial (e) substitution.

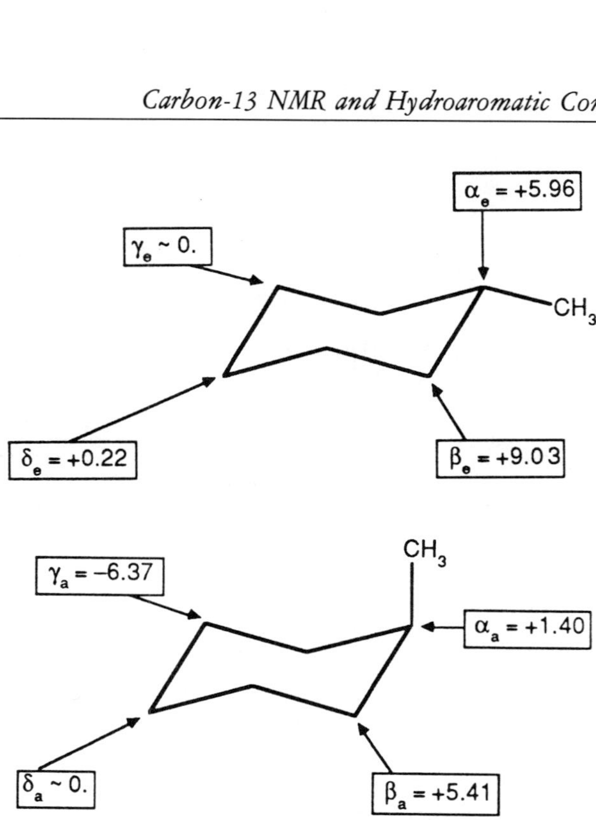

Figure 5.4 Illustration of the results of the regression analysis of the ^{13}C chemical shifts of the methylcyclohexanes. The shifts caused by equatorial substituents are uniformly more downfield than the corresponding axial ones.

perturbations for equatorial and axial methyls in the ring system. The β_e effect is found to be nearly the same as the β effect for the alkanes, but again β_a differs substantially from β_e. The γ_e parameter represents an effect that is expected to be similar to the addition of a methyl to propane to form s-*trans*-butane. This s-trans effect on the shift of the γ carbon is negligible, reflecting both the greater through-bond and through-space distance between the methyl group and the γ carbon. The replacement of an axial proton of cyclohexane by a methyl group produces structures that correspond to the gauche form of butane, and the γ_a parameter, as expected, is relatively large (-6.37 ppm). As the s-*trans* rotamer in butane is the dominant structure (ratio of trans to gauche is $\sim 0.8{:}0.2$ for an energy difference of ~ 0.9 kcal/mol), it is not surprising that the γ effect in the alkanes is closer in magnitude to the γ_e parameter of the cyclohexanes than γ_a. We finally note that in each case the equatorial substituents cause shifts that are uniformly more downfield than the corresponding axial ones.

This procedure for establishing shift parameters was performed on chemical shifts obtained at room temperature. Others have used variable temperature ^{13}C-NMR to freeze out the conformational inversion of methylcyclohexane,[11] *cis*-1,2-[12] and *cis*-1,4-dimethylcyclohexane,[12,13] and their findings have corroborated the shift parameters obtained on the rapidly converting systems. For instance, the observed shifts[12] of *cis*-1,2-dimethylcyclohexane at 158 K, where the conformational inversion is frozen out, compare well with the calculated ones as is also the case for *cis*-1,4-dimethylcyclohexane.[13] The agreement between the experimental and predicted shifts, considering the different solvents and the temperature dependence of chemical shifts, is very good.

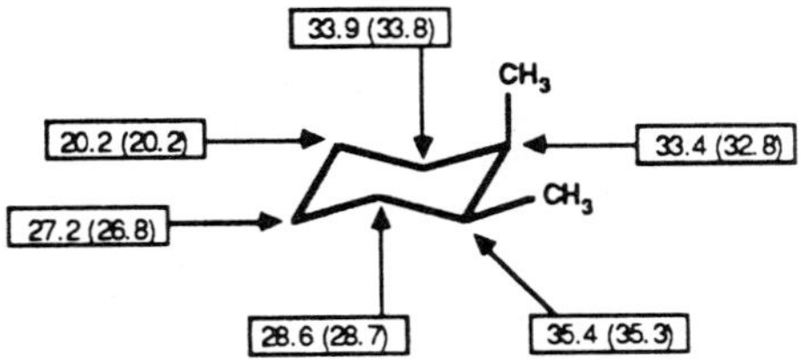

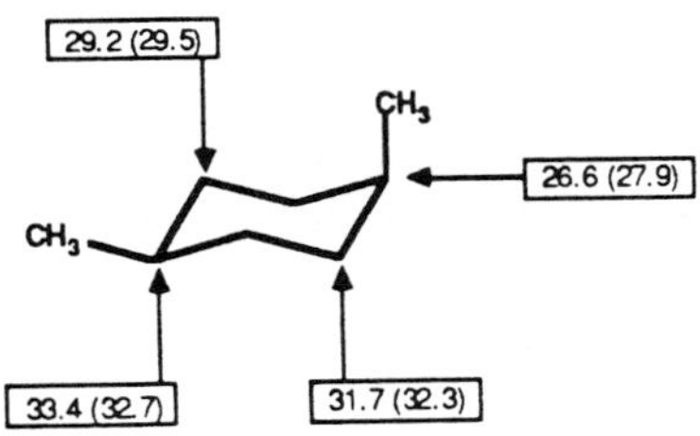

The generality of the methyl parameters was demonstrated by the successful application to the *cis*- and *trans*-decalins (decahydronaphthalenes),[14] and to perhydroanthracenes and perhydrophenanthrenes;[15] all four molecules are structurally related to cyclohexane and indicate the general applicability of these parameters to the cyclohexane moiety appearing in larger fused ring systems. Unfortunately, the choice of parameters used in any regression analysis is not unique and any logically chosen set of linearly independent parameters that describe the salient features of the system under study is valid. In the case of the decalins, a second parameter set which was less accurate but smaller in number than the methylcyclohexane substituent set was also employed for the sake of simplicity. The reader is referred to these investigations[14,15] for further details as well as to other parameterizations of the alkane data that have been formulated.[16]

5.3 HYDROAROMATICS

The investigation of conformation-dependent ^{13}C chemical shifts, using the multiple variable method outlined above, was continued with several subsequent studies of a series of hydroaromatics* such as tetralin (1,2,3,4-tetrahydronaphtha-

* The following abbreviations are used throughout this chapter: THA = 1,2,3,4-tetrahydroanthracene, DHA = 9,10-dihydroanthracene, THP = 1,2,3,4-tetrahydrophenanthrene, DHP = 9,10-dihydrophenanthrene.

lene) and the analogous 1,2,3,4-tetrahydroanthracenes[17] and the tetrahydrophenanthrenes.[18] The numbering schemes for naphthalene, anthracene, and phenanthrene are given below.

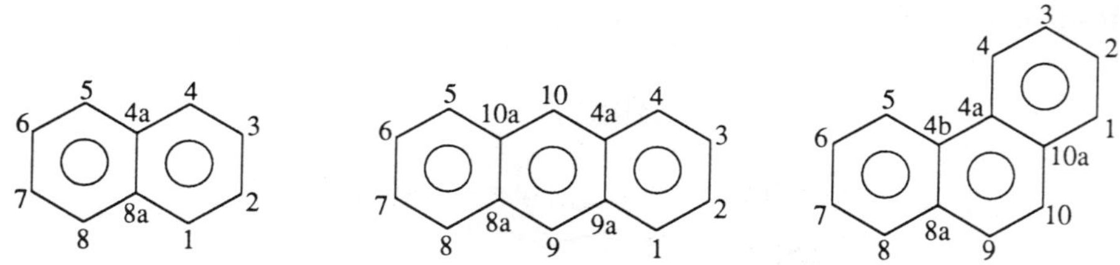

The aliphatic ring of hydroaromatics differs from cyclohexane in one important aspect in that there are two distinct sites, C-1 or C-2, for replacement of a hydrogen with a methyl group. This necessitates an increase in the number of substituent parameters required to characterize the structural features associated with the ^{13}C shifts of hydroaromatics. Fortunately, a very large number of molecules display such conformationally dependent shifts from which the parameter set may be determined. The α effect of the alkane parameter set, replaced by the α_e and the α_a parameters in the cyclohexane analysis, requires four parameters, α_{1e}, α_{1a}, α_{2e}, and α_{2a}, in tetralin. The four terms reflect, respectively, the presence of a methyl at C-1 (or C-4) oriented pseudoequatorially, a methyl at C-1 (C-4) oriented pseudoaxially, a methyl at C-2 (or C-3) oriented equatorially, and a methyl at C-2 (C-3) oriented axially.

As with the methylcyclohexanes, the methylhydroaromatics may undergo conformational mobility in the aliphatic ring at room temperature, exchanging the (pseudo)equatorial and (pseudo)axial positions and vice versa. This inversion usually occurs even faster for hydroaromatics than cyclohexanes due to a lower inversion barrier. However, depending on the nature of substitution, the methylhydroaromatics may be separated into three groups:

1. Those that possess a single highly favored conformer due to unfavorable steric interactions in the alternative half-chair structure.

2. Those that interconvert between identical forms of equal energy.

3. Those that have a significant but unequal contribution to the equilibrium form both half-chair conformations.

Only those molecules of groups (1) and (2) are suitable for inclusion in the regression analysis because of the need to identify the appropriate populational coefficients.

An important distinction between the hydroaromatics and cyclohexanes is the absence of axial hydrogens at the bridgehead position in the hydroaromatics. Thus, an axial substituent at C-2 encounters only a single *gauche*-butane interaction with the axial hydrogen at C-4 compared to two in the cyclohexane case. However, MM2 calculations by Schneider and Agrawal[19] have provided an estimate for the difference in steric energy of 1.4 kcal/mol between the axial and equatorial forms of 2-methyltetralin, larger than the 0.9 kcal/mol of a *gauche*-butane interaction. This energy difference of 1.4 kcal/mol limits the axial conformer to a contribution of $<10\%$ to the equilibrium at room temperature. Thus, as a good approximation, those compounds with a C-2 methyl are assumed to occupy only the form with the methyl equatorial. While an equatorial methyl is in a relatively unhindered position at C-2, a methyl at C-1 interacts with a single axial C—H group or the peri position of the aromatic ring so that a substituent at C-1 is expected a priori to have little preference for either orientation. Thus, 1-methyl and *trans*-1,4-dimethyltetralin were not included in the initial fitting.

Conclusions similar to those for tetralin can be made concerning the C-1 through C-3 positions of tetrahydrophenanthrene (THP), but the C-4 site is unique. A methyl group may occupy only the pseudoaxial orientation as the pseudoequatorial orientation is completely unattainable due to the steric interaction with C-5. Therefore, compounds such as *cis*- or *trans*-1,4-dimethyl-THP will be frozen into a single conformation, unlike the tetralin derivatives. Using these selection criteria, there are a total of 39 compounds with 144 chemical shifts available for the regression analysis in a set of reasonably diverse hydroaromatics.

In order to investigate the conformational-dependent ^{13}C shifts of both the tetralin and tetrahydrophenanthrene series of compounds as a group, the analysis was performed with the following parameters and assumptions. The substituent parameters at C-1 and C-4 were assumed to be identical. This is rigorously true for the tetralins, but not necessarily so for the tetrahydrophenanthrenes where the C-1 and C-4 sites are in quite different molecular environments. Nonetheless, this assumption will be shown to be justified (vide infra). The substituent parameters at C-2 and C-3 were also assumed to be identical, but different from those for C-1 and C-4 substitution. Again, this is rigorously true because of symmetry for the tetralins but also plausible for tetrahydrophenanthrene considering the similarities in the environments of C-2 and C-3 in that molecule. The number of parameters required to characterize the great variety of structural features in these interesting systems is unfortunately somewhat larger and requires the extensive terminology given in Figure 5.5. Thus, the initial set of 25 parameters consisted of the following:

1. B, the base value, here chosen to be C-2.

2. $P_{C\text{-}1}$, a parameter to account for the difference between the shift of the C-1,C-4 and C-2,C-3 carbons.

3. $P_{C\text{-}4}$, a parameter to account for the difference between the shift of the C-1 and C-4 carbons.

4. Parameters for both pseudoequatorial and pseudoaxial substitution at C-1: α_{1e}, α_{1a}, β_{1e}, β_{1a}, γ_{1e}, γ_{1a}, δ_{1e}, δ_{1a} (eight terms).

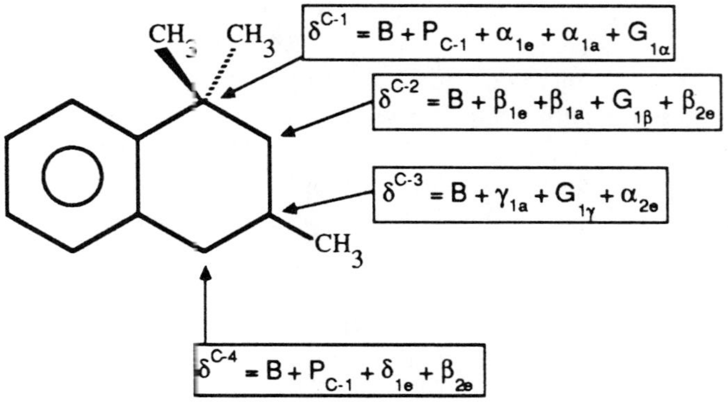

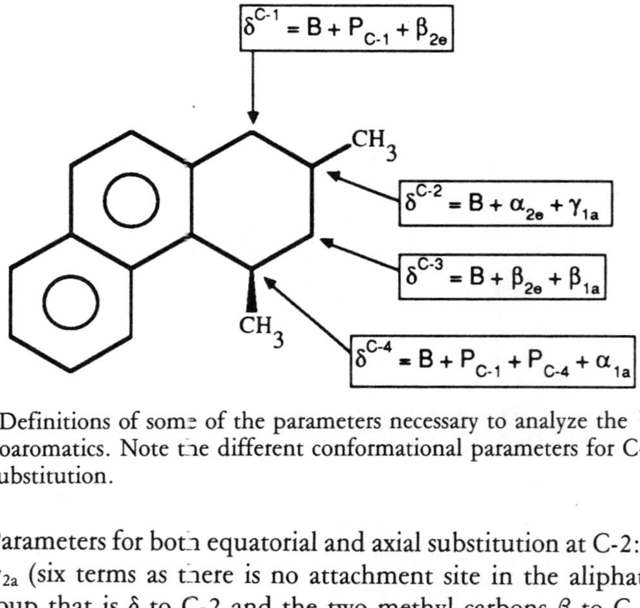

Figure 5.5 Definitions of some of the parameters necessary to analyze the ^{13}C shifts of methylhydroaromatics. Note the different conformational parameters for C-1 (C-2) and C-3 (C-4) substitution.

5. Parameters for both equatorial and axial substitution at C-2: α_{2e}, α_{2a}, β_{2e}, β_{2a}, γ_{2e}, γ_{2a} (six terms as there is no attachment site in the aliphatic ring for a methyl group that is δ to C-2 and the two methyl carbons β to C-2 are treated identically).

6. Parameters to account for geminal dimethyl groups at C-1: $G_{1\alpha}$, $G_{1\beta}$, $G_{1\gamma}$, $G_{1\delta}$ (four terms); and at C-2: $G_{2\alpha}$, $G_{2\beta}$, $G_{1\gamma}$ (three terms, again no δ term).

7. A term, V_{ea}, for interacting vicinal dimethyl groups.

8. A term, γ_5, for the presence of a γ methyl on the proximate aromatic carbon C-5 (or C-8) on C-4 (or C-1).

A preliminary calculation revealed that four of the above (γ_{1e}, δ_{1a}, γ_{2e}, and $G_{1\beta}$) were insignificant. They were removed and the calculation redone with 21

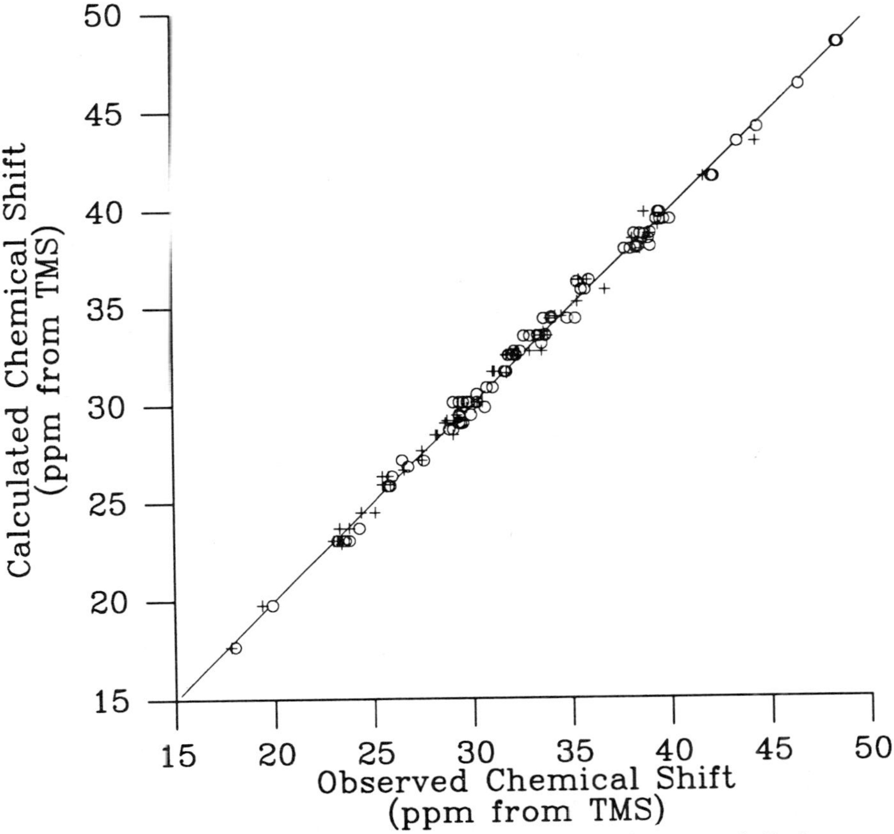

Figure 5.6 Plot of 144 calculated shifts versus observed shifts for the methylhydroaromatics. Tetralins are marked with circles and tetrahydrophenanthrenes with plus signs.

parameters. A plot of the calculated and observed shifts* is shown in Figure 5.6 with the shifts of the tetralins (and 1,2,3,4-tetrahydroanthracenes) marked with circles and the tetrahydrophenanthrenes with plus signs. The slope and intercept are 0.995 and 0.16, respectively, and the multiple correlation coefficient squared and standard deviation of the points are 0.9949 and 0.44 ppm. The agreement indicates that the structural features of unstrained compounds of unequivocal conformation are quite well described by the set of parameters and that the substituent parameters are generally applicable to either class of compounds.

The best-fit values of the parameters are graphically illustrated in Figure 5.7. The results for a few of the parameters are compared with the relevant parameters for the alkanes and cyclohexanes in Table 5.3. Several comments may be made: (1) as in the cyclohexanes, the (pseudo)equatorial parameters for the hydroaromatics are uniformly larger (downfield shift) than their (pseudo)axial counter-

* A complete tabulation of the ^{13}C chemical shifts of the hydroaromatics is provided in the Table 5.4.

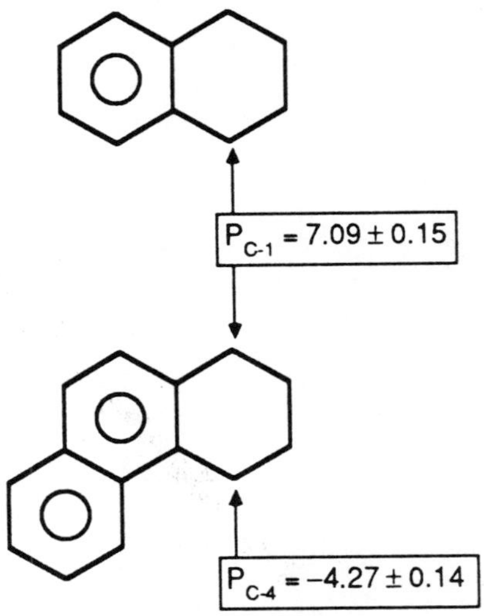

The base value is the shift of C-2 (C-3)

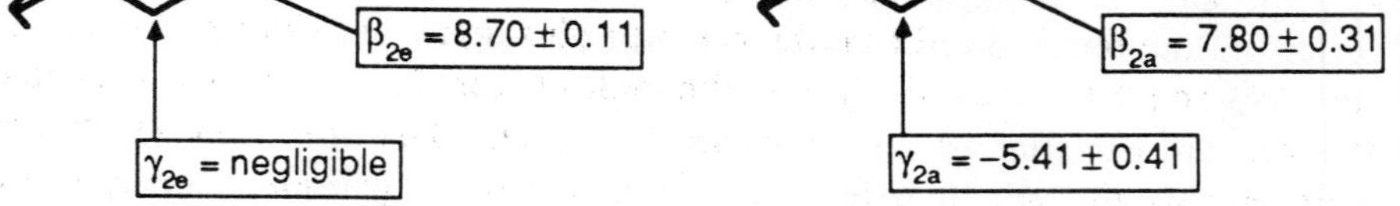

Substituent parameters for substitution at C-2 (C-3)

Figure 5.7 Illustration of the results of the regression analysis of the ^{13}C chemical shifts of the methylhydroaromatics. As with the cyclohexanes, the shifts caused by (pseudo)equatorial substituents are uniformly more downfield than the corresponding (pseudo)axial ones.

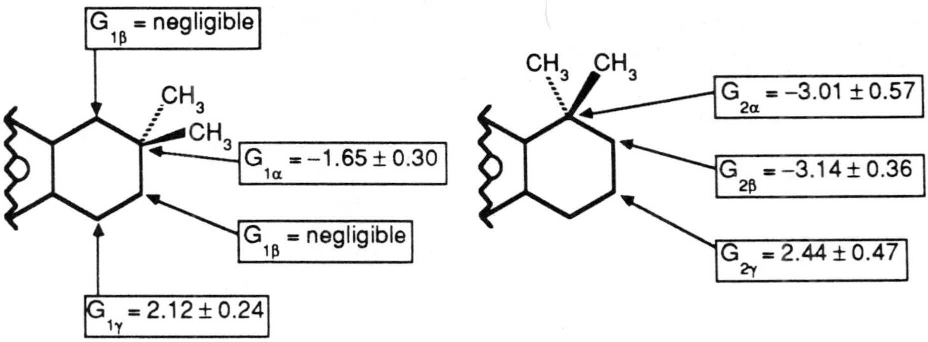

Substituent parameters for geminal dimethyl groups

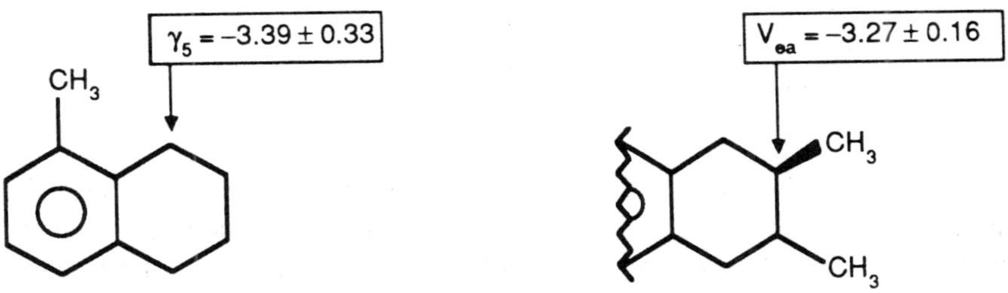

Miscellaneous substituent parameters

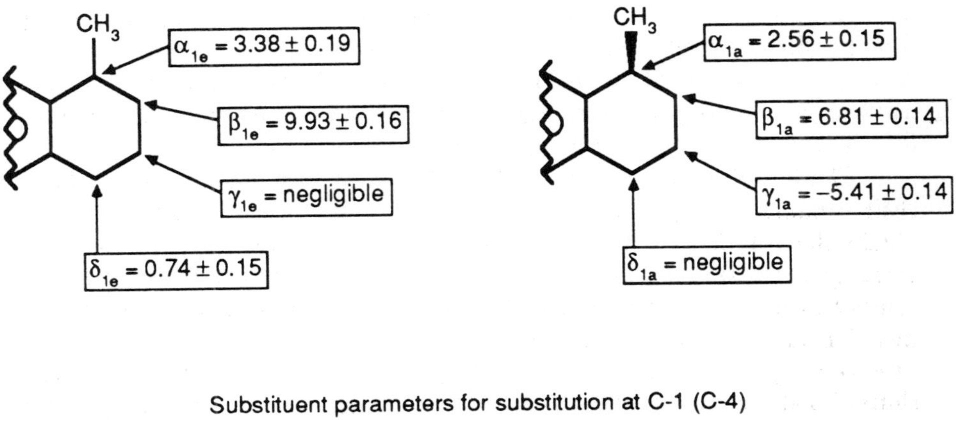

Substituent parameters for substitution at C-1 (C-4)

Table 5.3 Comparison of Some of the Substituent Effects from the Alkanes, Cyclohexanes, and Hydroaromatics

Alkanes	Cyclohexanes	Hydroaromatics
	$\alpha_e = +5.96$	$\alpha_{1e} = +3.38$
		$\alpha_{2e} = +6.07$
$\alpha = +9.12$		
	$\alpha_a = +1.40$	$\alpha_{1a} = +2.56$
		$\alpha_{2a} = +3.40$
	$\beta_e = +9.03$	$\beta_{1e} = +9.93$
		$\beta_{2e} = +8.70$
$\beta = +9.38$		
	$\beta_a = +5.41$	$\beta_{1a} = +6.81$
		$\beta_{2a} = +7.80$
	$\gamma_e =$ negligible	$\gamma_{1e} =$ negligible
		$\gamma_{2e} =$ negligible
$\gamma = -2.59$		
	$\gamma_a = -6.37$	$\gamma_{1a} = -5.41$
		$\gamma_{2a} = -5.41$

All values in ppm.

parts, (2) the parameters that pertain to a methyl group in the relatively unhindered C-2 equatorial position, α_{2e} and β_{2e}, are very close to the analogous parameters for cyclohexane, α_e and β_e, (3) the parameters pertaining to methyl substitution at C-1, for example, α_{1e} and α_{1a}, are substantially smaller than the parameters for substitution at C-2, (4) the parameters governing the effect of (pseudo)axial substitution at C-1 or C-2 upon a γ carbon, γ_{1a} and γ_{2a}, are similar to each other as well as to γ_a of the cyclohexane system (molecular models and geometries obtained from MM2 calculations show that these methyl groups are, in each case, approximately the same distance from the axial ring proton and this similarity is not surprising), and (5) in contrast to the alkanes and cyclohexanes, it was necessary to include a sizable δ_{1e} parameter in order to fit the data, indicating that the interaction of a pseudoequatorial methyl at C-1 (C-4) with H-8 (H-5) leads to a deformation of the ring geometry presumably due to the rather flexible nature of the tetralin aliphatic ring and thus provides a mechanism for transferring the effect of this otherwise remote substituent.

This flexible nature and the low barriers to interconversion between the half-chair conformations prevent the application of variable-temperature NMR to obtain the substituent effects directly from the chemical shifts in the slow exchange limit, as is possible for some of the methylcyclohexanes. This is also in contrast to the homologue of tetralin, benzocycloheptene, where the barrier to inversion of both the 2-methyl and 3-methyl derivatives is high enough that variable-temperature ^{13}C- and ^{1}H-NMR have been used to obtain the chemical shifts of both conformers in the slow exchange limit.[20] The substituent effects for both equatorial and axial substitution at either site, both of which are not directly bonded to the aromatic ring, were determined to be quite close to those for methylcyclohexane. Apparently, the aromatic ring is responsible for the differences in the C-1 and C-2 substituent parameters in hydroaromatics.

5.3.1 Tetralins with a Single Highly Preferred Conformation

The prediction with a high degree of accuracy (see Table 5.4) of the ^{13}C chemical shifts of the 2-methyl derivatives indicates that the assumption of a single contributor to the equilibrium of those derivatives is valid within the error limits of the regression analysis. The ^{13}C chemical shifts of *cis*-1,2-dimethyl derivative, which is rapidly equilibrating between two identical conformers in the analogous cyclohexane compound, are also predicted quite well under the assumption that only the 1a,2e conformer contributes to the equilibrium. The preference of a methyl group at C-2 to occupy the equatorial position overrides any preference a C-1 methyl may have for either orientation. For example, *trans*-1,3-dimethylcyclohexane undergoes a rapid equilibrium between two equivalent forms (one axial and one equatorial methyl in each) but the *trans*-1,3-dimethyltetralin is essentially locked with an equatorial methyl at C-3 and a pseudoaxial methyl at C-1.

It is possible to force a methyl at C-1 to occupy only the pseudoaxial site by placing a methyl group at C-8. In the 1,8-dimethyl compound, the ^{13}C chemical shifts are predicted reliably under the assumption that the methyl group occupies only the pseudoaxial position. The impossibility of orienting two methyl groups in a 1,3-diaxial relationship in the cyclohexane ring is unchanged in these hydroaromatic systems. Thus, to avoid a 1,3-diaxial interaction the *cis*-1,3-dimethyl and 1,3,3-trimethytetralins freeze into single conformations, those with a pseudoequatorial methyl group at C-1. This effect also results in a single conformer for the 1,1,3-trimethyl derivatives with the 3-methyl occupying the pseudoequatorial position.

5.3.2 Tetralins with Significant Contribution from Two Conformers

The substituent parameters obtained on well-characterized conformers may be used to estimate the relative contribution from those molecules that were not included in the fit (e.g., 1-methyltetralin) because they involve two possible half-chair conformations that are of comparable but unequal energies. The chemical shifts in such interconverting molecules with contributions from conformers with a pseudoaxial methyl and a pseudoequatorial methyl may be used with the appropriate substituent parameters to calculate the percentage of the two conformers from

$$\delta_{obs} = p_{psax}\delta_{psax} + p_{pseq}\delta_{pseq} \qquad [5.2]$$

where

$$p_{pseq} = 1 - p_{psax} \qquad [5.3]$$

and δ_{obs} and δ_{psax}, δ_{pseq} are the observed and calculated shifts, respectively. The mole fractions p_{psax} and p_{pseq}, of the molecules in the two forms sum to unity. For the C-2 carbon in 1-methyltetralin, a shift of 29.86 ± 0.14 ppm is predicted for a pseudoaxial methyl with 32.97 ± 0.16 ppm predicted for a pseudoequatorial methyl. From the observed shift of 31.8 ppm it may be estimated that this molecule exists as 62 ± 5% pseudoequatorial methyl. Similarly, the respective substituent effect of a 1-methyl in a pseudoequatorial or pseudoaxial conformation upon C-3 is either zero or −5.4 ppm which when added to a normal C-3 shift gives 23.6 and 18.2 ppm for the shifts of the alternative conformations. An observed shift of 20.8 ppm indicates that the molecule exists in a pseudoequatorial form 54 ± 4% of the time. The uncertainties are obtained from the error limits of the best-fit parameters. The predicted values for either C-1 or C-4 for the alternative conformations are so close together that the errors prevent one from determining conformational preference with any degree of reliability for data on these two carbons. These results suggest that the methyl group in 1-methyltetralin prefers only slightly, if at all, a pseudoequatorial position a C-1 and that both conformers contribute significantly at ambient temperatures.

Steric interactions involving a C-1 methyl are illustrated in Figure 5.8, where it is worth noting the similarity in the positions of the methyl groups in the two conformations relative to interactions with a proton. These considerations corroborate the conclusion that neither conformation is highly preferred. Using molecular mechanics calculations on 1-methyltetralin, Schneider and Agrawal[19] have estimated the strain energy difference between the two conformations to be only 0.1 kcal/mol with the pseudoaxial conformer being slightly more stable. The very small energy difference should be taken as an indication only of the closeness of the energies of the two conformers as the calculations within this level of accuracy lack the ability to state with certainty which conformer is the more stable.

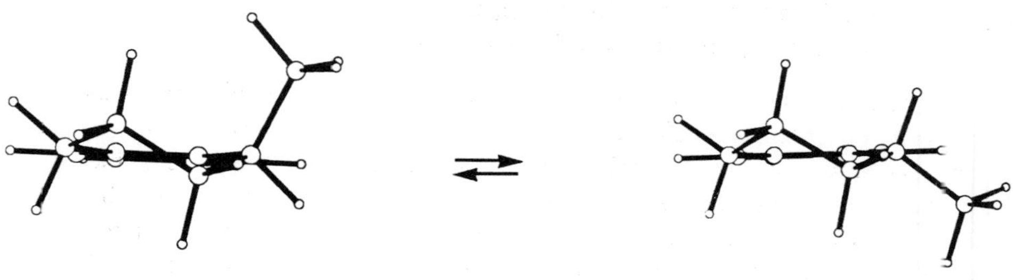

Figure 5.8 Illustration of the steric interactions involving a single pseudoaxial or pseudoequatorial methyl group at C-1. Note the similarity in the positions of the methyl groups relative to interactions with a proton (H-3_{ax} or H-8) in the two conformations.

Based on the 1-methyltetralin results, one might also expect *trans*-1,4-dimethyltetralin to have almost equally weighted conformers. The predicted shifts for C-2 are 24.45 ± 0.28 and 32.98 ± 0.16 ppm, respectively, for the pseudoaxial and pseudoequatorial conformers. The amount of *pseudoaxial* conformer determined from these predicted shifts and the observed shift of 28.4 ppm therefore is 54 ± 3%. Again, this estimate is close enough to 50% that a highly preferred conformation does not obtain for *trans*-1,4-dimethyltetralin. Similar conclusions concerning tetralins substituted at C-1 with other functional groups such as OH and NH_2 have been reached.[19]

Calculations for C-3 and C-4 of *trans*-1,2-dimethyltetralin give, respectively, 55 ± 4% and 57 ± 5% diequatorial conformation. Thus, a very substantial fraction of the time this molecule exists with diaxial methyl groups indicating that the interaction of a C-1 methyl group with the *peri* hydrogen must be sizable in the diequatorial conformation. In comparison, the equilibrium of *trans*-2,3-dimethyltetralin is strongly dominated by the diequatorial conformation. This latter molecule, like the analogous cyclohexyl compound *trans*-1,2-dimethylcyclohexane, is 99% diequatorial due to the $CH_3 \leftrightarrow$ axial C—H cross ring interactions in the diaxial conformation. In *trans*-1,2-dimethyltetralin, only two of these interactions are present in the diaxial form compared with four for the corresponding cyclohexane while another gauche-type interaction with the peri position is present in the diequatorial conformer. If *trans*-1,2-dimethyltetralin is included in the fitting procedure in either the 1e,2e or 1a,2a conformation, the average error in the predicted shifts is about 3.5–4.5 ppm.

Unfortunately, as the number of methyls on the ring increases, the accumulated errors in the predicted shifts limit the reliability of the estimates of conformational populations for interconverting molecules with comparable but unequal conformers. For this reason only general statements may be made about some of the remaining tetralins. Applying the substituent parameters to the 1,1,8-trimethyl derivative produces chemical shifts that typically differ by >3 ppm from the observed shifts. Severe crowding of methyls at C-1 from substituents in the aromatic ring obviously will cause distortion of the hydrogenated ring, and the fitting parameter set would not be expected to address such structural features. The lack of agreement in this highly sterically hindered system indicates that one cannot totally characterize major structural deformations which might be expected to arise. Conversely, where agreement is good in less strained molecules, one may make stronger arguments for the reliability of the structural assignments.

Predicted shifts for 1,1,4-trimethyltetralin have sufficient error to prevent them from being used in calculating conformational preferences. However, it has been demonstrated in 1-methyl and *trans*-1,2-dimethyltetralin, that an isolated methyl at C-1 or C-4 has no significantly preferred orientation, and one might also expect nearly equally populated conformations for 1,1,4-trimethyltetralin.

The three trisubstituted methyltetralins, 1,1,2-, 1,2,2-, and 2,2,3-, all may be expected to have a significant contribution from the conformer where the nongeminal methyl group is axial since this removes a gauche interaction between equatorial methyls. For instance, the C-3 shift of 1,2,2-trimethyltetralin is 3.6 ppm upfield of the same carbon of the 2,2-dimethyl compound, indicating a significant contribution from that conformation. When the 1,2,2- or 2,2,3-derivatives were included in the fit, the observed shifts were not reproduced very well (e.g., C-1 of 1,2,2- and C-4 of 2,2,3-, which would be most affected by an axial methyl group).

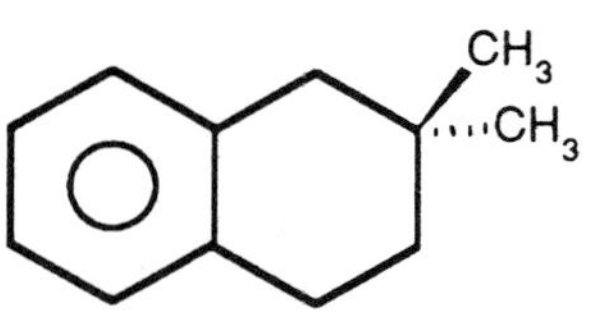

$\delta^{C\text{-}3} = 36.0$

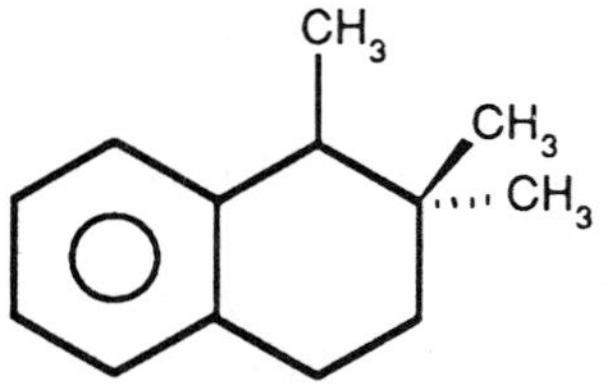

$\delta^{C\text{-}3} = 32.4$

It goes without saying that identical conformational characteristics for the 1,2,3,4-tetrahydroanthracenes were found to be the same as for the tetralins.

5.3.3 Methyl Carbons

Attempts to apply the same regression analysis technique to the methyl chemical shifts were considerably less successful, as it was not possible to find a relatively small parameter set that would adequately reproduce the observed shifts. The small range of shifts for the methyl groups also severely restricts the analysis. Several observations derived from the data provide an explanation for this dilemma. A reliable estimate of the shift of a pseudoequatorial methyl at C-1 may be obtained from **6** (21.8 ppm), **20** (21.9 ppm), and **24** (22.1 ppm), while the methyl shifts from **7** (24.4 ppm) and **25** (24.8 ppm) provide a value for the pseudoaxial methyl shift. We may predict then that an isolated C-1 methyl undergoing rapid interconversion between the two possible orientations in **8** (*cis*-1,4-dimethyltetralin) has an average shift of 23.2 ppm. The observed shift (22.7 ppm) for **8** slightly exceeds the experimental error for the deviation of the experimental value and the predicted average. Although this molecule was included among the fitted shifts, the deviation between predicted and observed is one of the largest of the set. Without a "locking" methyl group at the C-2 (or C-3) position, it may well be that the aliphatic ring in *cis*-1,4-dimethyltetralin distorts to relieve the steric strain between the pseudoequatorial methyl and either the C-5 or C-8 protons with only minor costs in aliphatic ring distortion energy. A similar distortion in *cis*-1,3-dimethyltetralin must result in an energy increase that is too large for the C-3 methyl in a pseudoaxial position and is thereby decreased in importance. Therefore, the pseudoequatorial methyl at C-1 in **8** probably experiences a slightly different environment and chemical shift than found in the otherwise similar methyl in **6**. Such distortions, of course, will increase the errors in the parametric fit. Numerous but unsuccessful attempts were made to predict with reasonable accuracy the methyl chemical shifts of **8** and the other conformationally characterizable compounds at the same time. The flexible nature of the saturated ring, in particular for groups in the C-1 position, is again indicated. The unusually large magnitude of δ_{1e} discussed briefly above probably also finds its origin in this type of ring deformation. As noted above, Figure 5.8 demonstrates that the environments of a pseudoaxial or pseudoequatorial methyl group are quite similar, and this is reflected in the relatively small difference in their chemical shifts (2.5 ppm) and in an inability to characterize clearly the system through a parametric fit of the methyl shifts.

In contrast to the variation in chemical shifts of methyl groups at C-1, those compounds that possess an isolated equatorial methyl at C-2, for example, **3**, **6**, and **7**, display a methyl chemical shift of 22.3 ± 0.2 ppm, not very different from the equatorial shift found in methylcyclohexanes.[6,10] Several interesting results are apparent in the methyl chemical shifts of tetrahydroanthracenes. Whereas the chemical shifts of C-2 methyls show little deviation from the analogous tetralins, C-1 methyls shift downfield by ~0.5 ppm when part of a geminal dimethyl moiety and upfield by as much as 0.7 ppm when isolated as in the case of the *cis*-1,4-dimethyl derivative. The extra benzene ring may increase the rigidity of the molecule and thereby limit deformations that can occur. Alternatively, these shifts may be produced by small variations in ring currents between the tetrahydroanthracenes and the corresponding tetralins.

5.3.4 Aromatic Carbons

Aromatic carbons 5 and 8 display chemical shifts that are dependent only on the presence or absence of a pseudoequatorial methyl at C-1 or C-4. When C-1 is unsubstituted or substituted with a pseudoaxial methyl, the aromatic carbon (i.e., C-5) chemical shift is 129.1 ± 0.2 ppm, while a pseudoequatorial methyl induces an upfield shift to 126.9 ± 0.2 ppm. Furthermore, a geminal dimethyl moiety at C-1 produces essentially the same shift at C-5 as a single pseudoequatorial methyl, indicating again that only the pseudoequatorial and not the pseudoaxial methyl produces a shift at C-8.

5.3.5 NMR Studies in the Solid State

Cross-polarization/magic-angle spinning NMR studies of the solid state,[21,22] where many motional processes cease, have been informative in some studies.[23] Since all the tetralins examined are liquids at room temperature, a solid carboxylic acid derivative, 5,8-dimethyl-5,6,7,8-tetrahydro-2-naphthoic acid, **27**, was synthesized. However, the aliphatic portion of the solid-state NMR spectrum of this molecule was identical to that of the solution spectrum, suggesting that the interconversion between conformers is not stopped in the solid state at ambient temperatures and that these motions continue to average the chemical shifts of the aliphatic carbons. Similar results may be encountered for other compounds in this study. Even at lowered temperatures ($< -30\,^\circ C$) there was no sign of freezing out of separate conformers. This result on a solid compound provides further corroborative evidence for the highly flexible nature of the saturated ring of the tetralin molecule and some molecular motion in the solid state may still be effective even at relatively low temperatures.

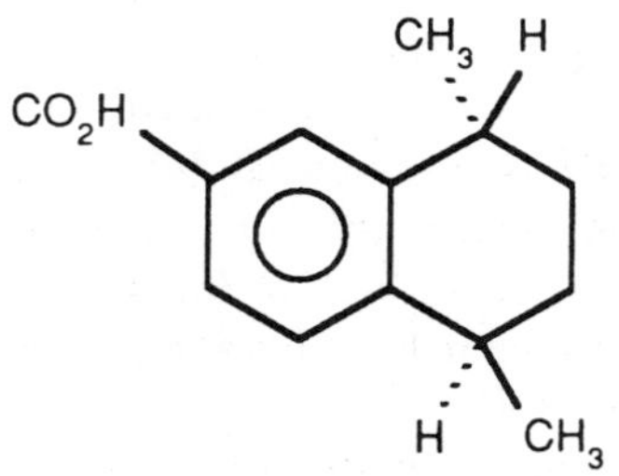

27

Of the tetrahydroanthracenes, only the parent and the 2-methyl derivative are solids at room temperature, due likely to the "planar" nature of these molecules which makes for efficient packing in the crystal. Not surprisingly, the solid-state NMR spectra of these conformationally pure compounds were identical to their solution spectra. None of the compounds that possess an axial or pseudoaxial methyl group is a solid at room temperature.

Wilson et al.[24] have reported the CP/MAS spectrum and X-ray crystallographic structure determination of 1,2,3,4,5,6,7,8-octahydroanthracene and deduced that of the two possible forms, syn and anti, the molecule takes up the syn form in the solid state. Small splittings of the resonances of the benzylic and nonprotonated aromatic carbons were postulated to arise from slight distortion from the idealized symmetry.

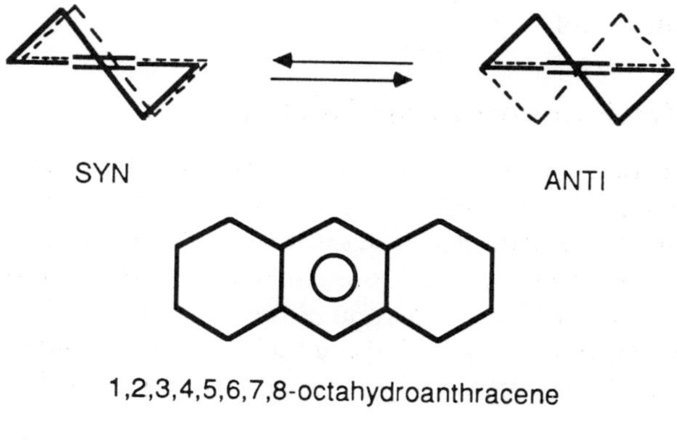

1,2,3,4,5,6,7,8-octahydroanthracene

5.3.6 Tetrahydrophenanthrenes

The general agreement, shown in Figure 5.6, between the predicted and observed ^{13}C chemical shifts for tetrahydrophenanthrene derivatives, is quite remarkable in light of the strong steric interaction between C-4 and C-5 and the anticipation that such a strong interaction would distort significantly the aliphatic ring. Indeed, an X-ray study[25] provides a distance from one of the C-4 hydrogens to the C-5 hydrogen of just 2.1 Å, well within two van der Waals' radii. Consequently, the C-4 shift of the parent (25.5 ppm) is 4.1 ppm upfield of that of tetralin itself (29.6 ppm) and a concomitant upfield shift of 4.3 ppm, from 127.3 ppm to 123.0 ppm, occurs for C-5. The remaining shifts for C-1, C-2, and C-3 are all rather modest (<1 ppm). Furthermore, all the changes in the methyl substituent parameters for THP relative to tetralin are relatively modest and apparently beyond an accurate experimental determination.

The results in Table 5.4 also validate the assumption that only a single conformation is needed in the fit for the THP series. Those THP derivatives with a C-4 methyl (**41**, **46**, **51**, and **56**) have, without doubt, been forced into a single conformation because of the proximity of the C-5 position. The 4-methyl derivative, **41**, clearly populates the pseudoaxial conformer only. The shifts of *cis*-1,4 (**46**) and *trans*-1,4 (**47**) were calculated assuming that they occur solely as the C-1e,C-4a and C-1a,C-4a conformers, respectively, and the success of the calculations corroborates these assumptions. In contrast, the analogous *trans*-1,4-tetralin had nearly equally populated conformations in both cases and the populations of the *cis*-1,4-tetralin are equal because of symmetry. It is important to note that it was imperative to include the δ_{1e} effect of +0.8 ppm if one is to calculate the shifts of C-4 of the tetrahydrophenanthrenes accurately.

As might be anticipated, substitution on the aliphatic ring at sites removed from C-4, that is, C-1, C-2, and C-3, have conformational consequences that are basically identical to the tetralins also. For instance, the 2- and 3-methyl derivatives are both shown to exist as a single equatorial conformer, whereas the 1-methyl derivative consists of a nearly 50:50 mixture of the pseudoequatorial and pseudoaxial forms.

An interesting result is found for the *cis*-2,4-THP isomer (**50**) not included in the regression. The two half-chair conformers in the tetralin analog are those with two equatorial or two axial methyls, with the former having been demonstrated above to be strongly preferred. In the tetrahydrophenanthrene compound, on the other hand, the 2e,4e form is strongly destabilized by the C-5 position. The MM2 strain energy of the boat form of THP has been calculated (F. G. Morin and D. M. Grant, unpublished results) to be 3–4 kcal/mol higher than the half-chair and, as the difference in energy for the 2e,4e and 2a,4a compounds is likely to be close to this value, significant distortion from the half-chair form may occur in this molecule. Nevertheless, for the 2e,4e conformation, the observed shifts are reproduced very well by the substituent parameters as indicated by the observed (calculated) shifts for C-1, 40.2 (39.5); C-2, 29.2 (29.1); C-3,

Table 5.4 Carbon-13 Chemical Shifts of Hydroaromatic Compounds and Their Methylated Derivatives[a]

	Tetralin										
Compound	1	2	3	4	4a	5	6	7	8	8a	CH_3 (position)
Parent (**1**)	29.6 (30.1)	23.6 (23.0)			137.1	129.4	125.8				
1-Methyl (**2**)	32.7	31.8	20.8	30.2	136.7	129.3	125.8	126.0	128.3	142.1	23.0
2-Methyl (**3**) (2e)	38.3 (38.8)	29.4 (29.1)	31.7 (31.7)	29.4 (30.1)	136.8[b]	129.2[b]	125.7	125.7	129.1[b]	136.5[b]	22.1
cis-1,2-Dimethyl (**4**) (1a,2e)	37.8 (38.0)	32.3 (32.5)	26.0 (26.3)	29.1 (30.1)	135.8	129.1	125.7	125.7	129.1	143.2	17.3 (1) 18.4 (2)
trans-1,2-Dimethyl (**5**)	40.1	35.8	28.9	28.2	136.5	129.1	125.5	126.0	128.7	141.5	22.1 (1) 20.4 (2)
cis-1,3-Dimethyl (**6**) (1e,3e)	33.5 (33.6)	42.3 (41.7)	29.5 (29.1)	39.4 (39.6)	136.9	129.0	125.7	126.0	126.9	141.3	21.8 (1) 22.5 (2)
trans-1,3-Dimethyl (**7**) (1a,3e)	32.5 (32.8)	39.0 (38.5)	24.3 (23.7)	38.6 (38.8)	136.2	129.1	125.7	125.9	128.9	141.8	24.4 (1) 22.1 (3)
cis-1,4-Dimethyl (**8**) (1a,4e ↔ 1e,4a)	33.0 (33.6)	28.9 (28.7)			141.9	128.0	125.9				22.7
trans-1,4-Dimethyl (**9**)	33.0	28.4			141.7	128.4	125.9				23.3
cis-2,3-Dimethyl (**10**) (2e,3a ↔ 2a,3e)	35.6 (35.9)	32.1 (32.6)			135.8	129.5	125.7				15.7
trans-2,3-Dimethyl (**11**)	38.6	35.5			137.1	128.8	125.7				19.6
1,1-Dimethyl (**12**)	33.7 (34.4)	39.6 (39.9)	19.9 (19.8)	30.8 (30.9)	135.9	129.3	125.5	126.1	126.7	145.7	31.9
2,2-Dimethyl (**13**)	43.5 (43.5)	29.4 (29.5)	36.0 (36.4)	26.5 (27.1)	136.4	129.7[b]	125.6[b]	125.7[b]	129.0[b]	135.7	28.1
5-Methyl (**14**)	30.3 (30.1)	23.8[b] (23.0)	23.2[b] (23.0)	26.8 (26.8)	136.5[b]	135.5	127.4[b]	125.5	127.3[b]	137.1[b]	19.4

1,8-dimethyl (**15**) (1a)	29.4 (29.3)	30.7 (30.1)	18.0 (17.9)	29.9 (30.0)	136.2	127.6	125.7	128.2	135.9	140.7	21.1 (1) 18.9 (8)
1,1,8-Trimethyl (**16**)	34.7	44.2	19.8	32.6	137.1[b]	128.1	125.5	130.9	136.8[b]	143.3	29.3 (1) 23.6 (8)
1,1,3-Trimethyl (**17**) (3e)	34.9 (34.4)	48.5 (48.5)	25.8 (25.8)	39.8 (39.6)	136.0	129.1	125.5	126.1	126.6	145.3	31.8 (1e) 32.8 (1a) 22.5 (3)
1,2,2-Trimethyl (**18**)	43.6	31.8	32.4	26.4	135.2	128.9[b]	125.5[b]	125.9[b]	129.3[b]	142.5	27.8 (2e) 25.9 (2a)
1,1,4-Trimethyl (**19**)	34.0	36.1	27.7	33.5	141.2	128.5	125.8	126.1	126.8	145.5	32.0 (1e) 32.0 (1a) 23.1 (4)
1,3,3-Trimethyl (**20**) (1e)	30.3 (30.6)	46.6 (46.4)	30.0 (29.5)	44.5 (44.3)	136.3	129.6	125.8	126.1	127.0	140.6	32.0 (3e) 25.2 (3a) 21.9 (1)
2,2,3-Trimethyl (**21**)	44.7	32.1	37.5	35.4	136.4[b]	129.4[b]	125.7[b]	125.6[b]	128.8[b]	136.1[b]	28.8 (2e) 20.0 (2a) 15.7 (3)
1α,2α,3α-Trimethyl (**22**) (1e,2a,3e)	38.4 (38.1)	39.1 (38.7)	33.8 (33.8)	34.1 (34.2)	136.5	128.7	125.6	126.1	127.0	140.0	19.7 (3) 17.7 (1) 6.1 (2)
1α,2α,3β-Trimethyl (**23**) (1a,2e,3e)	38.4[b] (38.1)	38.5[b] (38.1)	29.4 (29.4)	39.1 (38.8)	136.2	128.8	125.6	125.6	128.8	143.4	19.9 (3) 17.9 (1) 16.6 (2)
1,1,2-Trimethyl (**24**)	37.1	39.3	27.4	29.3	135.7	129.2	125.4	126.1	127.1	146.6	30.0 (1) 25.8 (1) 16.6 (2)
1α,2α,4α-Trimethyl (**25**) (1a,2e,4e)	38.8 (38.7)	31.9 (32.7)	35.4 (36.3)	33.4 (33.4)	143.8	127.3	125.7[b]	126.0[b]	129.3	141.0	17.4 (1) 19.4 (2) 22.1 (4)

(continued)

Table 5.4 Carbon-13 Chemical Shifts of Hydroaromatic Compounds and Their Methylated Derivatives[a] (*continued*)

Tetralin

Compound	1	2	3	4	4a	5	6	7	8	8a	CH_3 (position)
1α,2α,4β-Trimethyl (**26**) (1a,2e,4a)	38.1 (38.0)	27.6 (27.1)	33.6 (33.1)	32.2 (32.8)	*c*	129.1	125.8	125.8	129.1	*c*	17.4 (1) 18.5 (2) 24.8 (4)
5,8-Dimethyl-5,6,7,8-tetrahydro-2-naphtoic acid (**27**)	33.4[b] (33.5)	28.5 (28.7)	28.5 (28.7)	32.9[b] (33.5)	142.7	130.2	127.1	127.7	128.4	149.1	22.5 173.5 (carbonyl)

Tetrahydroanthracenes

Compound	1	2	3	4	4a	5	6	7	8	8a	9	9a	10	10a	CH_3 (position)
Parent (**28**)	29.8 (30.0)	23.5 (23.3)			136.3	127.3	125.1				127.0	136.5			
1-Methyl (**29**)	32.8	31.8	21.0	30.2	136.2	127.6[b]	125.1[b]	125.3[b]	127.2[b]	132.4[b]	126.0	141.5	126.8	132.7	22.7
2-Methyl (**30**) (2e)	38.6 (38.6)	29.5 (29.3)	31.8 (32.0)	29.6 (30.0)	136.2[b]	127.2	125.1	125.1	127.2	132.4[b]	126.9	135.9[b]	126.7[b]	132.5	22.1
cis-1,3-Dimethyl (**31**) (1e,3e)	33.8 (33.5)	42.2 (41.9)	29.6 (29.3)	39.6 (39.6)	136.5	127.7[b]	125.2[b]	125.4[b]	127.2[b]	132.7[b]	125.2	141.1	126.8	132.3	21.9 (1) 22.7 (3)
cis-1,4-Dimethyl (**32**) (1e,4a ↔ 1a,4e)	32.7 (33.4)	29.1 (28.9)			141.6	127.5	125.2				125.2			132.5	22.0
cis-2,3-Dimethyl (**33**)	35.8 (35.5)	32.3 (32.8)			135.1	127.3	125.0				127.2			132.4	15.8
trans-2,3-Dimethyl (**34**)	38.7	35.8			136.4	127.3	125.0				126.3			132.4	19.9
1,1-Dimethyl (**35**)	34.1 (34.6)	39.5 (39.9)	19.9 (19.9)	31.1 (31.0)	135.1	127.0	124.9[b]	125.1[b]	127.5	132.1[b]	125.2	145.0	127.0	132.8[b]	32.4
1,1,3-Dimethyl (**36**) (3e)	35.3 (34.6)	48.6 (48.5)	25.9 (25.8)	40.1 (39.6)	135.5	127.7[b]	125.3[b]	125.0[b]	127.0[b]	132.1[b]	125.2	144.8	126.7	132.7[b]	32.3 (1e) 33.6 (1a) 22.5 (3)

Compound	1	2	3	4	4a	4b	5	6	7	8	8a	9	10	10a	CH_3 (position)
								Tetrahydrophenanthrenes							
Parent (**37**)	30.4	23.2[b]	23.0[b]	25.5	132.5[b]	131.5	123.0	125.8	124.7	128.6	132.9[b]	125.9	128.3	134.2	
1-Methyl (**38**)	33.2	30.8	19.7	26.1	132.3[b]	131.4	123.3	126.0	125.0	128.6	132.8[b]	126.1	127.5	139.4	22.9
2-Methyl (**39**) (2e)	38.8 (39.1)	28.7 (29.1)	31.2 (31.7)	25.5 (25.5)	132.4[b]	131.0	123.0	125.8	124.7	128.5	132.7[b]	125.9	128.2	133.9	21.7
3-Methyl (**40**) (3e)	30.4 (30.4)	31.1 (31.9)	29.2 (28.9)	34.3 (34.2)	132.4[b]	131.4	123.0	125.8	124.8	128.6	132.7[b]	125.8	128.1	133.9	22.2
4-Methyl (**41**) (4e)	30.3 (30.4)	17.8 (17.7)	30.2 (29.7)	28.2 (27.8)	136.8	132.4[b]	123.5	125.8	124.6	129.0	133.1[b]	126.3	128.5	133.2	22.0
cis-1,2-Dimethyl (**42**) (1a,2e)	38.6 (38.2)	31.8 (32.6)	25.5 (26.2)	25.9 (25.5)	132.4[b]	130.6	123.3	125.9	125.0	128.6	132.7[b]	126.1	128.4	140.7	16.8 (1) 18.7 (2)
trans-1,2-Dimethyl (**43**)	40.7	34.5	27.2	23.5	132.2[b]	131.1	123.3	126.2	125.0	128.6	132.6[b]	126.2	128.2	138.6	22.7 (1) 19.8 (2)
cis-1,3-Dimethyl (**44**) (1e,3e)	34.1 (33.9)	41.8 (41.7)	29.2 (28.9)	35.4 (35.2)	132.1[b]	131.7	123.5	125.9	124.9	128.5	132.5[b]	125.8	126.1	138.7	22.2 (1) 22.6 (3)
trans-1,3-Dimethyl (**45**) (1a,3e)	33.6 (32.7)	38.2 (38.5)	23.8 (23.4)	34.6 (34.2)	132.4[b]	130.8	123.2	125.9	124.9	128.5	132.6[b]	126.1	127.9	138.9	23.8 (1) 22.5 (3)
cis-1,4-Dimethyl (**46**) (1e,4a)	33.6 (33.9)	27.5 (27.6)	29.8 (29.7)	28.8 (28.8)	137.2	132.1[b]	123.9	126.0	124.9	128.8	132.6[b]	126.5	126.6	138.4	23.0 (1) 22.1 (4)
trans-1,4-Dimethyl (**47**) (1a,4a)	33.0 (32.7)	25.1[b] (24.4)	24.4[b] (24.2)	28.3 (27.8)	136.4	133.0[b]	123.9	126.0	124.9	128.7	132.7[b]	126.4	128.9	138.7	24.1 (1) 22.0 (4)
cis-2,3-Dimethyl (**48**)	36.8	31.8[b]	32.1[b]	31.8	132.7	130.6	123.2	126.0	124.9	128.7	132.7	126.1	128.0	134.3	15.0[b] 16.7[b]
trans-2,3-Dimethyl (**49**)	39.5	35.0[b]	35.4[b]	35.0	132.7	131.5	123.2	126.0	124.9	128.7	132.7	126.0	128.0	134.3	19.4[b] 19.8[b]
cis-2,4-Dimethyl (**50**)	40.2	29.2	42.3	29.8	136.6	132.1	124.3	125.3	124.4	128.8	133.3	126.0	128.1	135.3	21.9 (2) 25.2 (4)

(*continued*)

Table 5.4 Carbon-13 Chemical Shifts of Hydroaromatic Compounds and Their Methylated Derivatives[a] (*continued*)

Tetrahydrophenanthrenes

Compound	1	2	3	4	4a	4b	5	6	7	8	8a	9	10	10a	CH_3 (position)
trans-2,4-Dimethyl (**51**) (2e,4a)	39.0 (39.1)	23.3 (23.6)	38.9 (38.3)	29.1 (27.8)	136.3	133.2	123.5	125.7	124.5	128.8	132.9[b]	126.2	128.2	132.0	22.4_9 (2) 22.5_3 (4)
1,1-Dimethyl (**52**)	34.0 (35.0)	38.8 (39.7)	19.4 (19.6)	26.6 (26.5)	130.5	132.0[b]	123.5	125.8	124.9	128.4	132.7[b]	126.4	125.1	142.4	31.3
2,2-Dimethyl (**53**)	44.4 (43.7)	29.3 (29.5)	35.9 (36.3)	23.4 (22.2)	132.4[b]	130.5[b]	123.2	126.0	124.9	128.7	132.6[b]	126.1	128.9	133.8[b]	28.0
3,3-Dimethyl (**54**)	27.5 (27.1)	35.5 (36.5)	29.3 (29.3)	39.5 (38.8)	130.9	132.9[b]	123.0	125.9	124.8	128.7	132.6[b]	125.8	128.1	133.1[b]	28.5
4,4-Dimethyl (**55**)	32.9	19.1	44.6	34.7	139.9	132.5[b]	126.8	124.6	124.0	129.5	134.1[b]	126.8	128.7	134.8[b]	30.3
cis-3,4-Dimethyl (**56**) (3e,4a)	30.6 (30.4)	24.8 (26.4)	32.4 (32.4)	33.7 (33.3)	138.4	132.8[b]	123.3	125.9	124.6	128.9	131.9[b]	126.2	128.4	132.8[b]	19.6 (3) 15.2 (4)

Dihydrophenanthrenes

Compound	1	2	3	4	4a	4b	5	6	7	8	8a	9	10	10a	CH_3 (position)
Parent (**57**)	128.3	127.5	127.1	123.9	134.7						137.4	29.0			
9-Methyl (**58**)	128.9	127.7[b]	127.1[b]	123.7	134.3[b]	133.9[b]	124.1	127.0[b]	127.9[b]	126.8	142.2	33.0	36.7	136.1	19.6
cis-9,10-Dimethyl (**59**)	126.8	127.9	126.9	124.0	134.0						141.8	37.8			14.6
trans-9,10-Dimethyl (**60**)	129.3	128.0	127.0	123.9	132.5						140.7	40.9			21.8
9,9-Dimethyl (**61**)	128.8	127.6	127.1[b]	123.7	134.4[b]	133.4[b]	124.3	126.7[b]	128.1	124.4	145.5	34.0	44.1	136.1	27.9
9,9,10-Trimethyl (**62**)	128.2	127.9[b]	126.9[b]	123.9	132.8[b]	133.0	124.0	126.6[b]	128.2[b]	125.3	143.7	37.1	46.4	142.3	24.6 (9e) 29.8 (9a) 17.1 (10)

Dihydroanthracenes

Compound	1	2	3	4	4a	5	6	7	8	8a	9	9a	10	10a	CH_3 (position)
Parent (**63**)	127.6	126.3			136.9						36.2				

9-Methyl (**64**)	127.1	126.2	126.2	127.9	136.0	142.0	41.1	35.1	23.6
9,9-Dimethyl (**65**)	124.4	126.7[b]	126.0[b]	128.0	135.8	145.2	39.1	35.8	28.8
cis-9,10-Dimethyl (**66**)	128.0	126.4			140.6		40.0		28.6
trans-9,10-Dimethyl (**67**)	126.0	126.4			141.5		38.5		36.3 (9a)
9,9,10-Trimethyl (**68**)	126.2	126.6[b]	126.4[b]	128.1	139.5	143.7	38.3	38.9	36.3 30.4 (9e) 29.0 (10)
9,9,10,10-Tetramethyl (**69**)	127.0	126.4			142.1		37.2		35.1
9-Ethyl (**70**)	127.1[b]	125.9[b]	125.8[b]	127.6[b]	135.5	139.9	48.8	31.3	12.1 35.3 (CH_2)
cis-9,10-Diethyl (**71**)	128.6	125.7			138.6		48.3		13.4 35.5 (CH_2)
trans-9,10-Diethyl (**72**)	126.7	125.8			138.6		43.8		10.0 27.0 (CH_2)

[a] Numbers beneath the aliphatic shifts are the calculated shifts using the parameters of Figure 5.7. When it is not obvious, the conformation assumed for each in the calculation is indicated below the compound name. Obtained as approximately 50% solutions in $CDCl_3$.

[b] Ambiguous assignments.

[c] Compound present as minor constituent with extensive overlap with signals of **25**.

42.3 (41.7); and C-4, 29.8 (29.2) ppm. The agreement is reasonable given the overall standard deviation of the least-squares fit. Alternatively, the conformer with diaxial methyl groups would be predicted to have the following respective shifts: (38.1), (20.7), (37.4), and (22.5) ppm. Obviously, from the predictions for C-2 and C-4 this diaxial conformer is unlikely to contribute to the equilibrium. A somewhat anomalous chemical shift for C-5 (see below) may also be taken as evidence that the C-4 methyl is not in a typical pseudoequatorial conformation and probably exhibits considerable distortion.

It should be recalled that the shifts of the *gem*-dimethyl derivatives in the tetralin series resulted in some of the larger deviations between the experimental and calculated values. Therefore, it is not surprising that the same parameters again lead to some of the more poorly predicted chemical shifts. Using data from a previous X-ray analysis[25] to construct Figure 5.9, one can see that the geminal dimethyl portion of 1,1-dimethyl-7-isopropyl-THP apparently flattens out the ring, resulting in very similar environments for the two methyl groups rather than producing distinctly pseudoaxial or pseudoequatorial positions. This is an indication that the high flexibility of the saturated ring may allow for significant distortions in the solid state.

Using the substituent parameters provided by the molecules of unambiguous conformation, one may make approximate determinations of the conformational preferences of the compounds that were not included in the fit. In the study of methylated tetralins and tetrahydroanthracenes, it was shown that the 1-methyl derivative populated the pseudoaxial conformer only to a slight extent over the pseudoequatorial conformer. If one assumes a 60:40 ratio for 1-methyl-THP (**38**) and uses the substituent parameters from the tetralin work, the observed (calculated) shifts are: C-1, 33.2 (33.0); C-2, 30.8 (31.1); C-3, 19.7 (19.8); C-4, 26.0 (26.1) ppm. A small change to 50:50 in the conformational ratio will still give reasonable agreement but the supposition of a slight preference for the pseudoaxial conformation would seem justified, as a result of the 1-methyltetralin results. A similar calculation for the *trans*-1,2 derivative, **43**, provides an estimate of 55% dipseudoequatorial conformer in the equilibrium, also essentially unchanged from the tetralin analog, confirming the assumption that the

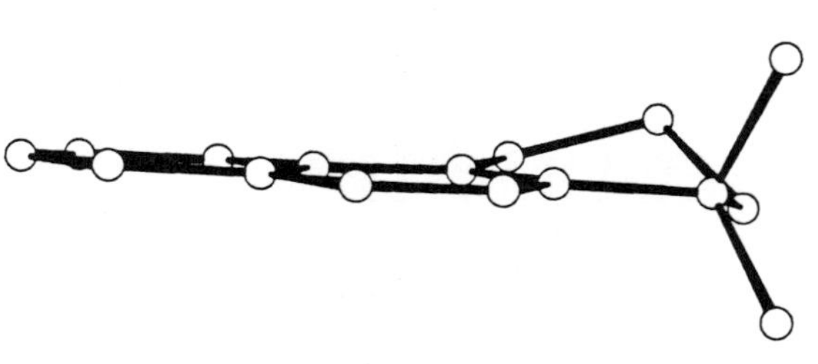

Figure 5.9 Illustration of the similarity of environments for the two methyls of a geminal dimethyl group in the solid state. Figure constructed from the X-ray structure determination[25] of 1,1-dimethyl-7-isopropyl-THP. Isopropyl group not shown.

tetrahydrophenanthrene ring system is very similar to the tetralins for methyl substitution at C-1, C-2, or C-3.

5.3.7 Aromatic Carbons

The aromatic chemical shifts in THP show little variation except for C-5 and C-10, which are the only carbons that can interact sterically with methyl groups in the aliphatic ring. The shift of C-5 is 123.3 ± 0.3 ppm in the absence of a pseudoequatorial methyl at C-4 but is moved downfield to 126.8 ppm in the 4,4-compound (**55**), where a severe interaction with at least one of the methyl groups is unavoidable. This downfield shift, opposite in direction to that normally encountered for nuclei involved in a γ interaction, is typical when the nuclei are situated in this δ (i.e., 1,4) manner. Grover et al.,[26] in a series of substituted cyclohexanes, norbornanes, and steroids, have discovered downfield shifts of 2.0–3.5 ppm when a proximate δ steric interaction is present. An even larger effect was found[27] in 1,8-dimethylnaphthalene for the methyl groups which resonate 6.7 ppm downfield of the methyl of 1-methylnaphthalene. An intermediate value of 124.3 ppm is found for the *cis*-2,4 derivative (**50**), indicating that the C-4 methyl is not occupying the pseudoequatorial position exclusively or that the molecule has distorted to some other undefined conformation rather than populating the alternative half-chair conformation with 1,3-diaxial methyls. This evidence supports the position that the conformation of this molecule is ill defined and that the surprisingly good agreement between predicted and observed shifts for the aliphatic carbons may be coincidental.

Conversely, C-10 resonates at 128.3 ± 0.4 ppm when C-1 is unsubstituted or substituted with a pseudoaxial methyl but is shifted upfield about 2 ppm to 126.1 or 126.6 ppm for compounds **44** and **46** which possess a pseudoequatorial methyl group.

5.3.8 Methyl Carbons

In the previous discussion of the tetralins it was indicated that a similar linear regressional analysis on the methyl carbon shifts was unsuccessful. The chemical shift range over which the methyls appeared was too small to give significant structural parameters and the same deficiency is manifested here for THP. However, the salient features of Table 5.4 will be noted. First, an isolated equatorial methyl group at C-2 (or C-3) (cf. **39**, **40**, **44**, **50**) is found at 22.1 ± 0.1 ppm, essentially the same as that found for a similar methyl on a cyclohexane ring.[10] Most of the remaining shifts are unchanged from those determined in the tetralin work, with minor deviations up to ~0.6 ppm. Those methyls that are oriented pseudoaxially, for example, **45** (23.8 ppm) and **47** (24.1 ppm), or pseudoequatorially, for example, **44** (22.2 ppm) and **46** (22.1 ppm), at C-1 are unchanged in their chemical shifts from the tetralins but those pseudoequatorial methyls at C-4 resonate 1.5–2.0 ppm upfield of the corresponding C-1 methyls, indicating that they detect the presence of the C-5 position to some extent. Even so, there is no effect on the methyl substituent parameters themselves.

The most interesting data are provided by the two 2,3-dimethyl-THP derivatives. Here the *trans*-2,3 isomer (**49**) undoubtedly exists in the diequatorial conformation,[28] but the methyl groups are nonequivalent to the extent of $\Delta\delta$ = 0.4 ppm, reflecting the small difference in environment of each site. The chemical shift difference between the equatorial methyls of 2-methyl and 3-methyl-THP is also only 0.5 ppm (21.7 versus 22.2 ppm). The shifts of the *cis*-2,3 isomer (**48**), which might be expected to be averaged between two rapidly interconverting conformations (2e,3a ↔ 2a,3e) that would result in the same time-averaged environments, are 15.0 and 16.7 ppm. However, MM2 calculations (F. G. Morin and D. M. Grant, unpublished results) that the 2a,3e conformer may be slightly favored (by ~0.4 kcal/mol) over the 2e,3a conformer. This biasing of the conformational equilibrium slightly toward one form is of the right magnitude to create the shifts difference, $\Delta\delta$ = 1.7 ppm, with the higher field resonance attributable to the C-2 methyl which spends more of its time in the axial position than does the C-3 methyl.

5.3.9 9,10-Dihydrophenanthrenes

There have been several investigations of alkyl-DHPs using ^{1}H-NMR. For instance, investigators have used variable-temperature ^{1}H-NMR to determine the free-energy barrier to inversion of the central ring of the parent[29] to be $<$ 9 kcal/mol and of the *cis*-9,10-dimethyl derivative[30] to be 10.9 kcal/mol at 250 K. Harvey et al.[31] have measured ^{1}H–^{1}H coupling constants which indicate that the 9-methyl derivative exists in the axial conformer 77% of the time.

The number of ^{13}C studies on DHP is also rather limited. One investigation[18] revealed a temperature dependence to the 75 MHz ^{13}C-NMR spectrum of *cis*-9,10-dimethyl-DHP, **59**. As the temperature was lowered the resonances became broadened and then separated into two peaks of equal intensity as the exchange between the two identical conformers becomes slow on the NMR time scale (about −25 to −35 °C). The lineshapes of the methyl and the aliphatic methine carbons were simulated as a function of temperature to obtain k_r, the exchange rate. The enthalpy and entropy of activation were found to be 10.3 ± 0.1 kcal/mol and −3.3 ± 0.4 cal/(mol·K), respectively, and the calculated $\Delta G^{\ddagger}$ is 11.1 kcal/mol at 250 K in agreement with the ^{1}H-NMR results.[30] The negative $\Delta S^{\ddagger}$ obtained may be ascribed to a change in rotational mobility of the methyl groups as the central ring becomes planar in the transition state where the hindrance to rotation is somewhat greater than in the ground state. A similar ^{13}C-NMR investigation of the conformational inversion rate of *cis*-1,2-dimethylcyclohexane[32] found $\Delta H^{\ddagger}$ = 9.3 kcal/mol and $\Delta S^{\ddagger}$ = −3.5 cal/(K·mol) ($\Delta G^{\ddagger}$ = 10.2 kcal/mol at 250 K). The similarity of the $\Delta S^{\ddagger}$ with that found for **59**, which closely resembles the dimethylcyclohexane in the vicinity of the methyl groups, suggests that the vicinal dimethyls are the source of the decreased entropy in the transition state.

The ^{13}C chemical shifts of **59** in the slow-exchange limit are themselves interesting. In this regime, the separate resonances for the methyl groups, one from the pseudoaxial and one from the pseudoequatorial carbons, differ by just 1.6 ppm in their chemical shifts while the aliphatic methine shifts differ by 4.8 ppm. It is expected that a pseudoequatorial methyl would be shifted upfield because of the interaction with the peri proton at C-1 (or C-8) compared to the axial position. Molecular models, however, demonstrate that an axial methyl is situated over the biphenyl ring moiety of the molecule, a region that is known to shift a nucleus in this position upfield.[33] As a consequence, the closeness of the shifts makes it difficult to assign the resonances in the slow-exchange limit.

Variable-temperature measurements were also obtained on the 9,9,10-trimethyl-DHP compound, **62**. Although the resonances are broadened between −20 and −40 °C due to slowing of the exchange between two conformations (those with the C-10 methyl axial or equatorial), the peaks due only to one conformer were observed in the slow-exchange limit (−60 °C). This indicates that the equilibrium strongly favors one of the conformers. It has been determined[34] previously that the conformational equilibrium of 1,1,2-trimethylcyclohexane favors the conformation with an equatorial 2-methyl group, which differs in an important way from **62**, however. The equatorial 2-methyl group of 1,1,2-trimethylcyclohexane is in a rather unhindered position relative to the other ring protons of the molecule, whereas an analogous equatorial methyl in **62** sterically interacts with the C-1 *peri* proton. Also, the destabilization of the conformer of 1,1,2-trimethylcyclohexane with an axial C-2 methyl axial arises from the γ interactions of the methyl with the axial ring protons at the C-4 and C-6 positions. The corresponding protons are absent in **62**. Conversely, the conformer of **62** with an equatorial C-10 methyl as compared to an axial C-10 experiences one more *gauche*-methyl interaction as well as the interaction with a peri proton in the adjacent aromatic ring. These effects result in a substantial preference for the axial conformation in direct contrast to 1,1,2-trimethylcyclohexane. The observation that the chemical shift of C-1 of **62** is the same as the parent confirms further that the methyl at C-10 is axial and not equatorial. An upper limit of ~5% can be placed on the population of the equatorial C-10 conformer corresponding to a free energy ($\Delta G°$) difference of >1.5 kcal/mol.

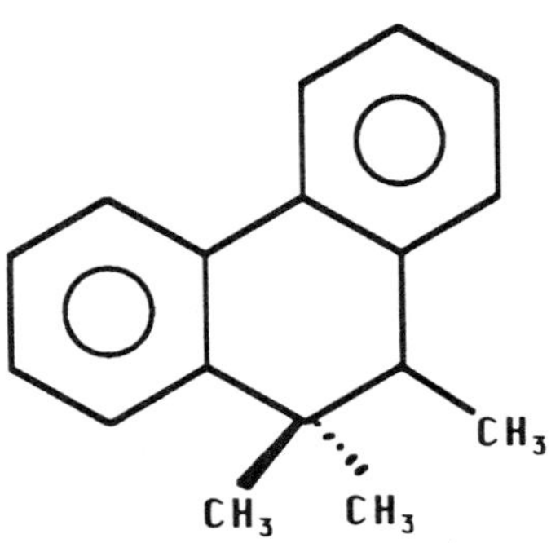

62

Insights into the conformational preference of 9-methyl-DHP (**58**) may be obtained by comparing the ^{13}C chemical shift of the methyl group (19.6 ppm) to that of the *trans*-9,10-dimethyl derivative (**60**) (21.8 ppm). If the methyl group of the former was essentially axial, one would expect the addition of the second *trans*-methyl to have a negligible effect on the chemical shift. The actual shift downfield of 2.2 ppm in the *trans*-9,10 compound would indicate that **58** has some contribution from the equatorial conformer and this presumably would result in an upfield shift due to the presence of the C-8 position which is not a factor in **60** when the methyl group is axial. Furthermore, the shift of C-8 of the *cis*-9,10 derivative (**59**), which has an equatorial methyl at C-9 half the time, is found to be identical to C-8 of **58**. While it may be that **58** fails to exist in the equatorial methyl conformation 50% of the time, these data corroborate that there is indeed a significant contribution from the equatorial conformer to the equilibrium.

The synthesis and ^{13}C-NMR spectrum of 9,9,10,10-tetramethyl-DHP have been reported.[35,36] The inversion of the central ring at room temperature is slowed in this tetramethyl molecule to the point where separate resonances are observed for axial and equatorial methyl groups, in contrast to the rapid inversion on the NMR time scale at room temperature in the *cis*-9,10-dimethyl derivative. At the transition state, two pairs of two methyl groups must pass each other compared to one pair in *cis*-9,10, and the barrier is raised to 15.7 kcal/mol at the coalescence temperature of 320 K.

5.3.10 9,10-Dihydroanthracenes

Dihydroanthracene and its alkyl derivatives have been investigated over the years through the application of ^{1}H- and ^{13}C-NMR. The parent was found by X-ray crystallography[37] to exist in the solid state with an angle of 145° between the planes of the aromatic rings. For this boat structure, substituents at the bridge carbons can assume either the pseudoaxial or pseudoequatorial orientations. The former structure is preferred for bulky substituents but contribution from the pseudoequatorial conformer is possible for small substituents, like methyl or phenyl.[38,39] Some flattening of the central ring was postulated[38] in *cis*-9,10-disubstituted compounds when bulky substituents were present.

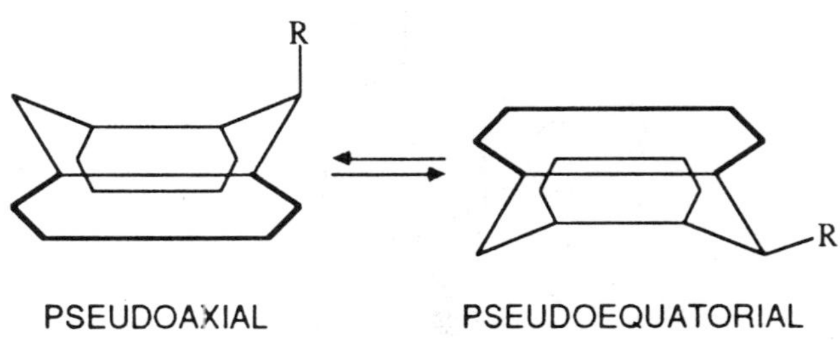

A combined ^{13}C-NMR study using both solution and solid-state NMR provided information on the conformations of various methyl derivatives.[23] The solution-state studies focused on the temperature dependence of ^{13}C shifts of methyls exchanging rapidly on the NMR time scale between pseudoaxial and pseudoequatorial environments. These values are governed by the following equation:

$$\delta_{obs} = \chi_a \delta_a + \chi_e \delta_e \tag{5.4}$$

where δ_a and δ_e are, respectively, the chemical shifts in the pseudoaxial and pseudoequatorial orientations and χ_a and χ_e are the corresponding mole fractions. The Boltzmann distribution relates the populations of the two states, p_a and p_e (assuming the e form is higher in energy):

$$p_e = p_a e^{-\Delta G/RT}. \tag{5.5}$$

Substitution of Equation (5.5) and (5.4) and rearrangement gives the temperature dependence as

$$\delta_{obs} = \frac{\delta_a + \delta_e e^{-\Delta H/RT} e^{\Delta S/R}}{1 + e^{-\Delta H/RT} e^{\Delta S/R}}$$

A nonlinear least-squares regression analysis of the methyl shifts of the 9-methyl, *cis*-9,10-dimethyl, and the 9,9,10-trimethyl compounds provided values for ΔH of 1.51, 1.25, and 1.35 kcal/mol, respectively, with ΔS values of 1.2, 0.2, and 0.7 cal/(K·mol). The calculated free-energy differences, ΔG, are ~1.5 kcal/mol. (The shifts of the methyl groups of the 9,9-dimethyl and *trans*-9,10-dimethyl derivatives were temperature independent as they should be if each interconverts between two equivalent boat conformers.) From the fit of the temperature data, the chemical shifts of an isolated methyl at C-9 were estimated to be δ_e = 12.6 ppm and δ_a = 25.2 ppm. The validity of these results is confirmed by the ^{13}C solid-state spectrum of the *trans*-9,10 derivative. The conformational inversion of the central ring is stopped in the solid and separate methyl resonances are found with shifts of 13.4 and 26.6 ppm, where intermolecular effects may account for the differences from the liquid values. The difference of (25.2 − 12.6) = 12.6 in the liquid (or 13.2 in the solid) is attributable to two upfield γ interactions of the equatorial methyl with the protons on the adjacent aromatic rings. Each γ interaction is then approximately equal to −6.5 ppm, a value comparable to the γ effects in the several different systems discussed above. From the ΔG values, the population of the pseudoaxial form of 9-methyl-DHA is 87%, which compares reasonably with the 75% approximated from homoallylic ^{1}H coupling constants and NOEs.[38,39]

From the temperature dependence of the methyl shifts of the *cis*-9,10 compound, it was determined also that the diaxial form contributes 87% to the equilibrium and the diequatorial 13%. The predicted methyl shift for the former is 30.8 ppm and the solid-state spectrum revealed a methyl resonance at 31.0 ppm, so that clearly the diaxial form is found in the solid as has been found in other cases of *cis*-9,10 disubstitution.

The solid-state NMR spectrum of 9,9-dimethyl-DHA displayed two methyl resonances that are separated by only 7.5 ppm. This decrease from the 13.2 ppm difference in the axial and equatorial shifts recorded for 9-methyl derivatives suggests considerable flattening of the central ring. The 7.5 ppm difference is larger than can be explained from intermolecular effects, however, and the ring apparently continues to pucker in the solid.

The separation of methyl shifts in the solid-state spectrum of the 9,9,10,10-tetramethyl compound is reduced to 2.5 ppm, a value small enough to be explained solely by intermolecular effects. An X-ray analysis (R. Neustadt et al., unpublished results) of this molecule revealed it to be indeed planar but with adjacent molecules located in such a manner that they could split the NMR resonances for the four methyls into two peaks as observed. As the liquid shift in this compound is reasonably close to the average of the two solid shifts, there is also no need to propose significant structural differences between the liquid and solid states, and it is reasonable to assume that the tetramethyl compound is planar in both cases.

Rabideau et al.[40] have used a combination of ^{13}C-NMR and molecular mechanics calculations to investigate a number of alkyl 9,10-dihydroanthracenes to address the question of preferred conformations in solution. They compared the ^{13}C spectra of *cis*-9,10-diethyl-DHA, **71**, and an analogous compound **73**, where the latter is fixed in a puckered conformation, and found the shifts of C-9 to be identical, signifying that **65** is also in the boat conformation. To model the planar conformation of DHA, the ^{13}C shifts of 9-ethyl-DHA (**70**), a boat, and **74**, a planar molecule, were compared. The effect of flattening the ring is a 3 ppm upfield shift. The difference of 1.6 ppm between the unsubstituted compounds, **63** and **71**, indicates some contribution from the planar form in the parent DHA.

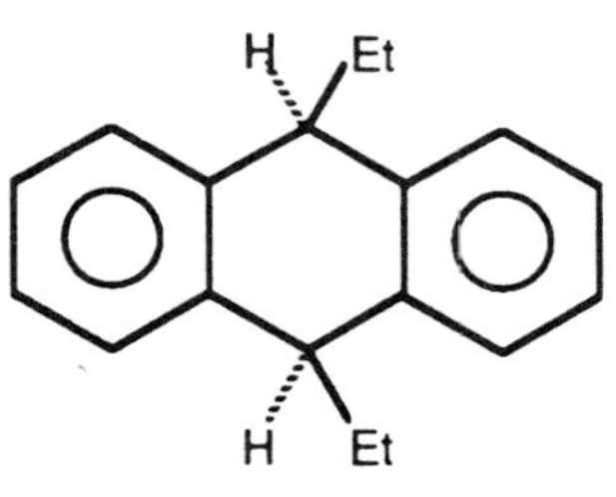

71

$\delta^{C\text{-}9}$ = 48.3 ppm

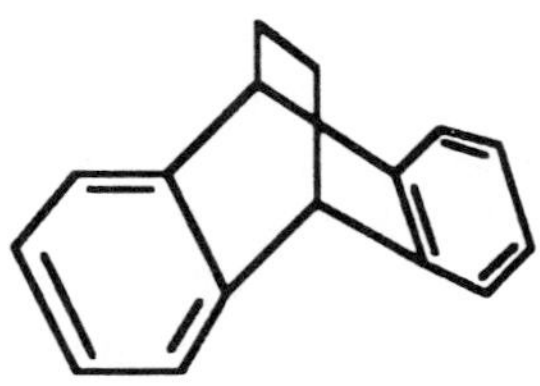

73

$\delta^{C\text{-}9}$ = 48.4 ppm

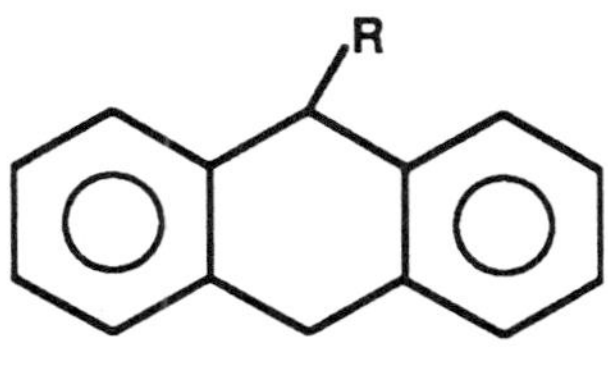

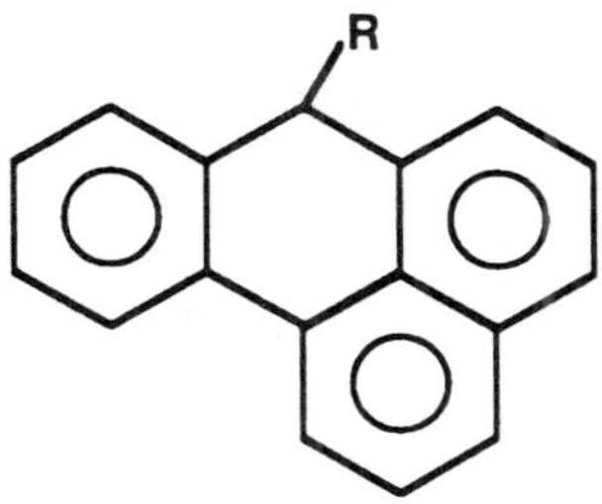

63, R=H $\delta^{C\text{-}9} = 36.9$ ppm **71**, R=H $\delta^{C\text{-}9} = 34.5$ ppm

70, R=Et $\delta^{C\text{-}9} = 48.7$ ppm **74**, R=Et $\delta^{C\text{-}9} = 45.7$ ppm

It was also shown[40] that one-bond $^{13}C-^{1}H$ coupling constants may be used to gain insight into the conformations of these molecules. Using a formula[41] for calculating J_{CH} as a function of the C—C_9—C angle and employing model compounds, a difference between puckered and planar values for J_{C_9H} of 1.2–1.4 Hz was found. The coupling constants for the *cis*-9,10- and *trans*-9,10-diethyl derivatives, **71** and **72**, are 128.5 and 126.9 Hz (ΔJ = 1.6 Hz), suggestive of different conformations of the central ring for these two molecules, a puckered form for the cis compound and a planar one for the trans. Molecular mechanics calculations (MM2) corroborated the finding of a planar structure for **74**.

The conclusions of Dalling et al.[23] that *trans*-9,10-dimethyl-DHA is puckered in the liquid is not contradicted with these results for the diethyl compound, **66**. Rabideau[40] has also performed MM2 and MMPI calculations on the dimethyl derivative and found the puckered conformer to be only slightly lower in energy (≤ 1 kcal/mol) than the planar and that both may contribute substantially to the equilibrium. Subtle changes in conformation depending on the type of substituents present are evident in significant differences between the calculated energy profiles for the dimethyl and diethyl derivatives.[40]

The molecular mechanics calculations on 9-methyl and *cis*-9,10-dimethyl-DHA appear to disagree with the experimental results, [23,38,39] however, which were interpreted on the basis of a significant contribution from the less-stable conformer. The pseudoequatorial conformer (or the dipseudoequatorial in the *cis*-9,10-dimethyl case) is calculated to be too high in energy, by 3–5 kcal/mol, to contribute to the ^{13}C chemical shift. The temperature dependence of the ^{13}C shifts of the methyl substituted DHAs discussed above is explained with wide amplitude oscillations of the planar conformer, as occurs in 1,4-cyclohexadiene, rather than with a change in the conformational populations. Conversely, the force-field calculations suggest that 9,9,10,10-tetramethyl-DHA assumes only the planar form and no energy minimum is found for the boat conformers, in agreement with the solution- and solid-state ^{13}C-NMR results and with X-ray crystallography findings that 9,9,10,10-tetrachloro-DHA is planar in the solid state.[42]

In conclusion, evidence from solid-state techniques (CP/MAS-NMR and X-ray crystallography) demonstrates unambiguously that substituted DHAs possess a boat structure when lightly substituted but become more planar as the substituents become increasingly bulky. While the *cis*- and *trans*-9,10-dimethyl derivatives are highly puckered (as is *cis*-9,10-diethyl-[43]), the 9,9-dimethyl compound is slightly less so and the tetramethyl compound is planar. The situation in solution is less clear, with some ^{13}C-NMR evidence in favor of boat-to-boat interconversions in methyl substituted compounds while molecular mechanics argue more strongly for planar or nearly planar structures. These disagreements emphasize the flexibility of these hydroaromatics and undoubtedly they will remain subjects of considerable interest for some time to come.

REFERENCES

1. Purcell, E. M.; Torrey, H. C.; Pound, R. V. *Phys. Rev.* **1946**, *69*, 37.
2. Bloch, F.; Hansen, W. W.; Packard, M. E. *Phys. Rev.* **1946**, *69*, 127.
3. Stothers, J. B. *Carbon-13 NMR Spectroscopy;* Academic Press: New York, 1972.
4. Levy, G. C.; Lichter, R. L.; Nelson G. L. *Carbon-13 Nuclear Magnetic Resonance Spectroscopy*, 2nd ed.; Wiley-Interscience: New York, 1980.
5. Levy, G. C., Ed. *Topics in Carbon-13 NMR Spectroscopy*; Wiley-Interscience: New York, 1979.
6. Dalling, D. K.; Grant, D. M. *J. Am. Chem. Soc.* **1972**, *94*, 5318.
7. Grant, D. M.; Paul, E. G. *J. Am. Chem. Soc.* **1964**, *86*, 2984.
8. Draper, N. R.; Smith, H. *Applied Regression Analysis*; Wiley: New York, 1966.
9. Neter, J.; Wasserman, W.; Kutner, M. H. *Applied Linear Statistical Models*, 2nd ed.; Richard D. Irwin: Homewood, IL, 1985.
10. Dalling, D. K.; Grant, D. M. *J. Am. Chem. Soc.* **1967**, *89*, 6612.
11. Anet, F. A. L.; Bradley, C. H.; Buchanan, G. W. *J. Am. Chem. Soc.* **1971**, *93*, 258.
12. Schneider, H.-J.; Price, R.; Keller, T. *Angew. Chem. Int. Ed. Engl.* **1971**, *10*, 730.
13. Booth, H.; Everett, J. R. *Can. J. Chem.* **1980**, *58*, 2709.
14. Dalling, D. K.; Grant, D. M.; Paul, E. G. *J. Am. Chem. Soc.* **1973**, *95*, 3718.
15. Dalling, D. K.; Grant, D. M. *J. Am. Chem. Soc.* **1974**, *96*, 1827.
16. Lindeman, L. P.; Adams, J. Q. *Anal. Chem.* **1971**, *43*, 1245.
17. Morin, F. G.; Horton, W. J.; Grant, D. M.; Dalling, D. K.; Pugmire, R. P. *J. Am. Chem. Soc.* **1983**, *105*, 3992.
18. Morin, F. G.; Horton, W. J.; Grant, D. M.; Pugmire, R. J.; Dalling, D. K. *J. Org. Chem.* **1985**, *50*, 3380.
19. Schneider, H.-J.; Agrawal, P. K. *Org. Magn. Reson.* **1984**, *22*, 180.
20. Menard, D.; St.-Jacques, M. *Tetrahedron* **1983**, *39*, 1041.
21. Fyfe, C. A. *Solid State NMR For Chemists*; C.F.C. Press: Guelph, Ontario, 1983.
22. Fukushima, E.; Roeder, S. B. W. *Experimental Pulse NMR—A Nuts and Bolts Approach*; Addison-Wesley: Reading, MA, 1981.
23. Dalling, D. K.; Zilm, K. W.; Grant, D. M.; Heeschen, W. A.; Horton, W. J.; Pugmire, R. J. *J. Am. Chem. Soc.* **1981**, *103*, 4817.
24. Wilson, M. A.; Vassallo, A. M.; Burgar, M. I.; Collin, P. J. *J. Phys. Chem.* **1986**, *90*, 3944.

25. Foresti, E.; Riva Di Sanseverio, L. *Att Accad. Naz. Lincei, Cl. Sci. Fis., Mat. Nat., Rend.* **1969**, *47*, 41.

26. Grover, S. H.; Guthrie, J. P.; Stothers, J. B.; Tan, C. T. *J. Magn. Res.* **1973**, *10*, 227.

27. Dalling, K. D.; Ladner, K. H.; Grant, D. M.; Woolfenden, W. R. *J. Am. Chem. Soc.* **1977**, *99*, 7142.

28. Peters, H.; Archer, A. R.; Mosher, H. S. *J. Org. Chem.* **1967**, *32*, 1382.

29. Oki, M.; Iwamura, H.; Hayakawa, N. *Bull. Chem. Soc. Jpn.* **1963**, *36*, 1542.

30. Rabideau, P. W.; Harvey, P. W.; Stothers, J. B. *Chem. Commun.* **1969**, 1005.

31. Harvey, R. G.; Fu, P. P.; Rabideau, P. W. *J. Org. Chem.* **1976**, *41*, 3722.

32. Dalling, D. K.; Grant, D. M.; Johnson, L. F. *J. Am. Chem. Soc.* **1971**, *93*, 3678.

33. Johnson, C. E.; Bovey, F. A. *J. Chem. Phys.* **1958**, *29*, 1012.

34. Eliel, E. L.; Chandrasekaran, S. *J. Org. Chem.* **1982**, *47*, 4783.

35. Mullins, D. F. Ph.D. Dissertation, McGill University, 1983.

36. Harpp, D. N.; Mullins, D. F. *Can. J. Chem.* **1983**, *61*, 757.

37. Ferrier, W. G.; Ball, J. I. *Chem. Ind. (London)* **1954**, 1296.

38. Rabideau, P. W. *Acc. Chem. Res.* **1978**, *11*, 141.

39. Brinkman, A. W.; Gordon, M.; Harvey, R. G.; Rabideau, P. W.; Stothers, J. B.; Ternay, A. L. *J. Am. Chem. Soc.* **1970**, *92*, 5912.

40. Rabideau, P. W.; Mooney, J. L.; Lipkowitz, K. B. *J. Am. Chem. Soc.* **1986**, *108*, 8130.

41. Baum, M. W.; Guenzi, A.; Johnson, C. A.; Mislow, K. *Tetrahedron Lett.* **1982**, *23*, 31.

42. Yannoni, N. F.; Silverman, J. *Acta Crystallogr.* **1966**, *21*, 390.

43. Ahmad, N-u-d.; Goddard, R. J.; Hatton, I. K.; Howard, A. K.; Lewis, N. J.; MacMillan, J. *J. Chem. Soc., Perkin Trans. 1* **1985**, 1859.

6

Conformational Analysis of Partially Unsaturated Six-Membered Rings Containing Heteroatoms

Slayton A. Evans, Jr.

6.1 INTRODUCTION

The considerable interest in the conformational and stereochemical considerations attending the heterologs (Figure 6.1) of the simplest ''6,6,6'' tricyclic system, 9,10-dihydroanthracene (**1**), is stimulated by the plethora of ''structure–reactivity'' information arising from the biological importance of these substances.[1] Generally, the bonding constraints within these systems allow for conformational flexibility about the *meso atoms* only, predicting a ground-state that is either planar[2,*] or puckered[†] into a steep- or shallow-boat conformation.[3] Various ring substitution patterns exert both electronic and steric effects which

X = Y = CH_2 ; 9,10-Dihydroanthracene (**1**)

X = Y = S; Thianthrene (**2**)

X = S; Y = NH Phenothiazine (**3**)

X = NH; Y = CH_2; Acridane (**4**)

X = O; Y = CH_2; Xanthene (**5**)

X = S; Y = CH_2; Thioxanthene (**6**)

X = S; Y = $C(CH_3)_2$; 9,9-Dimethylthioxanthene (**7**)

Figure 6.1

---

* The results of MNDO calculations demonstrate that several ''heterologs'' of 9,10-dihydroanthracene exhibit inherently planar conformations and respond to ring puckering induced by peri- and/or transannular interactions.

† The magnitude of the puckering translates into a dihedral angle, which is defined as the folding angle between the planes of the two aryl rings of the tricyclic array.

impact on the "limiting" conformations as well as the chemical reactivity of the heterocyclic system.[4] Both experimental and theoretical research on a variety of conformational descriptors have helped to identify the factors most responsible for controlling the reactivities of a wide variety of these substances.

This chapter focuses primarily on the conformational and stereochemical attributes of substituted thioxanthenes because of the availability of various NMR probes (^{1}H, ^{13}C) or "structural handles" for this particular system which have afforded an abundance of useful structural data. To maintain informative coverage on the thioxanthene theme, we have avoided substantive inclusion of structural and conformational data on substituted phenothiazines, thianthrenes, acridans, and so on. Nevertheless, a few brief comments on the conformational dynamics of several diaryl "6,6,6" heterocyclic or "butterfly" compounds seem appropriate to provide a proper introduction for the thioxanthenes.

First, thianthrene (**2**) is nonplanar in the solid[5] or the solution[‡] phase[6] and low barriers to ring inversion (3–6 kcal/mol)[7] preclude isolation of enantiomeric thianthrenes.[8] In fact, current *ab initio* (Gaussian 80) and energy-weighted

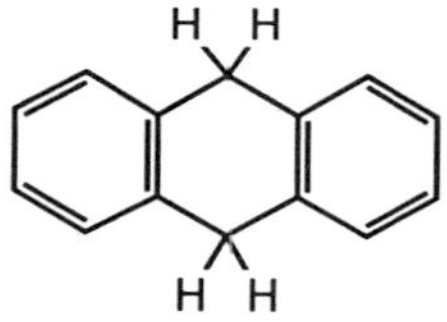

1

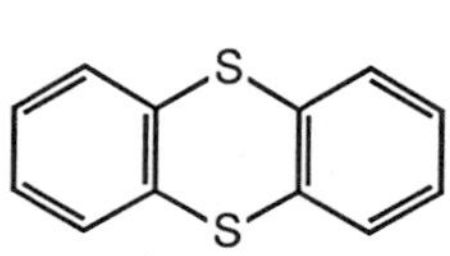

2

maximum overlap (EWMO) calculations predict a dihedral angle of 130° for the minimum energy[§] boat conformation.[9] In addition, *ab initio* calculations with the extended basis set (4–31G) suggest an inversion barrier of ~10 kcal/mol for 2.[9] Phenothiazine (**3**) is also nonplanar with a dihedral angle of 153.3° in the monoclinic form[10] and 158.5° in the orthorhombic crystalline form.[11] Caldwell et al.,[12] and more recently, Jovanovic and Biehl,[4] have described ^{13}C-NMR shift/dihedral angle correlations for phenoxathiins and phenothiazine derivatives. For phenothiazine derivatives, a correlation between dihedral angles and the internal S—C—C bond angles implies that "flattening" of the phenothiazine heterocyclic ring is compensated for by the vertical displacement of the sulfenyl sulfur and by changes in the hybridization of the nitrogen atom.[4] Interestingly, ^{13}C-NMR shift data for **3** suggest that *less* delocalization of electron density from nitrogen to the

‡ The angle of fold between the benzo rings is 144 ± 8° in benzene solution as determined from dipole moment analysis.

§ In fact, the total energy derived from *ab initio* and EWMO methods for the cation radical of **2** also suggest a dihedral angle of 130° and an inversion barrier of only 6 kcal/mol.

aryl rings occurs relative to acridane (**4**); consequently, **3** is less puckered than **4**.[13] "Torsional angles" defined in this study as the angles between a phenyl ring and the C—X—C planes and calculated from magneto-optical rotational data are ~30° for both xanthene (**5**) and acridane.[14] These values translate to shallow-boat conformations for both **4** and **5**. The dihedral angle of xanthene has been estimated to be 160° by Aroney et al.[15] on the basis of molecular polarizability measurements in solution. This flattened central ring of **5** is consistent with its ^{17}O-NMR shift (δ 100 ppm relative to $^{17}OH_2$) which reflects substantial oxygen 2p → π electron delocalization into the aryl rings.[16]

Finally, the X-ray crystallographic data on an extensive variety of 9,10-dihydro-9,10-dihetero anthracenes have been published.[17,18] The 9,10-dihetero anthracenes substituted at the meso positions with Si, Ge, and Sn atoms as electron acceptors and O, S, and N as electron donors are folded about the imaginary meso axis.[17] Significant evidence suggest that *p*-π and *d*-π conjugation involving the donor and acceptor atoms is substantial,[17] although some exceptions have been noted.[18] Replacement of the C-9 and C-10 atoms in **1** by identical Group 4 atoms causes an increase in the dihedral angle affording substantial flattening of the central ring.[19]

6.2 THE CONFORMATIONS OF THIOXANTHENE AND RELATED DIARYL HETEROCYCLES

6.2.1 Thioxanthene

The ^{1}H-NMR spectrum of thioxanthene (**6**) consists of a complex aromatic absorption and a singlet for the C-9 methylene hydrogens. 9,9-Dimethylthioxanthene (**7**) displays a similar NMR property.[20] This single resonance in **6** persists over a 125°C temperature range (+35 to −90°C), suggesting that either (1) thioxanthene exists primarily as a rapidly equilibrating, shallow-boat, or (2) that this restricted diaryl heterocycle prefers the planar conformation. Either explanation would give rise to isochronous methylene hydrogens in the ^{1}H-NMR. Dihydroanthracene **1** and thianthrene (**2**) exhibit dipole moments implying that they also possess nonplanar conformations in solution.[21] In fact, their X-ray crystal structures indicate that their central rings are boat shaped with "angles of fold" between the diaryl rings of 143.8°[22] and 128°,[23] respectively. Single "meso atom transposition" (i.e., S/CH_2) should give the structural composite, **6**, a hybrid of both **1** and **2** with respect to the degree of fold between the two phenyl rings. The X-ray crystal structure of **6** reveals that the C—S bond lengths are 1.781 and 1.759 Å, the C—S—C bond angle is 99.2°, and the angle between the phenylene planes is 135.3°.[24] Interestingly, Canselier and Cassoux[14] also conclude that **6** is folded based on molecular magneto-optical rotation data (i.e., the Faraday effect) employed to assess the extent of conjugative interactions between bridging biphenylic systems. Here, the magneto-optical rotation of compounds containing covalent σ or π localized bonds are calculable with a convincing degree of accuracy; consequently, deviations from additivity are interpreted in terms of unique structural features. Based on assumptions relating magneto-optical rotations to "torsional angles" between the phenyl ring and the C—S—C plane, the torsional angle in **6** is calculated as the $\theta = 37°$. This describes a shallow-boat conformation for the central ring in harmony with the report by Aroney et al.[15] which reveals a folded conformation (dihedral angle $= 135° \pm 8°$) for **6** in tetrachloromethane and benzene solution as derived from molecular polarizability data.

Finally, NMR support for the nonplanar conformation of **6** in solution is qualitatively estimated from an evaluation of the geminal coupling constant ($^2J_{HH}$) between the methylene hydrogens.[25] Barfield and Grant[26] have advanced a theoretical treatment which suggests that ^{1}H-NMR geminal couplings are influenced by the inclination of the H—H internuclear axis of the methylene group with the nodal plane of an adjacent π orbital. When the internuclear axis is perpendicular to this nodal plane, i.e., the optimum configuration, a large negative contribution to $^2J_{HH}$ is observed relative to the geminal coupling in methane ($^2J_{HH} = -12.4$ Hz).[27] For example, the geminal coupling for fluorene, a planar molecule[28] whose internuclear axis is of the optimum configuration, is ~ -22.3 Hz.[29] If **6** were a planar structure in solution, it would be expected to display a geminal coupling constant analogous to fluorene, assuming that sulfenyl sulfur makes no contribution via 3p-π conjugative interactions. However, the observed geminal coupling of -16.2 Hz for **6** suggest an absence of a large π-electron enhancement of the geminal coupling constant.[30] Nevertheless, the larger negative coupling observed for **6**, relative to that of methane, is taken as evidence supporting a conformation in which *one* of the methylene hydrogens of **6** resides in the nodal plane of the adjacent π orbitals.[30] Such a rationale is easily accommodated by a central ring, boat conformation for **6**.

Because **6** occupies a shallow-boat conformation, the activation energy required to attain the planar transition state is expectedly low. In fact, the major source of activation energy for inversion can be attributed to angle strain;[2a,7,9] molecular models indicate that torsional strain may not be significant since the C-9—H and C-1,8—H bonds can avoid one another during the interconversion process.

As a direct consequence of this folded geometry, substituents attached to the meso positions (i.e., C-9 and S-10) in **6** may occupy either of the *two* pseudoaxial (a′) or the pseudoequatorial (e′) conformations (Scheme 1). The a′-e′ positions at C-9 or S-10 are conformationally unique (but isoenergetic when R = R′), exhibiting different spectral characteristics, chemical reactivities, and so on. Substituted thioxanthenes undergo rapid ring inversion through a planar transition state at ambient temperatures, a process that interconverts a′ and e′ substituents.

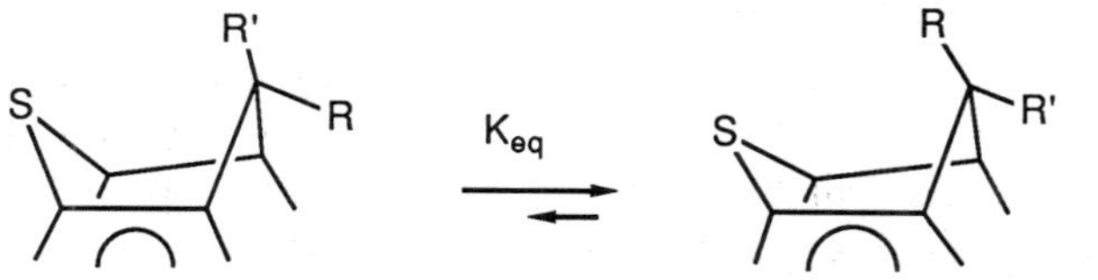

R' = H; R = Me, Et, i-Pr, i-Bu, t-Bu

R' = Me; R = Me, Et, i-Pr, i-Bu

Scheme 1

Several analogous diaryl heterocycles where at least one of the meso atoms is sp^3 hybridized (a carbonyl group in this case) exhibit both planar and folded geometries, reflecting the subtle sensitivities of substituents. Some representative examples are described in Figure 6.2.

2-Chlorothioxanthone[31] and Miracil D[32] have central ring geometries that are nearly planar, resulting (presumably) from delocalization of the sulfenyl sulfur electrons into the C-9 carbonyl group. Also, X-ray crystal data indicate that 4-methyl-9-oxothioxanthene 10,10-dioxide is planar[33] steming from the inability of

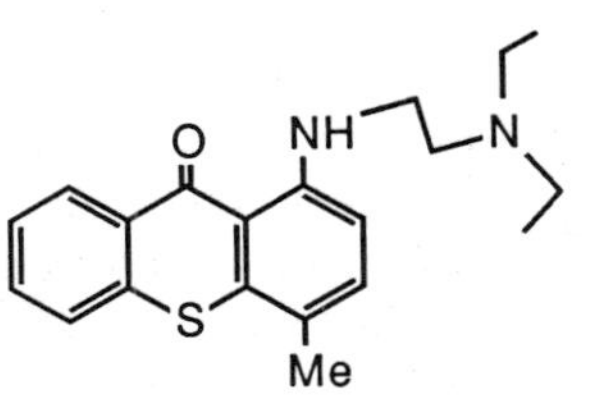

Miracil D

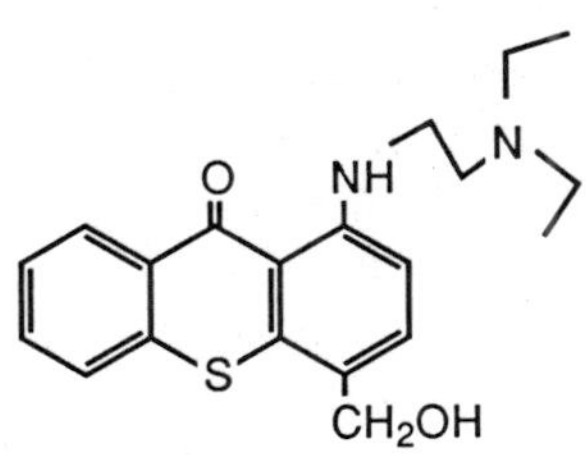

Hycanthone

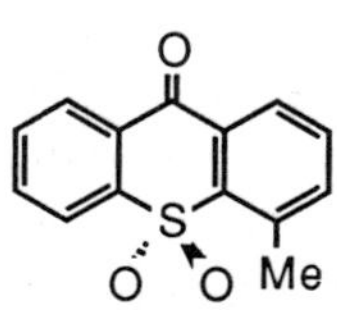

4-Methyl-9-oxo-
thioxanthene 10,10-dioxide

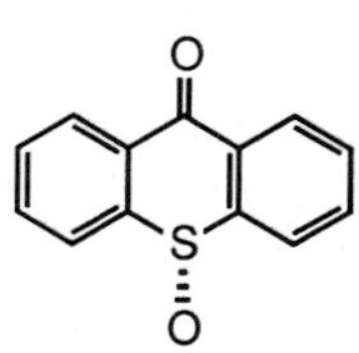

Thioxanthone 10-oxide

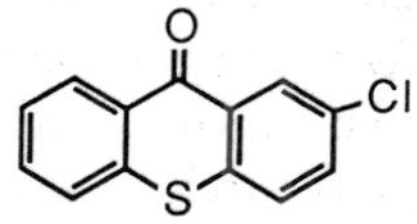

2-Chlorothioxanthone

Figure 6.2

the sulfonyl oxygens to easily rotate past the C-4 methyl group. However, thioxanthone 10-oxide[34] and hycanthone methanesulfonate[35] possess a "folded" thioxanthonyl ring skeleton. For thioxanthone 10-oxide, X-ray crystallographic analysis[36] indicates a dihedral angle of 155.7°, which is interpreted as a reduction in the aromaticity of the central ring resulting from a decrease in sulfur's electron-donating propensity compared to thioxanthone.

6.2.2 9-Alkylated Thioxanthenes

The interconversion of 9-substituted thioxanthenes through a planar transition state affords two conformational isomers: pseudoaxial (a′) and pseudoequatorial (e′) (see Scheme 1 where R ≠ R′). The barrier to interconversion is predictably low,[2a,7,9] such that even low-temperature NMR experiments do not allow for separate observation of each conformer; consequently, a temperature dependence of an intrinsic property (i.e., chemical shift, coupling constant, etc.) would only be indicative of a shift in the *position* of the conformational equilibrium (K_{eq}) and ultimately serving as a qualtitative barometer of the factors affecting the conformational equilibrium.

Johnson and Bovey calculations[37] confirm that the diamagnetic anisotropy of the aryl ring is the predominant if not the exclusive factor responsible for the observed chemical shift differences between a′ and e′ alkyl substituents.[38] The results of these calculations predict that a′ substituents will experience more diamagnetic shielding than the e′ substituents affording higher field ^{1}H-NMR shifts for the a′ substituents. Because of severe van der Waals interactions between e′ alkyl groups and the peri hydrogens (H-1, H-8), the a′ position is also predicted to be the thermodynamically more stable. The magnitude of the steric interactions between the 9a′ alkyl group and the relatively "soft", polarizable sulfenyl sulfur is not expected to be severe. These two points are dramatized in the variable temperature ^{1}H-NMR spectra of 9-methylthioxanthene (R = Me; R′ = H in Scheme 1) where increased shielding of the C-9 methyl doublet occurs as the temperature is decreased (+35 to −90 °C). These observations indicate that a methyl group in the a′ orientation is more stable than in the e′ array.[39]

For solution studies involving thioxanthenes, ^{1}H-NMR spectroscopy has been used extensively because hydrogen nuclei at C-9 and the alkyl fragments are easily discernible. If the ^{1}H-NMR chemical shift differences for C-9—H and the C-9—alkyl groups in the a′ or e′ orientations are sufficiently large, their chemical shifts compared to those for conformationally homogeneous (anancomeric[40]) a′ and e′ groups could serve as indicators of conformational preferences and ultimately stereochemical assignments.

While both conformational homogeneous a′ and e′ analogs are desirable, it has been synthetically more expeditious to create homogeneous a′ analogs through the technique of peri substitution.[41] In fact, the conclusion that the C-9 alkyls are largely a′ in 9-alkylthioxanthenes (9-alkyl-Ts) was confirmed by comparing the ^{1}H-NMR shifts of their C-9 alkyl groups with those in the 9-alkyl-

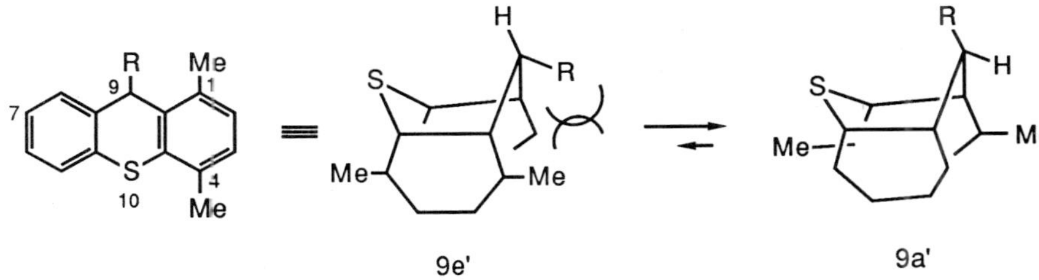

Scheme 2

1,4-dimethyl-Ts[30] (Scheme 2). In the latter compounds, the methyl group attached to C-1 in 9-alkyl-1,4-dimethyl-T is designed to cause severe steric interactions with a 9e′ alkyl group and steric relief, through ring inversion, forces the C-9 methyl into the a′ conformation. While the C-9 alkyl group in 1,4-dimethyl-T functions well as a sensor for the 9a′ orientation, the C-9 hydrogens ^{1}H-NMR chemical shifts are less reliable. Here, the C-9 hydrogen chemical shifts of the 9-alkyl-1,4-dimethyl-Ts are not superimposable on the shifts for C-9 hydrogens in the 9-alkyl-Ts. A van der Waals compression effect[42] amounting to ~0.42 ppm is responsible for this deshielding of the C-9 hydrogens in these conformational homogeneous systems (See Table 6.1)

By and large, monoalkylated 9-alkylthioxanthenes prefer the a′ orientation where the driving force for this conformational preference is the relaxation of the severe nonbonding interactions between the C-9e′ alkyl group and the peri hydrogens (H-1, H-8). The C-9 hydrogen ^{1}H-NMR chemical shift is far too sensitive to environmental variables to be employed as an unequivocal conformational probe. For example, the ^{1}H-NMR chemical shifts of C-9 hydrogen for a homogeneous series of 9-alkylated thioxanthenes is consistently shielded increasing steric bulk of the alkyl groups (at least, from methyl to isopropyl). Since with the increased alkyl steric encumbrance favors the 9a′ array (i.e, Me → Et → i—Pr), it is unreasonable to assume that the progressive *upfield shifts* of C-9 hydrogens arise from an increasing 9a′ contribution with subsequent exposure to the shielding component of the aryl ring diamagnetic anisotropy. At least three factors may influence the C-9 hydrogen chemical shifts in these thioxanthene derivatives: (1) the conformational equilibria reflecting the preference of the alkyl group for a′ or e′ at C-9, (2) the magnitude of the magnetic anisotropy resulting from the number of the C—C bonds disposed about C-1′, and (3) the hybridization index of the C-9—H bond reflecting internal angle strain. Consequently, a change solely in the chemical shift of C-9—H may not accurately reflect a change in preference for the a′ or e′ conformation. While C—H and C—C bond anisotropies arising from alkyl groups bonded to C-9 and changes in the hybridization at C-9 render the C-9—H absorption of dubious value in making conformational assignments, they should not affect the long-range coupling between the C-9—H and the aryl protons (especially the couplings to the H-1, H-8 hydrogens). It has been previously observed in similar systems[7b] that there exists

Table 6.1 ^{1}H NMR Data for Substituted Thioxanthenes[a]

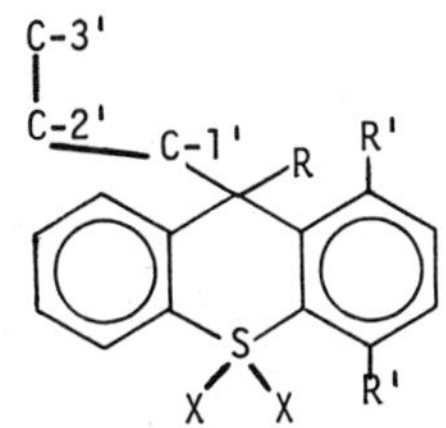

Compound	δ C-9 R(H or Me)	δ C-1′-H	δ C-2′-H	δ C-3′-H
Thioxanthene	3.80			
9-Methyl-T[b]	3.97 (7.0)	1.44 (7.0)		
9-Ethyl-T[b]	3.81 (7.2)	1.73	0.80 (7.2)	
9-isoPropyl-T[b]	3.50 (10.0)	2.22	0.76 (6.4)	
9-tert-Butyl-T[b]	3.71		0.90	
9-isoButyl-T[b]	4.06 (7.8)	1.59	1.29	0.86 (6.0)
9,9-Dimethyl-T[c]	1.66			
9-Ethyl-9-methyl-T[c]	1.82	1.87 (7.6)	0.54 (7.6)	
9-isoPropyl-9-methyl-T[c]	1.75	2.85 (7.0)	0.41 (7.0)	
9-isoButyl-9-methyl-T[c]	1.95	1.72 (5.4)	1.37	0.52 (6.5)
1,4,9-Trimethyl-T[d]	4.47 (7.1)	1.30 (7.1)		
1,4-Dimethyl-9-Ethyl-T[d]	4.16	1.43-2.04	0.79	
1,4-Dimethyl-9-isopropyl-T[d]	3.93 (10.2)	~2.33	0.74	
1,4-Dimethyl-9-tert-butyl-T[d]	4.18		0.90	
9-Ethyl-T 10,10-Dioxide[e]	3.88 (7.6)	2.08 (7.6)	0.95 (7.6)	
9-isoPropyl-9-methyl-T 10,10-dioxide[e]	1.77	2.84	0.70	
9-isoPropyl-T 10,10-dioxide[e]	3.53 (10.0)	2.48	0.88 (6.4)	

[a] ^{1}H NMR shifts (δ) are reported in parts per million (ppm) downfield from internal tetramethylsilane (Me_4Si) in deuteriochlorogrom ($CDCl_3$) solvent at *ca.* 30°C. The ^{1}H NMR shifts are followed by coupling constants (J) in hertz (Hz). See reference 30 for additional details.

[b] R = R′ = H; X = lone pair electrons. [c] R = Me; R′ = H; X = lone pair electrons. [d] R = H; R′ = Me; X = lone pair electrons. [e] X = 0; R = H, Me; R′ = H.

Table 6.2 ^{1}H-NMR Data for 9-Alkylated Thioxanthenes

Compound	C-9—H[a]	^{13}C—H (Hz)[b]	Percent Change[c]
Thioxanthene	3.80	129.0	—
9-Methylthioxanthene	3.97	129.0	21
9-Ethylthioxanthene	3.81	130.4	16
9-Isopropylthioxanthene	3.50	128.8	14
9-*tert*-Butylthioxanthene	3.79	127.3	22
9-Isobutylthioxanthene	4.06	—	—

[a] The chemical shifts (δ) are relative to tetramethylsilane (Me_4Si) and are recorded in deuteriochloroform ($CDCl_3$) solvent at ambient temperature.
[b] The ^{13}C—H coupling constants were obtained directly from the ^{13}C—H satellites.
[c] Represents the percent change in the C-9 hydrogen bandwidth with and without irradiation of the aryl hydrogens: see Ref. 30.

a relatively small coupling (<1 Hz) between the benzylic hydrogen (C-9 hydrogens) and the aromatic hydrogens. A benzylic hydrogen essentially perpendicular to the plane of the aromatic π system (C-9a′ in the thioxanthenes) is positioned for maximum orbital overlap and efficient nuclear spin–spin transmission compared to C-9e′ hydrogen. Exploitation of this angular-dependent coupling is manifested by observing the C-9 hydrogen absorption in any C-9 alkyl substituted thioxanthene while simultaneously decoupling the aryl hydrogens. Certainly, the trend in the data (column 3, Table 6.2) suggests that as the alkyl group increases in steric bulk the preference for the a′ conformation is more pronounced and the C-9 hydrogen assumes the e′ orientation with diminished benzylic coupling. The exception, the C-9 hydrogen in 9-*tert*-butyl-T, is strongly coupled, indicating an efficient overlap with the aryl π electrons. Here, transannular repulsions between the *tert*-butyl methyls and the sulfur atom result in a flattening of the central ring. A related phenomenon has been observed in the 9-alkyl-9,10-dihydroanthracene series.[43] This flattening process causes a displacement of C-9 hydrogen to a position that is more 9a′ in character with an apparent concomitant increase in its bandwidth ($W_{1/2}$) from the stronger "allylic" coupling interaction.

The ^{13}C—H coupling constants provide only marginal support for this view. Since 9-*tert*-butyl-T should be more flattened, an increase in the internal C—C-9—C angle is expected, requiring an increase in s character as well. Consequently, a reduction in s character in the C-9—H bond is expected. The relatively small reduction in the ^{13}C—H coupling constant is not inconsistent with the need for more p character in the C-9—H bond (column 2, Table 6.1) and the central ring flattening rationale.

The results of an X-ray study on 9-isobutyl-T (**8**) indicate that in the solid state the isobutyl group occupies the 9a′ conformation and the torsional angle about C-1′—C-9 of 51° is consistent with assignments employing ^{1}H-NMR solution data.[44]

Finally, Ternay and Chasar[45] have successfully demonstrated that thioxanthen-9-ol (**9**) and 9-trimethylsiloxythioxanthene (**10**) prefer that conformation in which the substituent occupies the e′ position. Apparently, the close proximity of

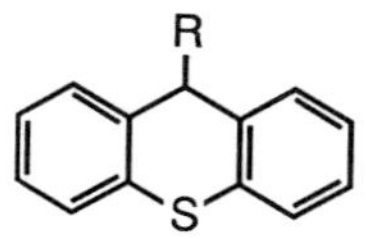

R = $CH_2CH(CH_3)_2$: **8**

R = OH: **9**

R = $OSiMe_3$: **10**

the lone pair electrons on both sulfur and the a′ oxygen creates repulsive coulombic interactions that are significantly diminished when the C-9 Me_3SiO group occupies the e′ position.[46] Recent studies show that the preferred conformer, 9e′ or 9a′, for **9** is solvent dependent.[47]

6.2.3 9-Alkyl-9-Methylthioxanthenes

The conformational interplay between alkyl and methyl groups attached to C-9 and their role in affecting the conformational equilibrium, e′ ⇌ a′, in 9-alkyl-9-methylthioxanthenes is particularly interesting. Some insight is necessary regarding (1) which of two different alkyl groups will prefer the 9a′ conformation, and (2) if the rotameric distributions of particular alkyl groups are altered significantly when compared to their rotomeric distributions in 9-alkyl-T's.

As previously mentioned,[20,30] the ^{1}H-NMR spectrum of 9,9-dimethylthioxanthene (7) at 30 °C exhibits a sharp singlet for the methyl groups consistent with a static planar structure or a rapid conformational interchange between two isoenergetic flattened-boat conformers. Although little data are available, one might anticipate that the manifestations of steric perturbations between the methyl groups and the peri hydrogens would translate into a more flattened ring and a significantly lower barrier to inversion.

The ^{1}H-NMR spectrum of 9-ethyl-9-methylthioxanthene (**11**) exhibits a singlet (δ 1.82) for the C-9 methyl group in $CDCl_3$ solvent which is deshielded by 0.5 ppm when compared to the C-9 methyls in *trans*-9-methylthioxanthene 10-oxide (9a′,10e′)[48] and other substituted thioxanthenes which are known to possess an a′ methyl group. However, when the ^{1}H-NMR chemical shift of the C-9 methyl in 9-methyl-9-ethyl-T (**11**) was compared to other substituted thioxanthenes known to have an e′ methyl group, a satisfactory chemical shift analogy is realized. Interestingly, the chemical shift of the methyl triplet of 9-methyl-9-ethyl-T occurs 0.25 ppm to higher field than the methyl triplet in 9-ethyl-T. A reasonable rationale requires that the ethyl group (1) occupy the 9a′ conformation and (2) orient its methyl group over the face of the aryl π cloud to relieve the van der Waals nonbonding interactions between the gauche methyls. The methyl group should then experience more of the anisotropic shielding exerted by the aryl rings of thioxanthenyl moiety. In addition, the CH_2 hydrogens in 9-methyl-9-ethyl-T are shifted downfield (δ 1.86) relative to those in 9-ethyl-T (δ 1.73) resulting from a removal of those hydrogens from the diamagnetic shielding region of the aryl rings. Overall, the smaller CH_3 group occupies the e′ array and the nonbonding interactions between the vicinal methyls encourage rotation of the ethyl group to a heavily weighted, time-averaged residence over the aryl rings.

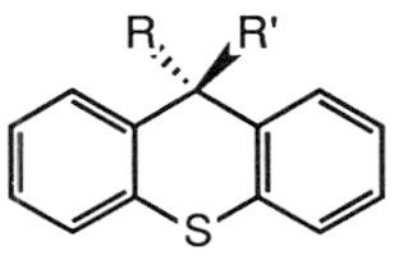

R = R′ = CH_3: **7**

R = CH_3; R′ = CH_2CH_3: **11**

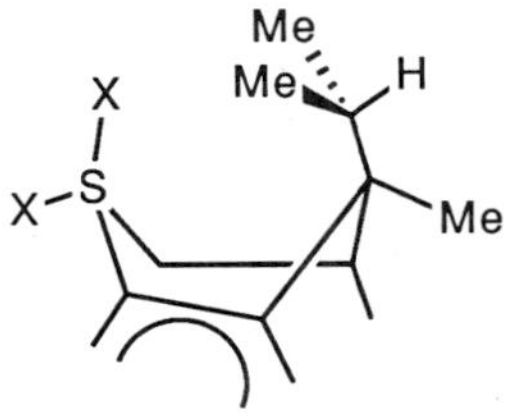

X = lone pair e⁻; **12**

X = O; **12-SO$_2$**

The 9-isopropyl methyl resonance in 9-isopropyl-9-methylthioxanthene (**12**) is ~0.35 ppm to higher field than that of 9-isopropyl-T. In both compounds, the C-9-isopropyl group is a′. This shielding effect is adequately explained if the *C-1′ hydrogen* of the C-9a′ isopropyl group in 9-isopropyl-T is antiperiplanar to C-9-He′. The magnitude of the vicinal coupling between C-9e′-H and C-1′-H as well as the lack of a substantial temperature dependence supports the view that the anti rotamer is favored (Figures 6.3 and 6.4). Replacement of C-9e′-H with a methyl group would introduce considerable nonbonding interactions between the C-9a′ isopropyl methyls and the single C-9e′ methyl. Relief from this steric congestion occurs only by rotation to the least sterically inhibited environment where the methyls are nearer the sulfenyl sulfur. The ^{1}H-NMR spectrum of 9-isopropyl-9-methylthioxanthene 10,10-dioxide (**12**-SO_2) provides substantial support for these suggestions. The C-9e′ methyl singlet and the multiplet arising from the isopropyl methine (C-1′) exhibit shifts that are practically superimposable on the corresponding resonances of 9-iso-propyl-9-methyl-T. However, the isopropyl methyl resonances in sulfone **12**-SO_2 are deshielded 0.29 ppm relative to **12**. These findings strongly suggest that both iso propyl groups in **12** and **12**-SO_2 not only occupy the 9a′ conformations but share the same preferred rotamers as well. (See Table 6-1.)

A final observation on the ^{1}H-NMR spectra of 9-alkyl-9-methyl-T's is in order. In the 9-alkylthioxanthenes, the ^{1}H-NMR chemical shifts of the aryl hydrogens peri to sulfur (C-4, C-5) are deshielded by ~0.20 ppm relative to the remaining aryl hydrogens.[49] This deshielding effect probably results from the combined effects of nonbonding interactions involving the sulfur lone pair electrons and the electronegativity/inductive effects of sulfur on these peri hydrogens. However, when a second alkyl substituent (methyl) is introduced at C-

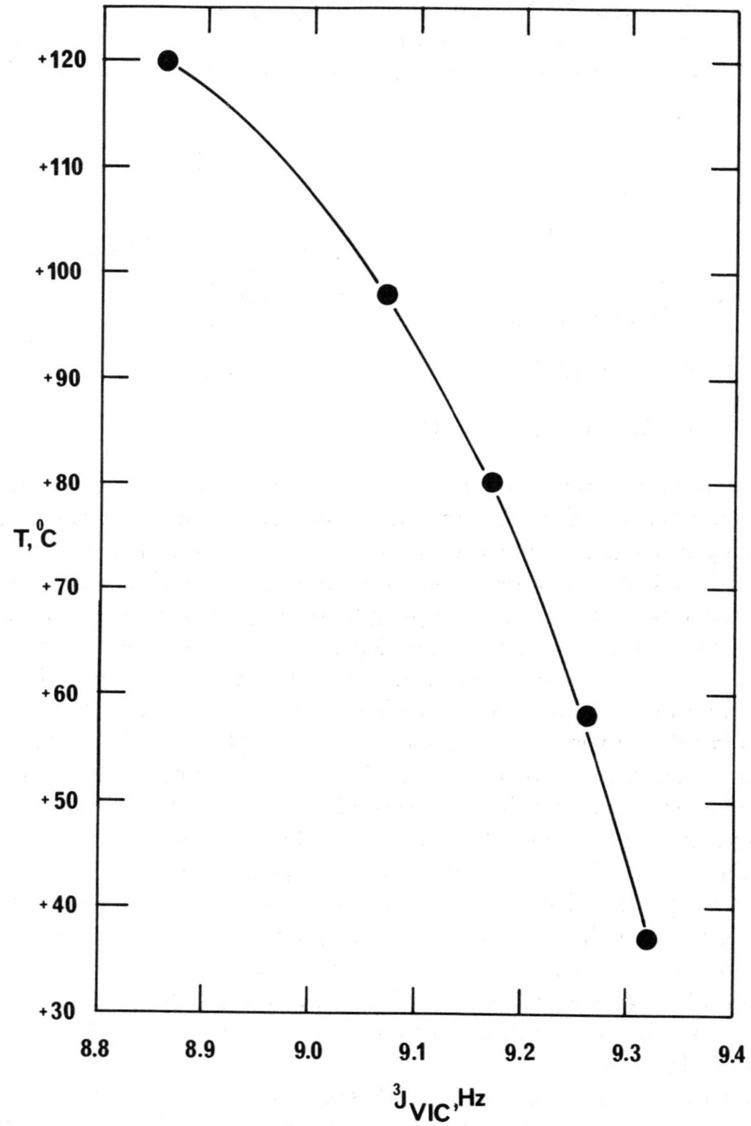

Figure 6.3 The temperature dependence (above ambient) of the vicinal coupling constant of 9-isopropylthioxanthene.

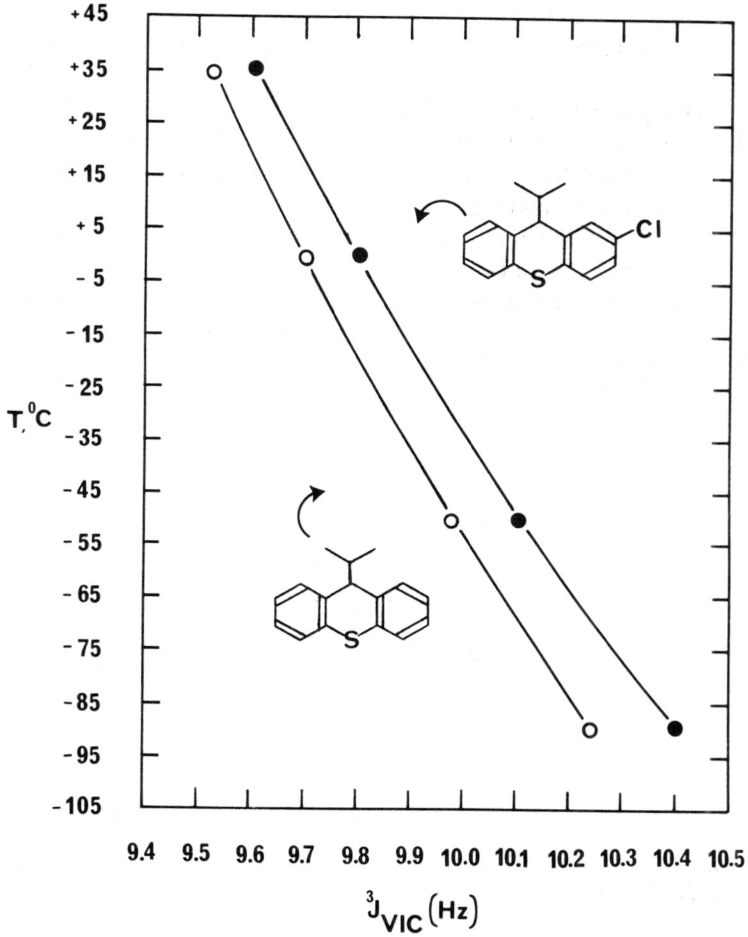

Figure 6.4 Temperature dependence (below ambient) of the vicinal coupling constants of 2-chloro-9-isopropylthioxanthene (●) and 9-isopropylthioxanthene (○).

9, the aromatic region exhibits *two* multiplets of equal intensity where the low-field region has doubled its intensity compared to the ^{1}H-NMR characteristics of 9-alkyl-T's. This means that the presence of a C-9e′ methyl group deshields the *o*-aryl hydrogens (H-1 and H-8) through a van der Waals compression effect.[42] A closer examination of this concept of deshielding suggests that in order for such an interaction to exist the C-9 methyl group must certainly be e′ and its hydrogens must be within at least 2.5 Å of H-1 and H-8. The magnitude of cross-relaxation between two nuclei through intramolecular direct dipole–dipole coupling is related to the inverse sixth power of the distance separation between the nuclei.[50] If two nuclei are efficiently cross-relaxed through such a mechanism, then they must be proximal and complete saturation of one nucleus should result in an enhancement in the integrated intensity of the other (i.e., the nuclear Overhauser enhancement).

The intensity enhancements of the aryl absorptions arising from irradiation of the C-9 methyl groups for the three 9-alkyl-9-methyl-Ts clearly establishes the C-9 methyl as proximal (i.e., within contact of their combined van der Waals radii) to the C-1,8 peri hydrogens and, therefore, e′. For example, while saturating the C-9e′ methyl resonance and integrating the aromatic region of 9-ethyl-9-methyl-T (**11**), 21% increase in the integrated intensity of the low-field aryl hydrogens was observed. In similar experiments with 9-isopropyl-9-methyl- (**12**) and 9-isobutyl-9-methyl-T (**13**), a 19 and 25% increase, respectively, in the integrated intensity was observed.[30]

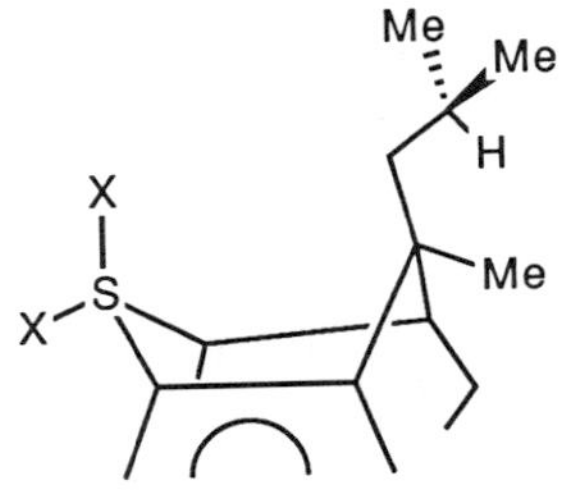

X = lone pair e⁻; **13**

X = O; **13-SO$_2$**

6.2.4 Sulfoxides

6.2.4.1 Thioxanthene 10-Oxide

Chemical oxidation of thioxanthene affords thioxanthene 10-oxide (**14**), which can exist in two distinctive boat conformations exhibiting either a 10a′ or 10e′ sulfinyl oxygen. The ^{1}H-NMR spectrum of sulfoxide **14** shows that the diastereotopic C-9 hydrogens appear as an unsymmetrical AB spin pattern where the broadened doublet is more shielded (e.g., upfield) than the sharper doublet.[41a] A decoupling experiment involving saturation of the aryl resonances causes a dramatic sharpening of the broadened absorptions, indicating the presence of a strong coupling interaction with the aryl protons. The C-9 axial, benzylic hydrogen is the best candidate for large coupling interactions with the aryl hydrogens because of the favorable σ to π orbital overlap which should encourage strong spin–spin interactions.[41] The chemical shift assignment of the axial C-9 hydrogen in **14** was confirmed by examining the ^{1}H-NMR spectrum of 1,4-dimethylthioxanthene 10-oxide (**15**).[41a] Here, the C-4 methyl forces the

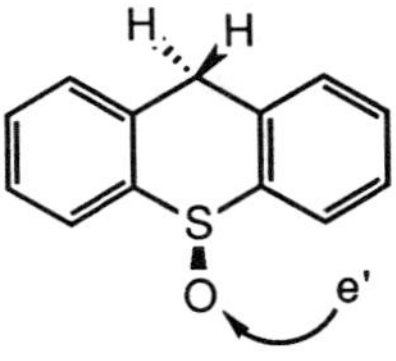

14

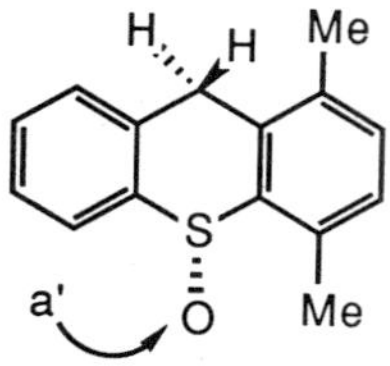

15

sulfinyl oxygen into the 10a′ conformation. Consequently, the proximity of the 1,4-synaxial C-9 hydrogen and sulfinyl oxygen results in substantial deshielding. In fact, the broadened segment of the AB pattern in **15** is *now* deshielded relative to the more sharpened doublet. The X-ray crystal structure of sulfoxide **15** confirms the conformational assignment of the sulfinyl group as 10a′.[51] The e′ S—O preference in **14** is in contrast to that exhibited by phenoxathiin 10-oxide[52] and phenothiazine 10-oxide[53] where both have an a′ sulfinyl oxygen in the solid and solution state (Scheme 3).

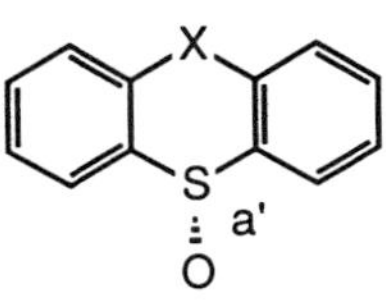

X = O; phenoxathiin 10-oxide

X = NH; phenothiazine 10-oxide

Scheme 3

Perhaps the most interesting conformational feature attending sulfoxide **14** is the demonstration that the a′ and e′ preference of the sulfinyl group can be controlled by hydrogen bonding solvents and complexation with halogens. For example, in $CDCl_3$ or C_6D_6 solvents, the S=O group in thioxanthene 10-oxide prefers the e′ orientation, but in trifluoracetic acid (CF_3COOH), a strongly hydrogen-bonding medium, the a′ conformation is preferred.[54] Apparently, the hydrogen-bonding solvent creates an effective sterically larger sulfinyl group that experiences less steric crowding in the a′ array.[54] With iodine monochloride, the 1041 cm^{-1} sulfinyl oxygen stretching frequency is diminished in intensity and affords two new absorptions at 992 and 953 cm^{-1}. These observations are interpreted in terms of a mixture of a′ and e′ complexed conformers.[55]

In passing, Ternay et al.[56] have pointed out that particularly for the thioxanthonyl system the C-2,7 ^{13}C-NMR shifts rather than the C-9 carbonyl shifts are an efficient moniter of the oxidation state and mesomeric potential at the sulfur functionality.

6.2.4.2 cis- and trans-9-Substituted Thioxanthene 10-Oxides

Direct oxidation (e.g., $NaIO_4$, H_2O_2, or *m*-CPBA) of a 9-alkylthioxanthene affords two diastereomers (cis and trans) each possessing two rapidly equilibrating conformational isomers (Scheme 4). Control of a particular conformation by a substituent group reflects the magnitude of the relative energy differences associated with the a′ and e′ orientations.

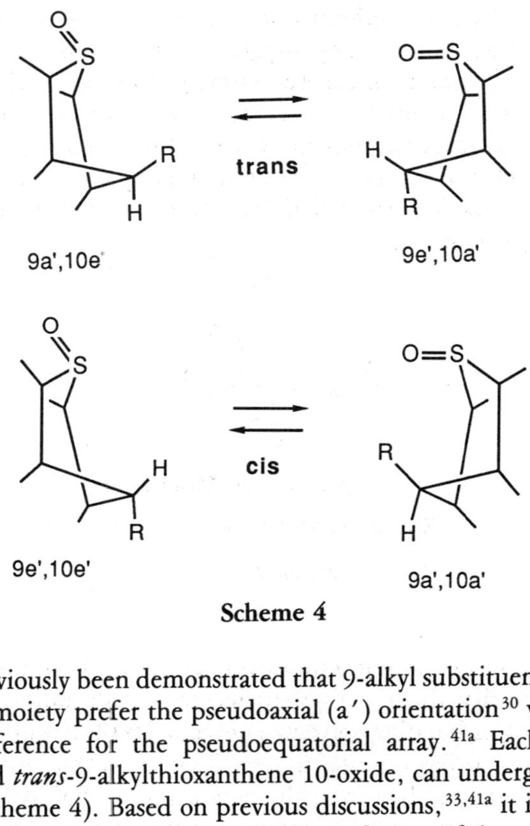

Scheme 4

It has previously been demonstrated that 9-alkyl substituents attached to the thioxanthenyl moiety prefer the pseudoaxial (a′) orientation[30] while 10-oxide **14** displays a preference for the pseudoequatorial array.[41a] Each configurational isomer, *cis*- and *trans*-9-alkylthioxanthene 10-oxide, can undergo conformational interchange (Scheme 4). Based on previous discussions,[33,41a] it is anticipated that conformer 9a′,10e′ will be the more stable conformer of the trans diastereomer. However, predicting the more stable conformer of the cis diastereomer is less straightforward. For example, when R = CH_3, the 9e′,10e′ conformer is preferred, indicating that the e′ sulfinyl oxygen is conformationally demanding. However, for R groups larger than methyl, the 9a′,10a′ conformer is preferred, demonstrating that the interplay of C-9 and S-10 group preferences is controlled exclusively by the larger C-9 alkyl group.[57]

In 1963, Schindler et al.[58] described the oxidation of 9-[(*N*-methyl-3′-piperidyl)methyl]thioxanthene (**16**) with hydrogen peroxide to a mixture of two diastereomeric sulfoxides (*cis*- and *trans*-**17**). Configurational analyses of these

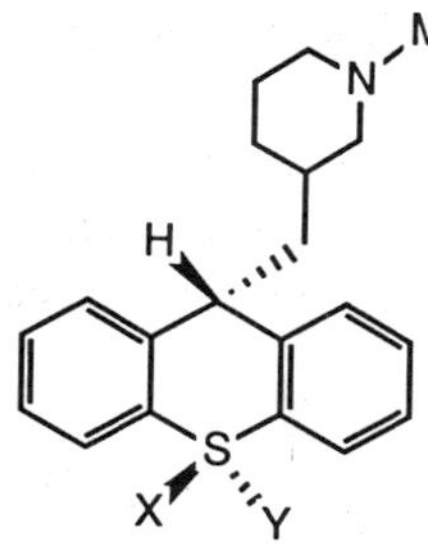

X = Y = lone pair e$^-$: **16**

X = lone pair e$^-$; Y = O; **cis-17**

Y = lone pair e$^-$; X = O; **trans-17**

substances, employing IR, UV, and ^{1}H-NMR spectroscopic techniques, supported the assignment of *cis*-**17** to the diaxial conformation (9a′,10a′) and *trans*-**17** to the 9a′,10e′ array.[59] For each diastereomer, the large alkyl group prefers the occupancy of the least hindered conformation (a′).[59]

The configurations and preferred conformations of the isomeric sulfoxides of thioxanthen-9-ol (**9**) and appropriate derivatives were ascertained from ^{1}H-NMR chemical shift data.[46] For example, *trans*-thioxanthenol 10-oxide (*trans*-**18**), *trans*-9-trimethylsiloxythioxanthene 10-oxide (*trans*-**19**), and *trans*-2-chloro-9-trimethylsiloxythioxanthene 10-oxide (*trans*-**20**) were assigned the 9a′,10e′ conformation

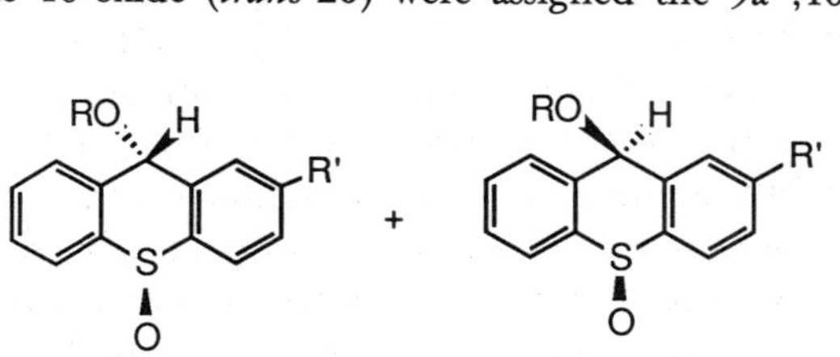

trans-(9a',10e')   +   cis-(9e',10e')

R = R′ = H; **18**

R = $SiMe_3$; R′ = H; **19**

R = $SiMe_3$; R′ = Cl; **20**

based on the similarity in the C-9 hydrogen chemical shifts. In addition, X-ray structural data for *trans*-**18** confirmed the presence of a slightly flattened boat conformation with the sulfinyl oxygen occupying the e′ array and the C-9-OH residing in the a′ orientation.[60] The cis diastereomers of the above were assigned the 9e′,10e′ conformation based on the ^{1}H-NMR chemical shifts of the C-9 hydrogens.[46]

6.2.4.3 1,4,9- and 2,4,9-Trimethylthioxanthenes 10-Oxides

Interesting conformational features attend 2,4,9-trimethylthioxanthene 10-oxides in that *three* unique forms have been isolated. Oxidation of 2,4,9-trimethylthioxanthene (**21**) with *m*-chloroperoxybenzoic acid affords a 4:1 ratio of

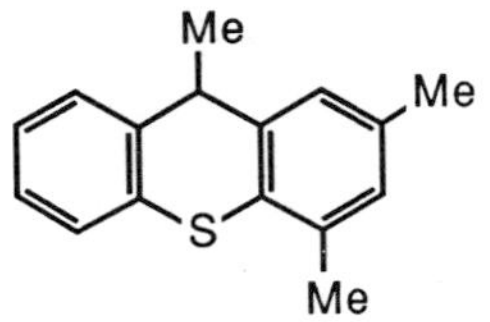

21

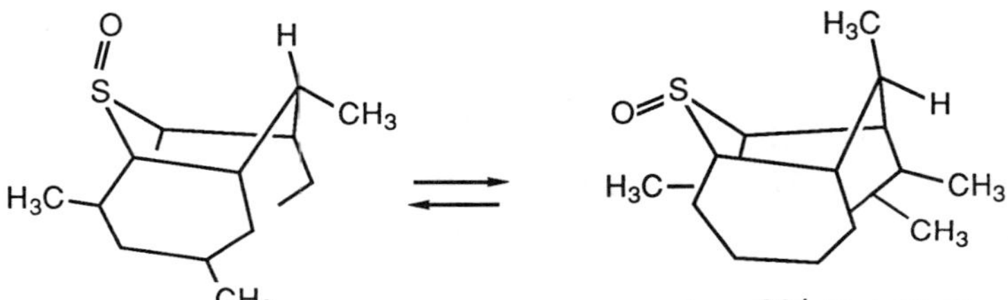

trans-**22** (9e′,10a′) *trans*-**22′** (9a′,10a′)

cis:trans diastereomers. X-ray analysis of *cis*-sulfoxide[61] **22** shows that both the C-9 methyl and the 10-oxide exist in a shallow-boat 1,4-diaxial conformation (9a′,10a′) with an angle of fold about the C-9 and S-10 atoms of 147.3°. On the other hand, trans isomer is found to exist in two different crystalline forms,[62] *trans*-**22** (9e′,10a′; mp 374–377 K) and *trans*-**22′** (9a′,10e′; mp 397–400 K) with identical ^{1}H-NMR solution spectra and consequently the same conformation in solution (presumably, *trans*-**22**; 9e′,10a′). Interestingly, recrystallization of *trans*-**22′** from methanol solvent gave the thermodynamically more stable *trans*-**22** conformer whose X-ray structure confirmed the sterically congested C-9e′ methyl and the a′ sulfinyl oxygen which seeks steric relief from the perturbation caused by the C-4 methyl. The folding angle about the meso atoms of the central ring for *trans*-**22** is 134.2°. Since the folding angle between the planes of the two aryl rings is correlated with the bond angles involving the meso atoms of the central ring, the larger folding angle in the *cis*-**22** is due presumably to the nonbonded interactions between the axial 9-methyl substituent and the axial 10-oxide group. *cis*-9-Methylthioxanthene 10-oxide with both substituents in equatorial orientations exhibits folding angle of 127.2° and reflects the absence of severe steric perturbations.[63] A subtle proximity effect of a methyl group is seen in *cis*-1,4,9-trimethylthioxanthene 10-oxide (**23**) where the angle of fold is 140.7°.[64] A measure of the degree of central ring flattening can be appreciated by comparing the angle of fold between *cis*-2,4,9-trimethyl-T 10-oxide (147.3°) and *cis*-1,4,9-trimethyl-T 10-oxide (140.7°). In sulfoxide **23**, the 1,4-diaxial interac-

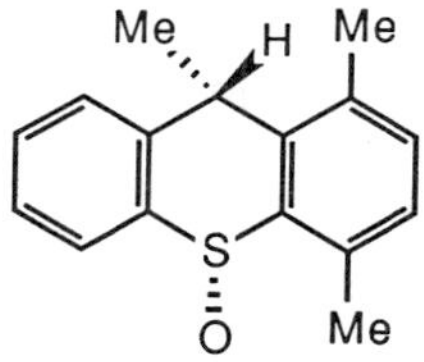

cis-**23** (9a′, 10e′)

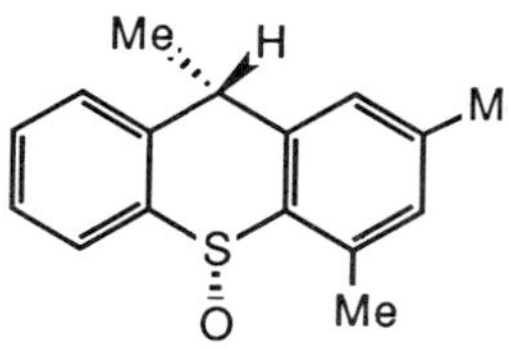

cis-**22** (9a′,10a′)

tions between the axial C-9 methyl and the a′ sulfinyl oxygen giving rise to ring deformation is apparently buffered by the close approach of the C-9 and S-10 substituents to the 1,4-dimethyl groups.

6.2.4.4 9-Ethyl-1,4-Dimethyl- and 9-Ethyl-2,4-Dimethylthioxanthene 10-Oxides

Direct oxidation of 9-ethyl-2,4-dimethylthioxanthene (**24**) affords a 4:1 ratio of cis:trans diastereoisomeric sulfoxides. The X-ray crystal structure of *cis*-9-ethyl-2,4-dimethylthioxanthene 10-oxide (**24**-SO)[65] has both substituents in the a′ position and the folding angle between the planes of the two benzene rings is 141.5°. The magnitude of this folding angle is initially surprising when compared to the 147.3° angle found in *cis*-2,4,9-trimethylthioxanthene 10-oxide (*cis*-**22**) (vide supra). While it seems apparent that some flattening of the central ring results from repulsive van der Waals steric interactions between the a′ sulfinyl oxygen and the C-9a′ ethyl, "total energetic relaxation" through this bending mode must be hampered by interactions between the CH_3 of the ethyl group and the peri C-1,8 hydrogens. This rationale is consistent with the expected smaller C—S—C and C—C—C internal bond angles involving the meso atoms.

The X-ray structural data for *cis*-9-ethyl-1,4-dimethylthioxanthene 10-oxide (*cis*-**25**) has been published[66] and while the 9a′,10a′ diaxial array for the meso substituents is expected, the preferred rotamer of the 9a′ ethyl group in *cis*-9-ethyl sulfoxide **25** is different from that in *cis*-9-ethyl 10-oxide (*cis*-**24**-SO). The torsional angles defining the γ gauche and γ anti rotamers involving the CH_3 of the ethyl group are $-169.8°$ for 1,4-dimethyl-T 10 oxide (*cis*-**25**) and 65° for *cis*-9-ethyl 10-oxide **24**-SO. This indicates that the conformation of the ethyl group in the 1,4-dimethylthioxanthenyl moiety is significantly influenced by the nonbonding steric interactions of the attached C-1 peri methyl.

The X-ray single crystal data[67] for *trans*-9-ethyl-2,4-dimethyl-T 10-oxide (*trans*-**24**-SO) indicates that the 9-ethyl group is a′ and the 10-oxide is e′. This is in contrast to the trans isomer of 2,4,9-trimethyl-T 10-oxide (*trans*-**22**) in which the C-9 methyl group is in a boat-equatorial array while the 10-oxide is axial. These observations are interpreted to mean that the 10-oxide group governs the stereochemistry of the sulfoxide isomers when the size of the 9-alkyl substituent is small (i.e., methyl). However, when the size of the alkyl group increases to ethyl, the larger alkyl group overrides the conformational driving force of the sulfinyl

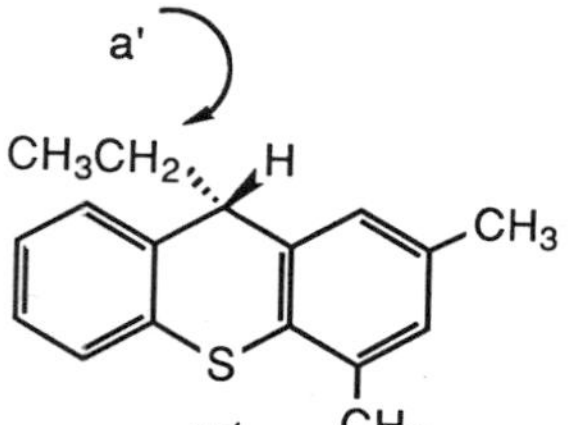

24

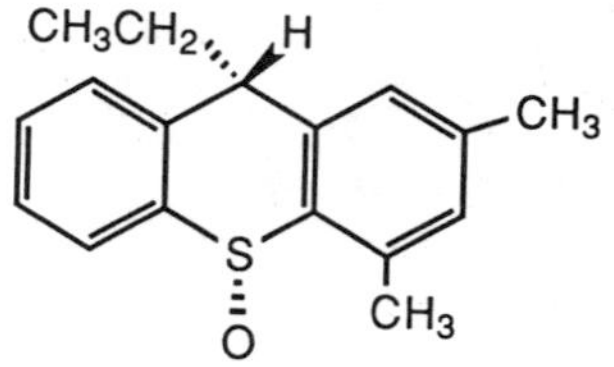

cis-**24**-SO (9a′,10a′)

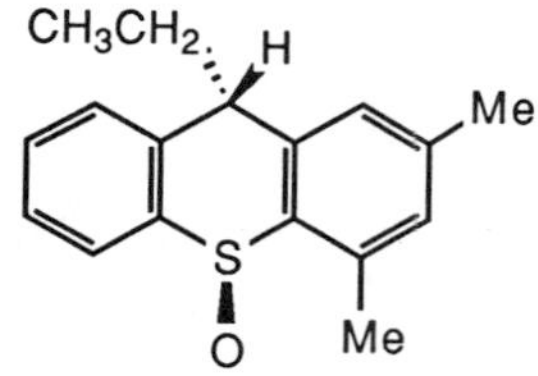

trans-**24**-SO (9a′,10e′)

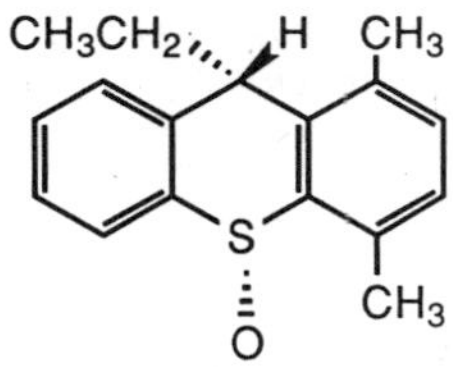

cis-**25** (9a′,10a′)

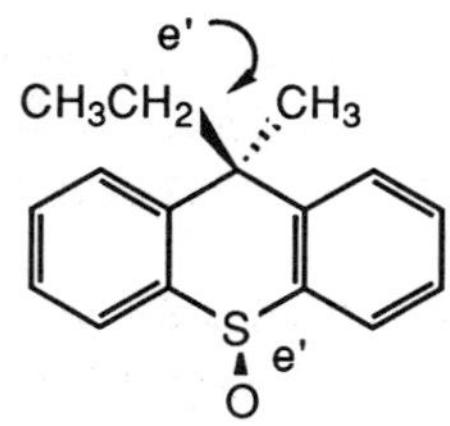

26

group. The folding angle between the aryl rings in *trans*-**24**-SO is 135.4° and in *cis*-**24**-SO is 141.5° demonstrating that van der Waals steric interactions induce significant flattening of the central ring in the cis isomer.[67]

Chu and Rosenstein[68] have demonstrated through X-ray analysis that *cis*-9-ethyl-9-methylthioxanthene 10-oxide (**26**) possess a central ring boat conformation and the C-9 ethyl and sulfinyl oxygen are cis and both occupy the e′ orientation. The folding angle between the planes of the two mirror-related phenyl rings is 135.0° which is surprisingly analogous to that observed in *cis*-9-isopropyl-T 10-oxide.[69]

6.2.4.5 1,4-Dimethyl-9-isopropylthioxanthene 10-Oxide

The orientation of the C-9 isopropyl group and the sulfinyl oxygen in 1,4-dimethyl-9-isopropylthioxanthene 10-oxide (*cis*-**27**) is 9a′,10a′ as determined

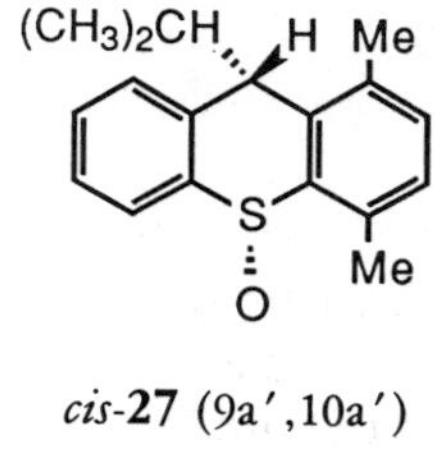

cis-**27** (9a′,10a′)

from X-ray analysis.[70] Interestingly, the conformation of the isopropyl group here mirrors that found in *cis*-9-isopropylthioxanthene 10-oxide where the isopropyl group is nearly symmetrical with respect to a plane passing through the meso atoms. Apparently, the "near" symmetrical orientation of the isopropyl group in the 1,4-dimethylthioxanthenyl system provides minimization of the nonbonded interactions between the isopropyl methyls and the a′ sulfinyl oxygen as well as the C-1 methyl group.

6.2.4.6 9-Arylthioxanthene 10-Oxides

In these systems "spectroscopic protocols" for determining the preferred conformation of the C-9 aryl group are interesting and worthy of comment. When a phenyl group occupies the 9e′ position, it experiences severe rotational constraints around C-9, C-1′ single bond which places the peri hydrogens (H-1, H-8) in the diamagnetic shielding region of the phenyl group. When the C-9 aryl group is a′ its rotational freedom about the C-9–C-1′ bond is sterically hampered by the C-2′ and C-6′ hydrogen interactions with both the C-9-e′ hydrogen and the sulfenyl sulfur. Consequently, the time-average conformation of the aryl (or phenyl) ring is the one that is perpendicular to the imaginary meso axis. Here, the magnetic anisotropy of the thioxanthenyl ring shields the C-2′, C-6′ hydrogens or substituents when the phenyl occupies the pseudoaxial (a′) orientation.

The preferred conformation of the sulfinyl oxygen in the thioxanthenyl system is easily assessed by (1) aromatic solvent-induced shifts (ASIS)[71] and (2) taking advantage of the strong anistropic effect of a′ sulfinyl group which causes a sizeable downfield shift of the C-9a′ hydrogen or substituent compared to its C-9e′ counterpart.

Direct oxidation of 9-phenylthioxanthene (**28**) with hydrogen peroxide [35% H_2O_2 in HOAC-CH_2Cl_2, 25°] gives a 3:1 trans:cis ratio of diastereomers whose conformations are assigned as *trans*-**29**: 9a′,10e′ and *cis*-**29**: 9e′,10e′.[72] If

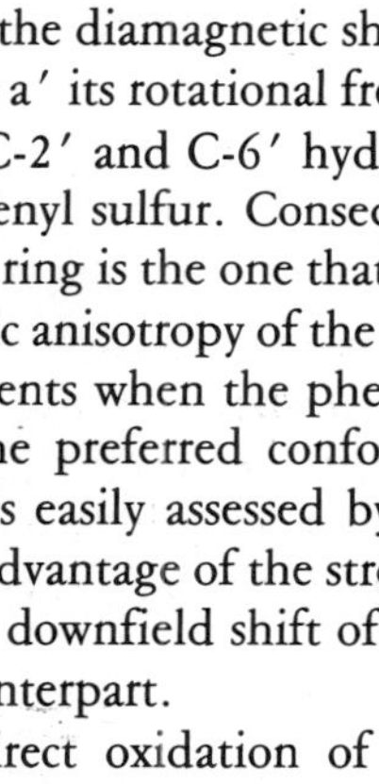

28

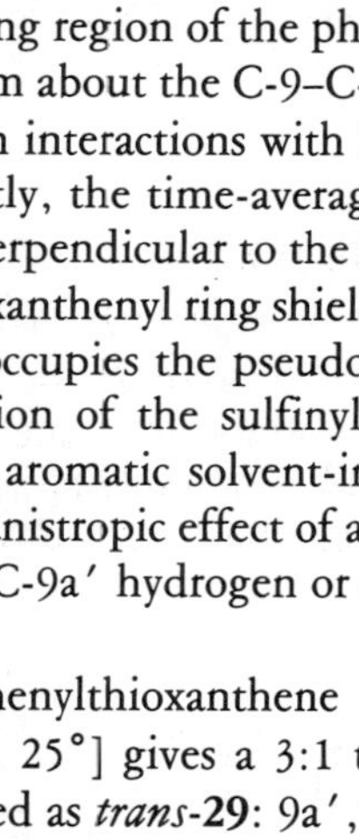

trans-**29** (9a′,10e′)

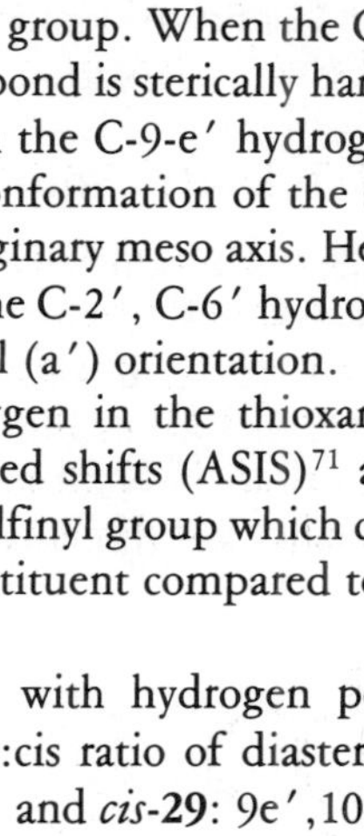

cis-**29** (9e′,10e′)

the C-9 phenyl group prefers the a′ conformation in **28** then these results imply a kinetic controlled oxidation favoring oxidant approach from the least hindered 10e′ side.[72]

The C-9, C_6H_5-, C_6D_5-, and C_6F_5-substituent groups surrender to the conformational demands of the sulfinyl moiety; in the trans isomer, the preferred conformer is 9a′,10e′ and the preferred conformer in the cis diastereomer is 9e′,10e′. When the aryl group is mesityl (i.e., 2,4,6-trimethylphenyl), the

9a′,10e′ conformer of the trans disastereomer is no longer preferred but gives way to the 9e′,10a′ conformer where the bulky mesityl group occupies the 9e′ conformation. This is also true for the 2,3,5,6-tetramethylphenyl C-9 substituent as well. Apparently, the C-2′,C-6′ methyl groups of mesityl and duryl create severe repulsive van der Waals interactions with the thioxanthenyl skeleton in the C-9a′ position.[72]

When both an alkyl and a phenyl group are attached to the C-9 position, the larger alkyl group prefers the 9a′ position with the sulfinyl group in the 10e′ position. Even when very large groups like isopropyl and phenyl are attached to C-9, the two diastereomeric sulfoxides both have 10e′ sulfinyl groups. In one isomer, the isopropyl group prefers the 9a′ position and in the other, the phenyl prefers the 9a′ orientation.[73]

6.2.5 Diaryl Sulfones

6.2.5.1 Thioxanthene 10,10-Dioxides

While thioxanthone 10,10-dioxide (**30**) may be nearly planar,[33] various thioxanthene 10,10-dioxides have central rings that are boat shaped. For example, the structure of 3,6-bis(dimethyl amino)thioxanthene 10,10-dioxide (**31**) has been described using ^{1}H-NMR and HMO methodology as having a "tub" conformation.[75] Finally, 2,4-dimethylthioxanthene, 10,10-dioxide (**32**) is boat shaped by

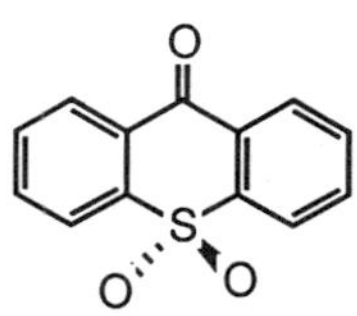

30

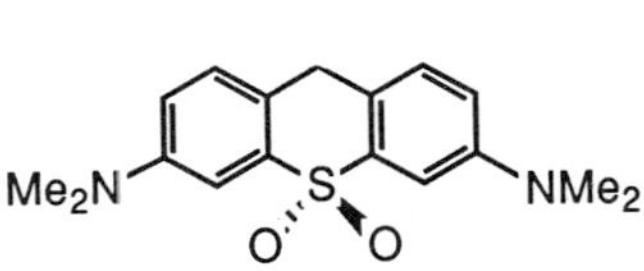

31

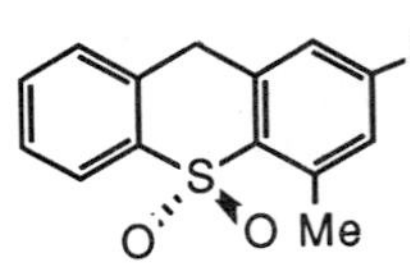

32

X-ray diffraction with a folding angle between the planes of the two benzo rings of 136.6°.[76]

6.2.5.2 9-Substituted Thioxanthene 10,10-Dioxides

A brief comment is appropriate regarding the preferred conformations of several 9-substituted thioxanthene 10,10-dioxides. First, the ^{1}H-NMR spectrum of thioxanthene-9-ol 10,10-dioxide (**33**) exhibits two doublets attributable to the C-

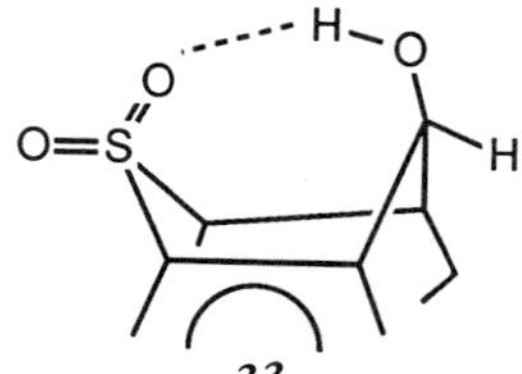

33

9-H and the C-9-OH with a coupling constant $^3J_{H,OH}$ = 8.0 Hz in $CDCl_3$ solvent. The magnitude of the coupling is consistent with a transoid or antiperiplanar arrangement of the H—C-9—O—H bonds and the apparent resistence to rapid exchange characterizes a relatively strong intramolecular hydrogen bond.* By contrast, 9-trimethylsiloxy- and 9-acetoxythioxanthene 10,10-dioxide prefer the 9e′ conformation as revealed from ^{1}H-NMR chemical shift data. Presumably, the 9e′ preference is controlled by the magnitude of the 1,4-synaxial oxygen–oxygen repulsive interactions which are probably coulombic in origin.

The ^{1}H-NMR spectral data for 9-alkylthioxanthene 10,10-dioxides is consistent with a C-9a′ conformation for the alkyl group. Confirmation of these assignments is dramatically demonstrated by comparison of the ^{1}H-NMR chemical shifts of 9-ethyl-T and the 10,10-dioxide as well as 9-isopropyl-T and its 10,10-dioxide. From an examination of the ^{1}H-NMR shifts in Table 6-1, it is clear that the presence of the 10a′ sulfonyl oxygen induces a downfield shift on all of the hydrogens attached to the 9a′ alkyl group (compared to similar hydrogens in 9-alkyl-T where alkyl = ethyl and isopropyl. In the solid state, C-9 methyl and isopropyl groups prefer the C-9a′ conformation in 2,4-dimethylthioxanthene 10,10-dioxide where the folding angle between the planes of the two benzo rings are 142.9° and 134.3°, respectively.[76] The folding angle is smaller when the size of the C-9 substituent is increased from methyl to isopropyl and this apparently results from an increase in the nonbonded interactions between the 9-isopropyl substituent and the thioxanthenyl ring system. The X-ray structure of 1,4-dimethyl-9-isobutylthioxanthene 10,10-dioxide[77] exhibits a 9a′ isobutyl group.

6.2.5.3 9-Arylthioxanthene 10,10-Dioxides

In contrast to the diastereomeric sulfoxides where two isomers give rise to four possible conformers, only two conformational isomers are possible for 9-arylthioxanthene 10,10-dioxides. In conformer **34e′**, the ^{1}H-NMR resonances for H-1, H-8 should be shifted upfield because of the magnetic shielding anisotropic effect of the 9e′ aryl group. On the other hand, in conformer **34a′**, the 2′,6′ hydrogens should experience substantial shielding caused by the anisotropy of the

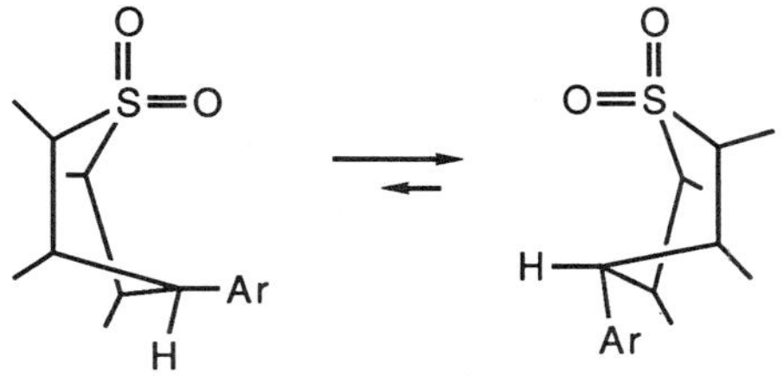

34a′ **34e′**

* Thioxanthene-9-ol 10,10-dioxide also displays a moderately intense absorption at 3508 cm^{-1} (1.3 × 10^{-5} M, CCl_4). See Ref. 46.

thioxanthenyl ring. Based on ^{1}H-NMR data, all the 9-phenylthioxanthene sulfones of the type represented above exist predominantly with the e′ phenyl group (**34e**).[78] This preference for the 9e′ conformation in the sulfones is in contrast with the 9a′ conformation exhibited for 9-phenylthioxanthene and probably reflects both steric and coulombic repulsive interactions between the 1,4-synaxial sulfonyl oxygen and the 9a′ phenyl group.

6.2.5.4 Thioxanthene 10,10-Dioxide Iron Complexes

The use of C-9 substituents as conformational probes for stereochemical assignments is not so straightforward. For example, it is a bit surprising that η^6-thioxanthene-η^5-cyclopentadienyl iron hexafluorophosphonate (**35**) displays isochronous or equivalent C-9 methylene absorptions [δ 4.56 in $(CD_3)_2C=O$] in the ^{1}H-NMR.[79] This is in contrast to the ^{1}H-NMR spectra of 9,9-dimethyl-T-(**36**) and 9,9-dimethyl-T 10,10-dioxide (**37**) cyclopentadienyl iron complexes where

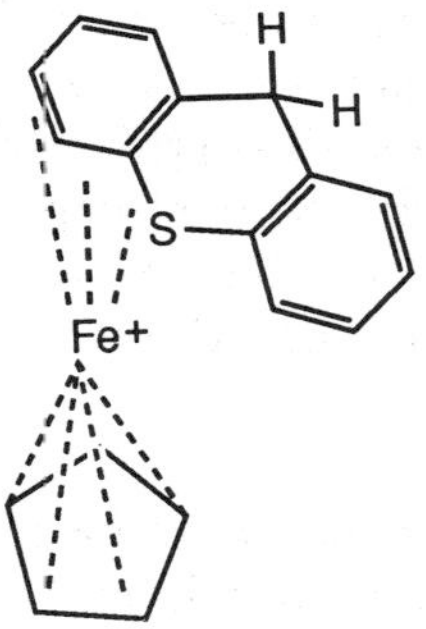

35

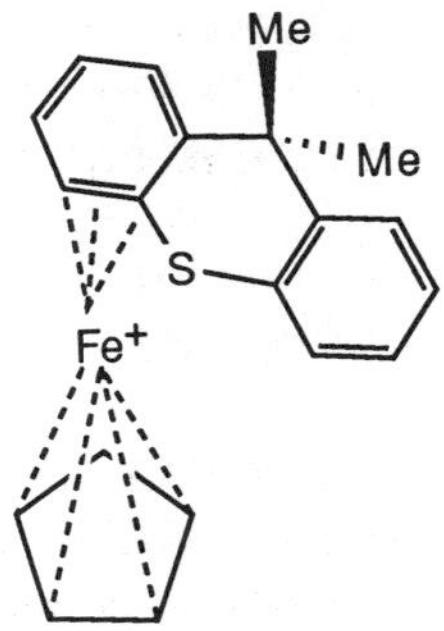

36

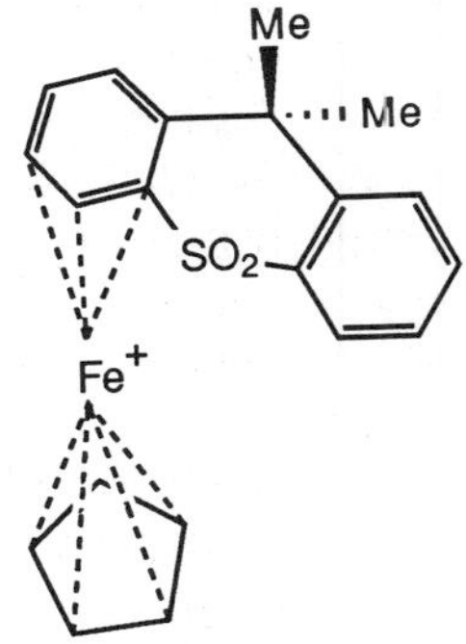

37

the C-9 dimethyl group is highly diastereotopic.[79] It seems apparent that the thioxanthenyl ring in **35** undergoes facile boat-to-boat conformational interchange or the methylene hydrogens are accidentally chemical shift coincident.

6.2.6 10-Alkyl-9-Arylthioxanthenium Sulfonium Salts

As previously discussed, the assignment of a preferred conformation of the 9 aryl group in thioxanthene is determined from (1) the magnitude of the anistropic shielding caused by the 9 aryl group in the 9e′ position on the proximal peri hydrogens (H-1, H-8) and (2) the magnitude of the shielding caused by the thioxanthene ring on the H-2′,H-6′ hydrogens of the C-9a′ aryl group. The conformational identity of the 10 alkyl groups are determined by the expected shielding difference within the 10a′ and 10e′ conformational arrays. The 10 a alkyl group is expected to be more shielded than the 10e′ due to the magnetic anisotropy of the thioxanthenyl ring.

The S-methyl and S-ethyl groups attached to 9-phenylthioxanthene (**28**) are conformationally demanding and prefer the 10a′ array. For example, the cis diastereomers favor the 9a′,10a′ conformer while the trans isomers prefer conformer, 9e′,10a′ (Scheme 5). In fact, an X-ray structural analysis of the cis

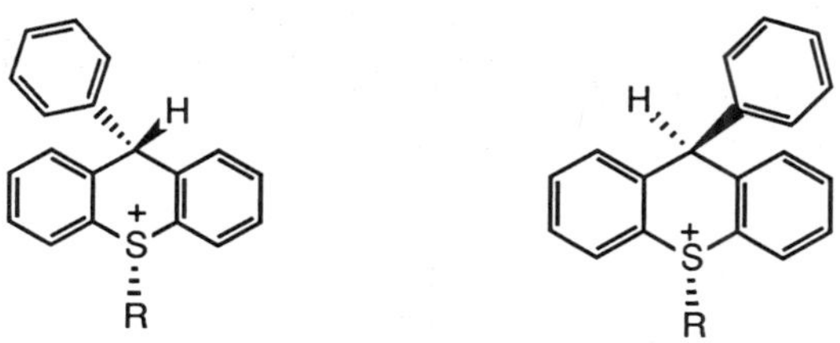

cis (9a',10a') R = Me, Et **trans** (9e',10a')

Scheme 5

isomer (R = Me) reveals that both the C-9 phenyl and S-10 methyl groups occupy pseudoaxial positions where the phenyl ring is perpendicular to the meso imaginary axis. The six-membered heterocyclic ring has a shallow-boat conformation with a dihedral angle between the planes of the two benzene rings of 139.5°.[80]

The interplay between *two* C-9 substituents (i.e., alkyl and aryl) projects an enhanced 9a′ conformational preference for the phenyl group which supercedes the S-Me preference for the 10a′ orientation. The balance between the relief of strain energy arising from van der Waals interactions between the C-9 geminal substituents is clearly nullified by the S-ethyl group since its preference for the 10a′ orientation is clearly dominant.[81]

It is interesting that treatment of thioxanthylium perchlorate[82] with KCN in $H_2O \cdot CH_2Cl_2$ affords 9-cyanothioxanthene[83] in 95%, which is methylated with MeI in the presence of $AgClO_4$ or $AgBF_4$ to give 9-cyano-10-methylthioxanthenium perchlorate or tetrafluoroborate (**38**). It is reported that both the perchlorate and tetrafluoroborate salts exist as the cis isomer having the 9e′10e′ conformation (i.e., **38**), which must arise from a kinetically controlled alkylation from the 10e′ side. This also implies that the 9-cyano group in 9-cyanothioxanthene probably prefers the 9e′ orientation. Finally, in trifluroacetic acid solution, some

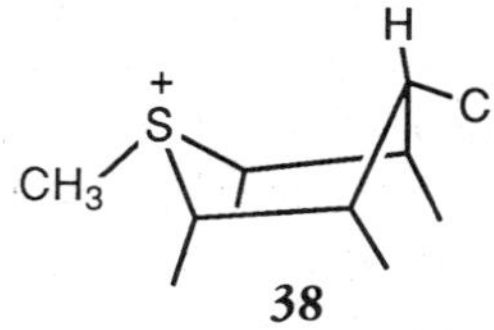

38

$X = ClO_4^-$, BF_4^-

equilibration of the cis isomers occurs to afford a cis:trans ratio of 1.55;[84] however, the mechanism of this important equilibration is not discussed.

An interesting reaction involving the rearrangement of a 10-aryl-thioxanthenium salt having conformational and stereochemical implications has been reported.[85] Maryanoff et al.[85] have demonstrated (Scheme 6) that deprotonation

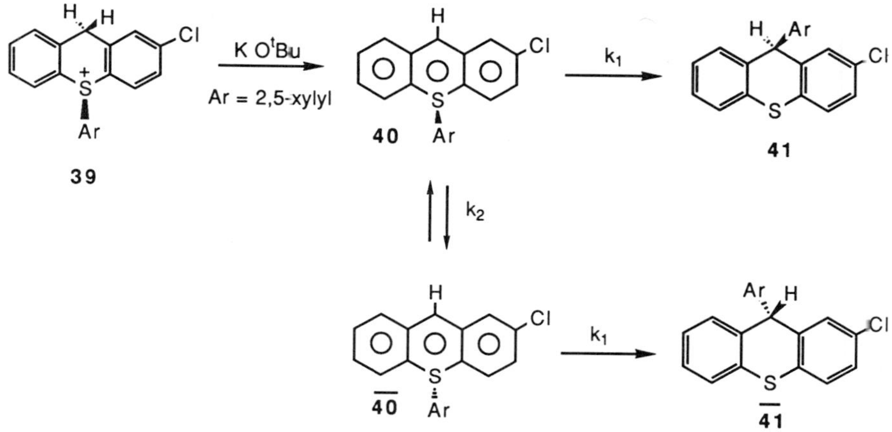

Scheme 6

of optically active 2-chloro-10-(2,5-xylyl)-10-thioxanthenium perchlorate (**39**), resolved via the (+)-camphor-10-sulfonate salt, afforded 2-chloro-10-(2,5-xylyl)-10-thiaanthracene (**40**), which rearranged to enantomerically enriched 2-chloro-9-(2,5-xylyl)-10-thioxanthene (**41**) (7% ee by ^{1}H-NMR). The first-order rate constant for racemization (**40** $\rightleftharpoons$ $\overline{\mathbf{40}}$) is estimated as $k_{rac} = 1.5\text{–}3.5 \times 10^{-5}\ s^{-1}$ (-15 °C) with $\Delta G^{\ddagger} = 19.1\text{–}19.5$ kcal/mol. Similarly, conversion of **40** to **41** gives a $k_1 = 1.3 \times 10^{-5}\ s^{-1}$ (-18 °C) with $\Delta G^{\ddagger} = 20.7 \pm 0.1$ kcal/mol.[85] The results of this study obviously indicate that the rearrangement proceeds through the 10a′ conformer of **39**; however, it also demonstrates the pyramidal lability of a tricoordinate and "neutral" tetravalent sulfur species in the thioxanthenes.

6.2.7 Assignment of Sulfinyl Oxygen Stereochemistry Employing Aromatic Solvent-Induced Shifts (ASIS)

Evans and Ternay[71] have demonstrated that the peri hydrogens (H-4,5) adjacent to the sulfinyl sulfur of thioxanthene are deshielded relative to the remaining aryl protons. This deshielding effect is considered to result from the combined inductive and anisotropic effect of the sulfinyl group. Interestingly, equatorial sulfinyl groups form dynamic complexes with benzene solvent in such a way as to deshield the peri hydrogens while an axial sulfinyl encourages complexation which ultimately shields the peri hydrogens. Evaluation of this effect on axial and equatorial sulfinyl oxygen model compounds provide definitive support for the usefulness of this stereochemical probe. As a consequence, the preference for the axial or equatorial conformation of diastereomeric sulfinyl groups can be adequately evaluated. In all cases, the Δ–δ values (ASIS = δ $CDCl_3$-δ C_6H_6) ranged from −0.20 ppm for equatorial sulfinyl oxygens to +0.22 ppm for axial sulfinyl oxygens.

6.2.8 N-Tosylsulfilimines

The *N*-tosylsulfilimines possess unique structural characteristics having a second-row heteroatom directly bound to sulfenyl sulfur and as a consequence they provide interesting structural analogies with sulfinyl groups. In addition, the facile base-catalyzed rearrangement of thioxanthene-*N*-(*p*-toluenesulfonyl)sulfilimines to 9-(*N*-*p*-toluenesulfonamido)thioxanthenes has generated considerable mechanistic interest (Scheme 7).[86]

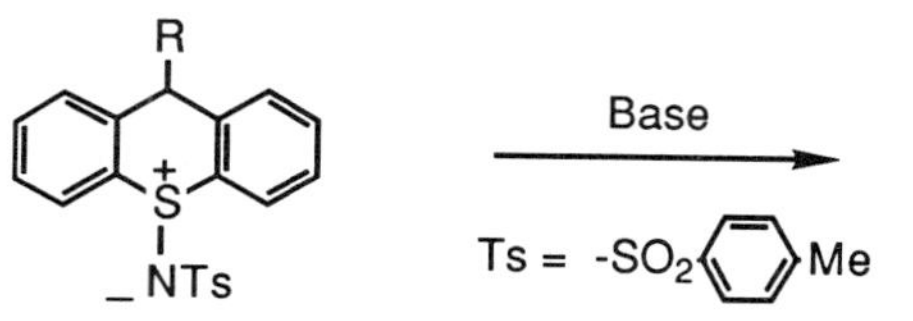

Scheme 7

As previously discussed, thioxanthene 10-oxide (**14**), 2-chlorothioxanthene 10-oxide, and related thioxanthene 10-oxides which lack a substituent at both C-4 and C-5 have preferred conformations in weakly interacting solvents (e.g., $CDCl_3$, C_6D_6) where the sulfinyl oxygen is pseudoequatorial (e′). This preference for the 10e′ conformation can be synthetically compromised by placement of a methyl group in the C-4 position. Consequently, attainment of the axial conformation is encouraged through relief of the nonbonding interactions between the sulfinyl oxygen or the S^+NTs and the peri methyl group in 1,4-dimethyl- and 2,4-dimethylthioxanthenes. If either the SO or SN group is equatorial, the equatorial C-9 hydrogen appears at lower field than the axial C-9 hydrogen mainly because the former is in the deshielding region of the thioxanthenyl moiety, and if the heteroatom attached to sulfenyl sulfur is axial, the axial C-9 hydrogen is shifted to lower field due to deshielding arising from the polarized functional group.

Treatment of thioxanthene (**6**) with chloramine-T affords thioxanthene *N*-*p*-toluenesulfonyl sulfilimine (**42**) having the nitrogen substituent group in the 10e′ conformation.[86a] The C-9 methylene group displays an AB quartet in the ^{1}H-NMR at δ 4.32 and 3.88 (J_{AB} = 17 Hz), where the upfield doublet is broader due to "allylic" coupling with the peri hydrogens at C-1 and C-8. These NMR shift trends and NMR signal coupling patterns parallel those observed for thioxanthene 10-oxide,[41a] implying that both S-10 substituents occupy the e′ conformation. The 1,4-dimethylthioxanthene *N*-tosylsulfilimine (**43**) is also prepared in 38% yield by reaction of 1,4-dimethylthioxanthene with chloramine-

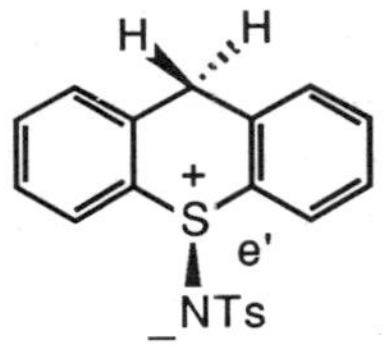

42

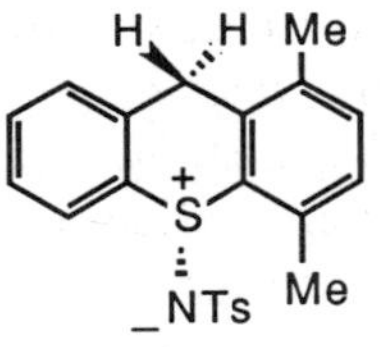

43

T.[87] Here, the C-9 AB quartet exhibits ^{1}H-NMR shifts at δ 4.10 and 4.58 (J_{AB} = 18 Hz) where the *low* field doublet (δ 4.58 ppm) exhibits broadening due to "allylic" coupling with the single CH-8 aryl hydrogen. This finding is consistent with an axial NTs group, which is in good agreement with that observed in the corresponding 2,4-dimethylthioxanthene derivative.[86c] Undoubtedly, the NTs group prefers the axial conformation to avoid the repulsive interactions between this functional group and the 4-methyl group.

The 9-alkylthioxanthene-*N*-(*p*-toluenesulfonyl)sulfilimines are particularly interesting since they afford a measure of (1) the influence of the 9-alkyl group on the diastereoselectivity of the reaction, (2) insight on the relative thermodynamic stability of the diastereomeric pair, and (3) the role of the C-9 alkyl group in controlling the conformational preferences of the C-9 and S-10 substituents. Two synthetic routes to the 9-alkylthioxanthene-*N*-(*p*-toluenesufonyl)sulfilimines have been described: (1) *method A* requires tosylation of the 10-aminothioxanthenium mesitylene sulfonates which are prepared by reaction of the thioxanthenes with *O*-mesitylene sulfonyl hydroxylamine[88] and (2) *method B* describes the reaction of the 9-alkylthioxanthene with chloramine-T.[89]

The reaction of 9-methyl-T with 1 equiv of chloramine T gives a mixture of two isomeric sulfilimines (**44**) in a cis:trans ratio of 1:5 (Scheme 8).

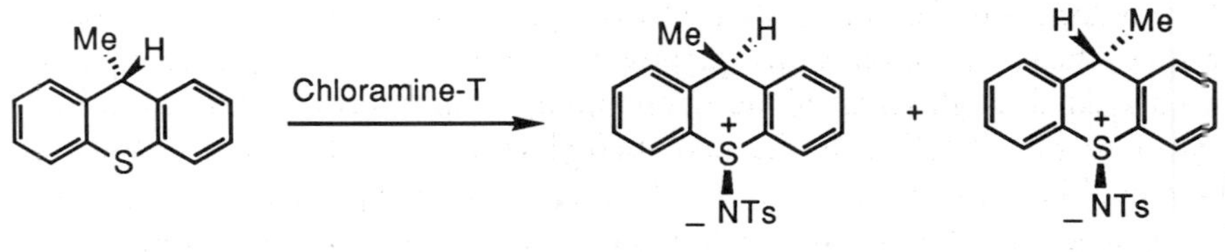

cis-**44** (9e′,10e′) *trans*-**44** (9a′,10e′)

Scheme 8

Similar reactions with 9-ethyl- and 9-isopropyl-T gave only the trans sulfilimine isomers, respectively (i.e., **45** and **46**).[89] Ternay and co-workers[57,71]

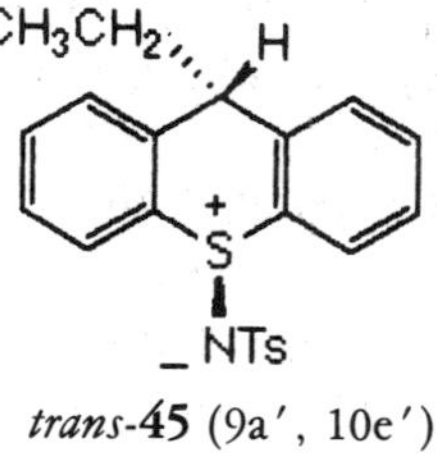

trans-**45** (9a′, 10e′)

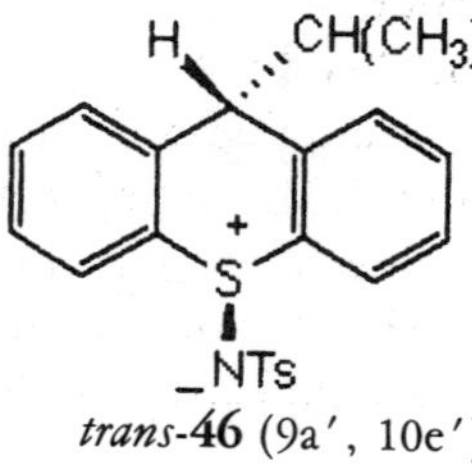

trans-**46** (9a′, 10e′)

had previously demonstrated that the sulfinyl group prefers the e′ conformation and tends to govern the conformation of the stereoisomers in 9-methyl-T. However, larger C-9 alkyl groups clearly prefer the a′ conformation and this preference tends to control the conformations of the diastereomers. This generalization has found merit and some validity in the sulfilimines. Thus, the the C-9 methylthioxanthene-*N*-(*p*-toluenesulfonyl)sulfilimine prefers the 9e′10e′ conformer in *cis*-**44** and the 9a′,10e′ conformer in *trans*-**44** (Scheme 8). For C-9 ethyl and isopropyl groups, the trans isomer (i.e., **45** and **46**) favors the 9a′10e′ conformer and the cis isomer opts for the 9a′10a′ conformer as the most stable array for *cis*-**45**.

It is particularly interesting to note that when refluxed in benzene solvent for 10 h, both *cis*-**44** and *trans*-**44** gave separately an equilibrium mixture consisting of cis:trans = 1:2, indicating that the C-9a′ methyl is preferred by 0.45 kcal/mol. *cis*-9-Ethyl-T sulfilimine under similar conditions isomerized completely to *trans*-9-ethyl-T diastereomer **45** suggesting that the driving force for complete isomerization to the trans isomer is derived from the relief of steric repulsions between the 1,4-diaxial substituents in the cis isomer. As expected, *trans*-9-isopropyl-T sulfilimine (**46**) (9a′,10e′) is both conformationally and configurationally stable and resistant to isomerization to the cis isomer.

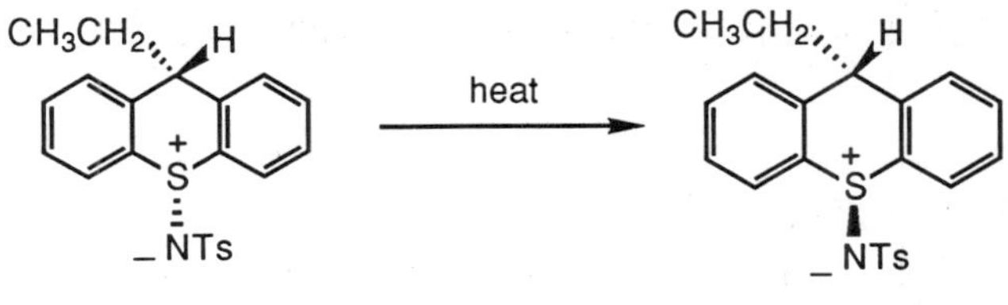

cis-**45** (9a′,10a′) *trans*-**45** (9a′,10e′)

The conformational preferences within the cis and trans diastereomers (**47**) of 9-phenylthioxanthene-*N*-(*p*-toluenesulfonyl)sulfilimine are not particularly surprising although their conformational and stereochemical assignments are insightful.[86b] From ^{1}H-NMR data, *cis*-**47** is assigned the 9e′10e′ conformer while

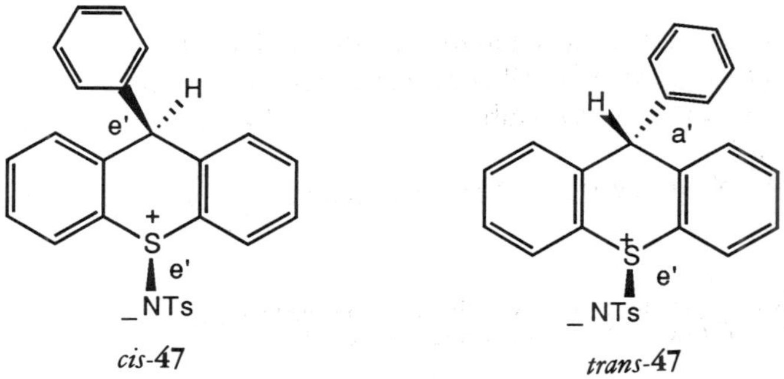

cis-**47** *trans*-**47**

trans-**47** is assigned the 9a′10e′ conformer. This assignment is accomplished by observing the shielding influence of the C-9 aryl ring anisotropy has on the C-1 and C-8 hydrogens. The upfield shift of these hydrogens is consistent with a C-9e′ phenyl group existing essentially perpendicular to the plane of the thioxanthene ring and bisecting the C-9 meso atom. When *cis*- or *trans*-sulfilimine **47** is refluxed in benzene for 10 h, an equilibrium mixture consisting of *cis:trans*-**47** of 1:3 is obtained. This preference is consistent with the notion that the $SNSO_2C_6H_4Me$ group favors the equatorial position whereas the C-9 phenyl prefers the axial position. These conclusions are based on results that are consistent with the findings of Hori and co-workers[90] who investigated the conformational preferences of 9-arylthioxanthene 10-oxides by NMR spectroscopy.

As one might anticipate, the presence of a C-4 methyl group on the thioxanthene skeleton peri to the substituent attached to sulfur should cause severe steric crowding resulting in a substituent preference for the a′ conformation. Interestingly, reaction of 9-methyl- and 9-ethyl-2,4-dimethylthioxanthenes with 2 equiv of chloramine T affords exclusively *trans*-2,4,9-trimethylthioxanthene-*N*-(*p*-toluenesulfonyl)sulfilimine* (*trans*-**48**)[91] and *trans*-9-ethyl-2,4-dimethylthioxanthene-*N*-(*p*-toluene sulfonyl)sulfilimine (*trans*-**49**) in 79 and 67% yields, respectively.[86c] Kinetically, the trans isomers are preferred, but thermodynamically the cis isomers are the most populous and can be obtained by thermal isomerization of the trans diastereomers. The complete equilibration of *trans*-**49** (R = Et) to *cis*-**49** is consistent with the need to relieve the steric interference

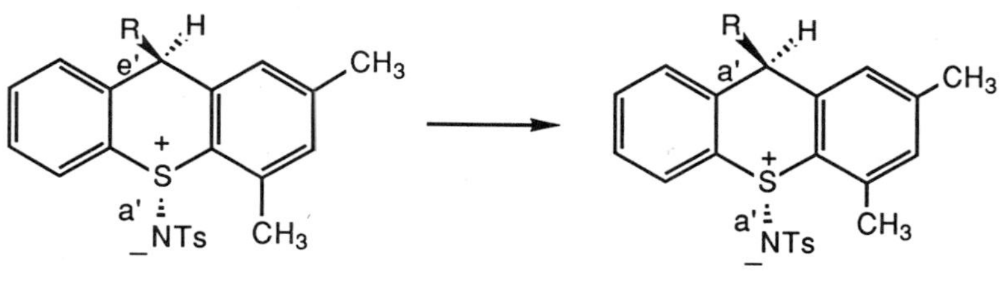

trans-**48**	R = CH_3	*cis*-**48**
trans-**49**	R = CH_2CH_3	*cis*-**49**

between the equatorial alkyl group and the peri hydrogens at C-1 and C-8. Similarly, the 1,4-dimethyl-9-alkylthioxanthene-*N*-tosylsulfilimines (alkyl = Me, Et) are thermodynamically stable as the cis diastereomer in the 9a′,10a′ conformation.[87]

The ^{13}C-NMR spectra of these diastereomers are particularly instructive.[86c]

---

* The X-ray crystal structure reveals that the trans diastereomer prefers the 9e′, 10a′ conformer in the solid state with a dihedral angle of 130.2°.

For example, the ^{13}C-NMR shifts attributable to C-9 in the trans isomers are 5–8 ppm to higher field than C-9 in the cis isomers. It seems unlikely that steric compression between the equatorial C-9 alkyl group and the peri hydrogens at C-1 and C-8 is responsible for this shift. On the other hand, the proximity of the SNTs group may impact on the shielding of C-9 when the alkyl group occupies the e′ conformation.

6.2.9 Thioxanthenium Bis(carboalkoxy) Methylides

The potential for base-promoted rearrangement of thioxanthene sulfonium methylides to 9-bis(methoxycarbonyl)methyl thioxanthenes have electronic similarities with the reactions involving the thioxanthene *N*-tosylsulfilimines. As a result, the conformational demands of the bis(methoxycarbonyl)methyl group have significant mechanistic implications (Scheme 9).

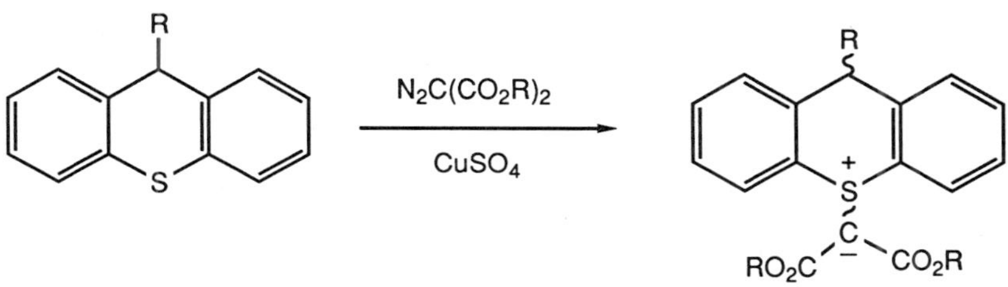

Scheme 9

Generally, the thioxanthenium bis(carbomethoxy)- and bis(carboethoxy)-methylides are prepared by sulfenyl sulfur "trapping" of a carbene or carbenoid species derived from the copper(II)-catalyzed decomposition of dimethyl- or diethyl diazomalonate[92] (Scheme 9). Specifically, both carboethoxy and carbomethoxy methylides of *un*substituted thioxanthenes exhibit their malonyl residues in the 10e′ conformation as determined by ^{1}H-NMR spectral characteristics of the C-9 CH_2 diastereotopic hydrogens.[93,94] For example, the C-9 AB quartet absorbs at δ 3.91 and 4.31 ppm (J_{AB} = 18 Hz) for thioxanthenium bis(carbomethoxy)methylide (**50**).[93] The more shielded doublet (9a′ hydrogen) exhibits a larger $W_{1/2}$ due to enhanced "allylic coupling" with H-1 and H-8 of the aryl ring. As corroborative evidence, the malonylide groups in 1,4-[93] and 2,4-dimethylthioxanthenium bis(carbomethoxy)methylide[94] (**51**, **52**) are expected, by virtue of steric perturbations caused by the C-4 CH_3, to favor the 10a′ conformation. In this orientation the C-9a′ hydrogen experiences significant deshielding relative to the C-9e′ hydrogen and the broadened doublet now appears downfield. These e′ conformational assignments are confirmed for the *C-9 unsubstituted* thioxanthenium bis(carbomethyl)methylides from X-ray analysis data.[95] Here, the crystallographic data for 2-chlorothioxanthenium (**53**), thioxanthenium (**50**), and thioxanthonium bis(carbomethoxy)methylides (**54**) indicate that the malonylide fragments are e′ planar, and oriented with their π systems perpendicular to the lone pair on the sulfonium sulfur. This orientation of the malonylide moiety brings two oxygen atoms of the anion in close proximity to the positively charged sulfur atom. The authors suggest that coulombic stabilization in this orientation may have reasonable merit.

Perhaps, the most compelling evidence for the suggestion of coulombic attraction arises from a determination of the kinetic parameters for carbomethoxy interchange in 2-chlorothioxanthonium bis(carbomethoxy)methylide (**55**). At 30

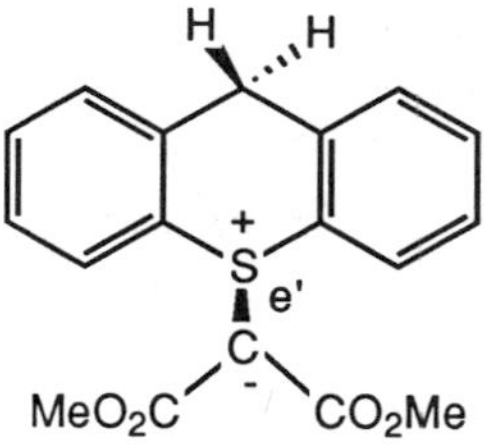

50

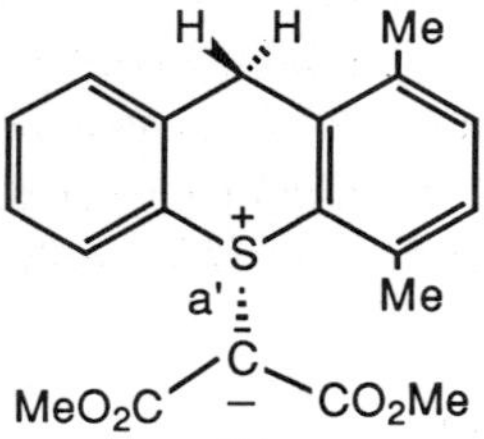

51

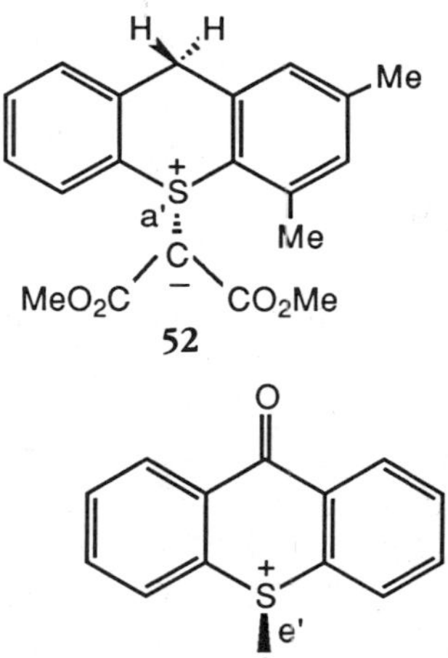

52

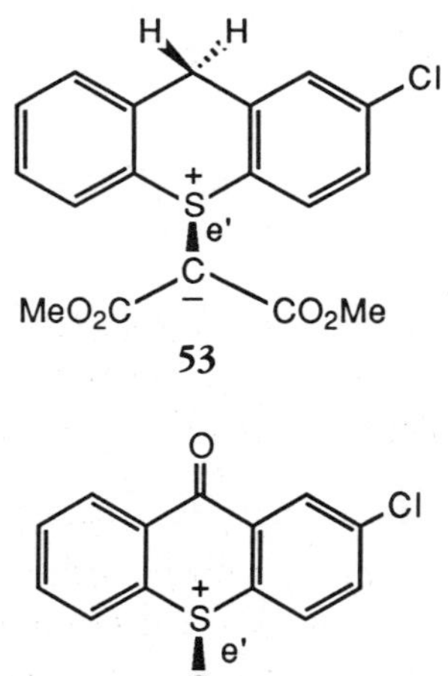

53

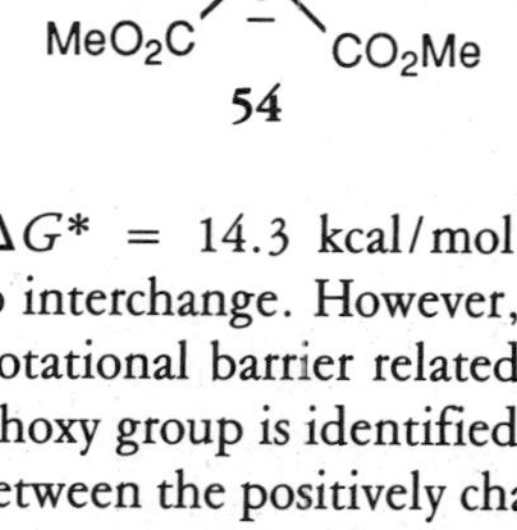

54

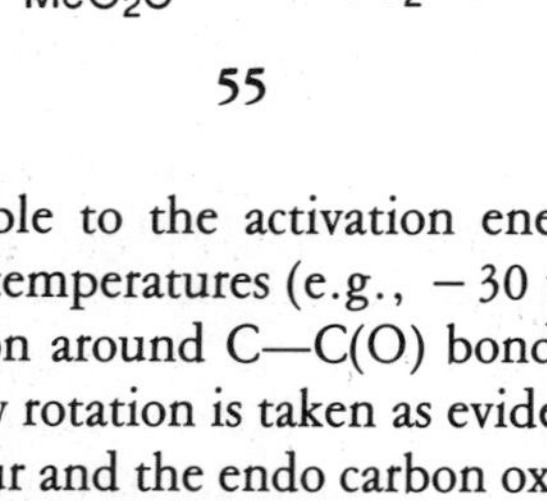

55

°C in C_6D_6, $\Delta G^* = 14.3$ kcal/mol attributable to the activation energy for methoxy group interchange. However, at lower temperatures (e.g., -30 to -90 °C) a second rotational barrier related to rotation around C—C(O) bond in the endo carbomethoxy group is identified. This slow rotation is taken as evidence for an attraction between the positively charged sulfur and the endo carbon oxygen.[95]

The analog, thianthrenium bis(carbethoxy)methylide (**56**), has been prepared by reacting thianthrene with diethyl diazomalonate with Cu(II) catalysis.[96] An X-ray analysis reveals that the central ring is a boat conformation (dihedral angle = 135.7°) with the methylide carbon in the e′ conformation. Since the angle of fold for **50** (138°) is similar to that found in **56**, it appears that repulsive 1,4-transannular heteroatom–heteroatom interactions are not significant enough to impact on the ring deformation.

On the other hand, the X-ray structural data for 9-hydroxy-1,4-dimethylthioxanthenium bis(carbomethoxy)methylide (**57**) indicates that the two phenyl

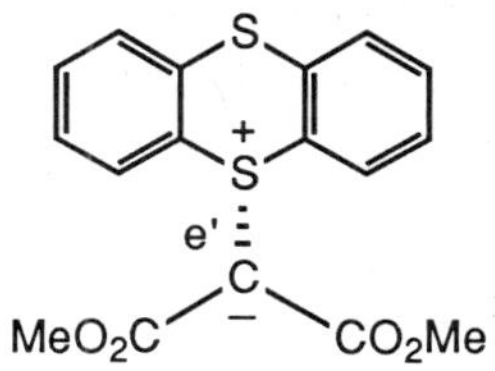

56

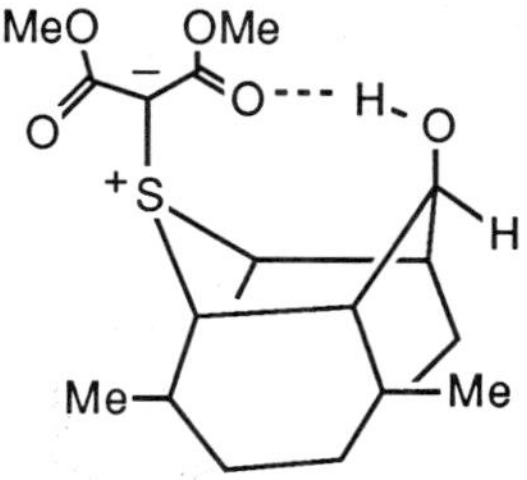

57

rings exhibit an interplanar angle of 158.7(2)°.[97] The six-membered ring containing the sulfur atom is in a boat conformation with the C-9 OH and the S-10 malonyl group occupying the axial sites. The carbanion moiety is planar and occupies the axial site because the methyl group at C-4 forces the carbanion into the sterically less-hindered axial conformation. An intramolecular hydrogen bond involving the hydroxyl group at C-9a′ and the carbonyl oxygen further stabilizes this particular conformation.[97]

As an aside, an X-ray crystal structure study of 1,4-oxathianium bis(carbomethoxy)methylide (**58**) has also been performed.[97] As perhaps expected, the six-

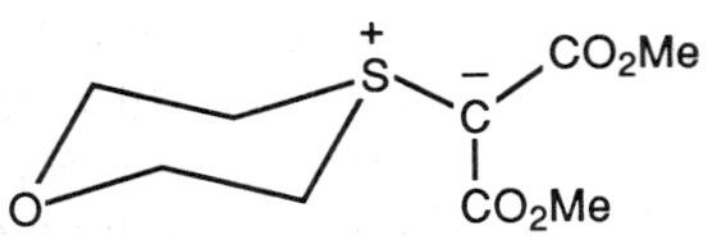

58

membered ring exhibits a chair conformation with the carbanion occupying an equatorial site.

In principle, the 9-alkylthioxanthene bis(methoxycarbonyl)methylides can exist in two configurational and four conformational arrays. Consistent with expectations from the sulfinyl derivatives, the preferred conformations within each configurational isomer is determined by a combination of factors including:

1. The steric interference between the equatorial malonyl group and the peri hydrogens (H-4, H-5) and/or a C-4 methyl group.

2. The steric hindrance caused by the interactions between the equatorial C-9 alkyl group and the peri hydrogens (H-1, H-8) and/or a C-1 methyl group

3. The nonbonding repulsive or electrostatic attractive interactions between two 1,4-diaxial (S-10a′, 9a′) substituents.

Reactions of 9-methyl-, 9-ethyl, 9-isopropylthioxanthenes and 9-ethyl-2,4-dimethylthioxanthene with dimethyl diazomalonate and $CuSO_4$ as a catalyst[92] affords exclusively the trans diastereomers of the requisite sulfonium ylides, all having the preferred 9a′,10e′ conformation (Scheme 10).[94] Here, the the C-9

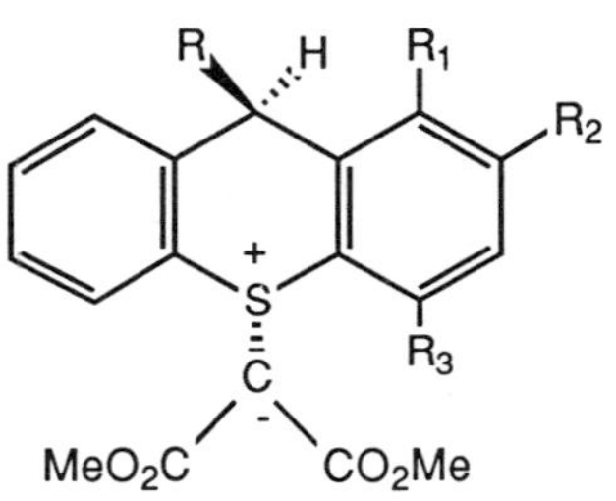

$R_1 = R_2 = R_3 = H$; R = Me, Et, i-Pr

$R_2 = R_3 = Me$; R = Me, Et

$R_1 = R_3 = Me$; R = Me, Et

Scheme 10

alkyl group clearly dictates conformational control, forcing the malonyl group into the e′ position despite the apparent repulsive interactions caused by the C-4 methyl group. This is further dramatized in the 9-alkyl (R = Me, Et, i-Pr), 2,4-dimethyl-, and 9-alkyl (R = Me, Et)1,4-dimethylthioxanthenium bis(carbomethoxy)methylides which afford only the trans diastereomers in the 9a′,10e′ array where the bis(methoxycarbonyl)methylide group is equatorial (Scheme 10).[87] Such a preference for the equatorial position for the $SC(CO_2Me)_2$ group has been rationalized by Ternay et al.[95,96] In the 9-alkyl-T methylides, the 9a′10a′ cis conformer is severely destabilized either by steric interactions between one of the ester groups and the axial C-9 substituent or by electron–electron repulsion between the sulfur lone pair and the *p*-orbital of the methylide carbon. Tamura et al.[87] comment that the longer S—C bond length (~1.74 Å) compared to that for S—O (~1.50 Å) and the S—N (~1.64 Å) may also contribute to the decrease in the unfavorable interaction between the equatorial $SC(CO_2Me)_2$ and the C-4 methyl group.

Finally, from ^{13}C-NMR data, it is notable that thioxanthonium bis(carbomethoxyl)methylide (**54**) prefers the 10e′ orientation while thioxanthonium bis(carbomethoxy)methylide 10-oxide (**59**) prefers to have its sulfoxium oxygen in the 10e′ array complemented by the 10a′ bis(carbomethoxy) group.[98]

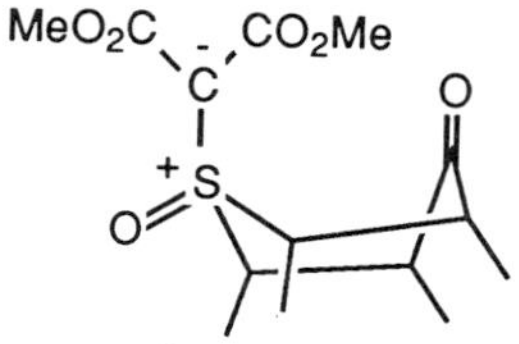

59

6.2.10 Sulfonium Ylides from 9-Alkylidenethioxanthenes

Numerous sulfonium ylides have been prepared by the reaction of 9-alkylidene thioxanthenes with dimethyl diazomalonate with anhydrous copper(II) sulfate as catalyst (Scheme 11).[92] The ^{13}C-NMR chemical shifts characterizing the

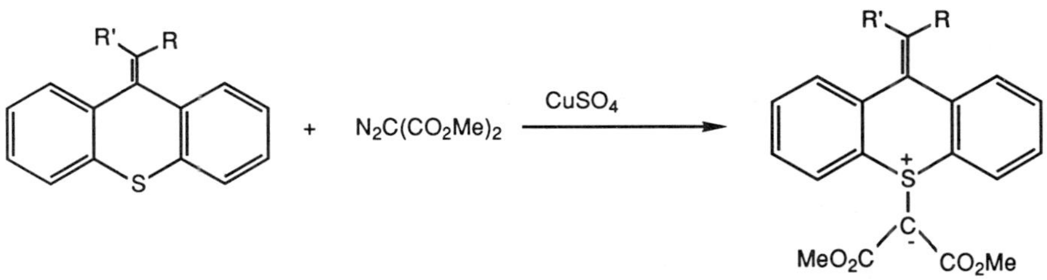

Scheme 11

methylide carbons of the malonylide fragments derived from thioxanthenes have been used to assign the a′ or e′ orientation of the malonylide fragment in the 9-alkylidene thioxanthenium ylides. Generally, the ^{13}C-NMR shifts of the methylide carbons for e′ malonylides occur near δ 50 ppm while the a′ malonylide fragments exhibit their methylide resonances near δ 60 ppm. The 9-alkylidene thioxanthenium malonylides have methylide resonances near δ 50–55, which lead to the 10e′ conformational assignments.[99] In fact, the high field shift for the methylide carbon of the e′ malonylide fragment suggest a sensitivity for steric compression.[100]

Perhaps one of the most intriguing features of this work[99] is the unraveling of the mystery surrounding the influence of substituent effects on the ^{1}H-NMR chemical shift equivalence/nonequivalence of the methoxy resonances in several 9-alkylidene thioxanthenium bis(carbomethoxy)methylides. For example, when

C-9 is attached as methylidene, ethylidene, or propylidene, the two methoxy groups are isochronous, appearing as a sharp singlet at δ 3.6 ppm. However, those ylides possessing an isopropylidene fragment at C-9 exhibit two methoxy resonances in the ^{1}H-NMR. In fact, the ^{13}C-NMR spectrum of the C—O resonances at δ 165 and 168 ppm and the methyl carbons of the carbomethoxy groups at δ 50 and 51 ppm support this contention. The question is *why do the lower alkylidenes show relatively free rotation and chemical shift equivalence of the carbomethoxy groups while the isopropylidine group serves to apparently restrict free rotation?* Ternay and co-workers[99] proposed that rapid interchange of the carbomethoxy groups is promoted by ring inversion coincident with bond rotation (Scheme 12).

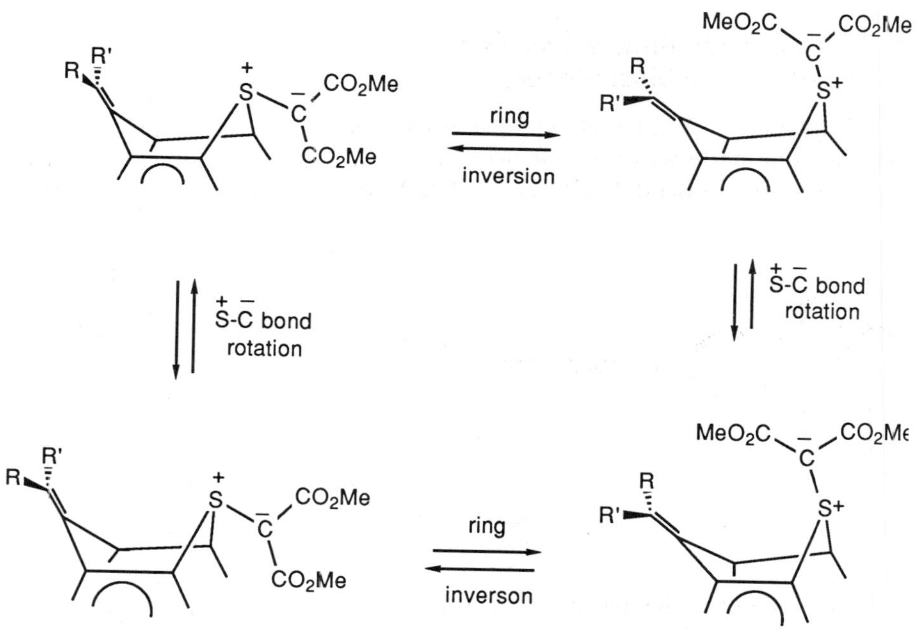

Scheme 12

Since direct interchange of carbomethoxy groups by rotation while occupying the thermodynamically preferred e′ orientation is disfavored by steric hindrance, ring inversion would place the bis(carbomethoxy) groups in the a′ orientation with considerably more rotational freedom. If ring inversion is discouraged, anisochronous carbomethoxy groups would result, as demonstrated by ^{1}H- and ^{13}C-NMR chemical shift nonequivalence of the isopropylidene derivatives. Consequently, the higher barrier to ring inversion in the isopropylidene examples results from steric repulsions between the methyl groups of the isopropylidene moiety and the hydrogens attached to C-1 and C-8. Rapid rotation

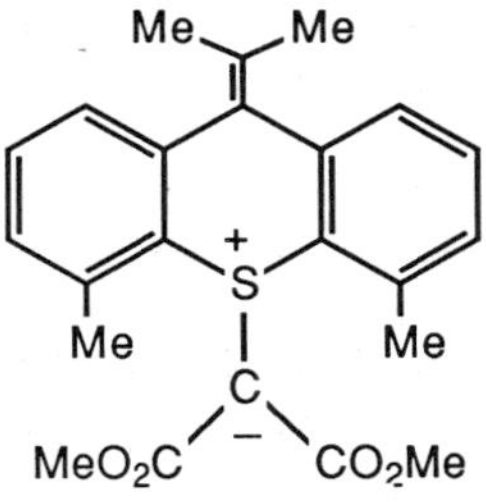

60

does, in fact, occur easily in the ylide derived from 4,5-dimethyl-9-isopropylidene thioxanthene (**60**). Here, the two methyl groups peri to the sulfur force the methylide carbon into the pseudo axial position where it experiences considerably more freedom to initiate facile C—S bond rotation.[99]

6.2.11 Summary

The various substitution patterns attending the thioxanthenyl skeleton influence both the stereochemical and conformational preferences of C-9 and S-10 substituents. Subtle changes in torsional angles are apparent from ^{1}H-NMR spectral data as well as X-ray crystallographic data. The general trends reflect a shallow-boat conformation for the central ring or a planar array depending on the substituents.

ACKNOWLEDGMENTS

Acknowledgment is made to the Ford Foundation administered through the National Research Council–National Academy of Sciences, National Science Foundation (CHE-8720270), and the University of North Carolina's Research Council for support of this effort.

REFERENCES

1. Kaiser, C.; Setler, P. E. In *Burger's Medicinal Chemistry*, Part III, 4th ed. M. E. Wolff, Ed.; Wiley Interscience: New York, 1981; Chapter 56, p. 859.
2. (a) Lipkowitz, K.; Burkett, A.; Landwer, J. *Heterocycles* **1986**, *24*, 2757. (b) Saratov, I. E.; Yakovler, I. P.; Reikhsfel'd, V. O.; Zakharov, V. I. *Zh. Obshch. Khim.* **1980**, *50*, 1090.
3. Aizenshtat, Z.; Klein, E.; Weiler-Feilchenfeld, H.; Bergmann, E. D. *Isr. J. Chem.* **1972**, *10*, 753.
4. For a pertinent example, see Jovanovic, M. V.; Biehl, E. R. *J. Heterocycl. Chem.* **1987**, *24*, 51.

5. (a) Rowe, I.; Poss, B. *Acta Crystallogr.* **1958**, *11*, 372. (b) Hosoya, S. *Acta Crystallogr.* **1963**, *16*, 310. (c) Hosoya, S. *Acta Crystallogr.* **1966**, *21*, 21. (d) Larson, S. B.; Simonsen, S. H. *Acta Crystallogr.*, **1984**, *40 Sect. C*, 103.
6. Aroney, M. J.; LeFevre, R. J. W.; Saxby, J. D. *J. Chem. Soc.* **1965**, 571.
7. (a) Chandra, A. K. *Tetrahedron* **1963**, *19*, 471. (b) Lansbury, P. T. *Acc. Chem. Res.* **1969**, *2*, 210.
8. Chickos, J.; Mislow, K. *J. Am. Chem. Soc.* **1967**, *89*, 4815.
9. Jorgensen, K. A. *Tetrahedron* **1986**, *42*, 3707.
10. Bell, J. D.; Blount, J. F.; Briscoe, O. V.; Freeman, H. C. *Chem. Commun.* **1968**, 1656.
11. McDowell, J. J. H. *Acta Crystallogr., Sect. B.* **1976**, *32*, 5.
12. Caldwell, S. R.; Turley, J. C.; Martin, G. E. *J. Heterocycl. Chem.* **1980**, *17*, 1145.
13. Ragg, R.; Fronza, G.; Mondelli, R.; Scapini, G. *J. Chem. Soc., Perkin Trans.* **1983**, 1289.
14. Canselier, J. P.; Cassoux, P. *J. Mol. Struct.* **1977**, *39*, 301.
15. Aroney, M. J.; Hoskins, G. M.; LeFevre, R. J. W. *J. Chem. Soc. B* **1969**, 980.
16. Kintzinger, J.-P.; Delseth, C.; Nguyen, T. T.-T. *Tetrahedron* **1980**, *36*, 3431.
17. Bel'skii, V. K.; Satatov, I. E.; Reikhsfel'd, V. O.; Simonenko, A. A. *J. Organomet. Chem.* **1983**, *258*, 283.
18. D'yachenko, O. A.; Sokolova, Yu. A.; Otormyan, L. O. *Zh. Strukt. Khim.* **1984**, *25*, 83. *Chem. Abstr.* **1985**, *102*, 2469e.
19. McCarthy, W. Z.; Corey, J. Y.; Corey, E. R. *Organometallics* **1984**, *3*, 255.
20. Ricci, A.; Pietropaolo, D.; Distefano, G.; Macciantelli, D.; Colonna, F. P. *J. Chem. Soc., Perkin Trans. 2* **1977**, 689.
21. Campbell, G. M.; LeFevre, C. G.; LeFevre, J. J. W.; Turner, E. E. *J. Chem. Soc.* **1938**, 404.
22. Ferrier, G.; Iball, J. *Chem. Ind. (London)* **1954**, 1296.
23. (a) Linton, H.; Cox, E. *J. Chem. Soc.* **1956**, 4886. (b) Rowe, I.; Post, B. *Acta Crystallogr.* **1958**, *11*, 372. (c) Wei, C. H. *Acta Crystallogr., Sect. B* **1971**, *27*, 1523. (d) Larson, J. B.; Simonsen, S. H.; Martin, G. E.; *Acta Crystallogr.* **1984**, *40c*, 103.
24. Gillean, J. A., III; Phelps, D W.; Cordes, A. W. *Acta. Crystallogr., Sect. B* **1973**, *29*, 2296-98.
25. Barfield, M.; Chakrabarti, B *J. Am. Chem. Soc.* **1969**, *91*, 4346.
26. (a) Barfield, M.; Grant, D. M. *J. Am. Chem. Soc.* **1961**, *83*, 4726. (b) Barfield, M.; Grant, D. M. *J. Am. Chem. Soc.* **1963**, *85*, 1899.
27. Karplus, M.; Anderson, D. H.; Farrar, T. C.; Gutowsky, H. S. *J. Chem. Phys.* **1957**, *27*, 597.
28. (a) Brown, C. M.; Bortner, M. *Acta Crystallogr.* **1954**, 7, 1. (b) Burns, D. M.; Iball, U.; *Proc. R. Soc. London Ser. A* **1955**, *227*, 200.
29. (a) Cookson, R. C.; Crabb, T. A.; Frankel, J. J.; Hudec, J. *Tetrahedron (Suppl. 7)* **1966**, 355. (b) Bothner-By, A. A. *Advances in Magnetic Resonance*, Vol. 1, Waugh, J. S., Ed.; Academic Press: New York, 1965.
30. Ternay, A. L., Jr.; Evans, S. A. *J. Org. Chem.* **1974**, *39*, 2941.
31. Chu, S. S. C.; Yang, H. T. *Acta. Crystallogr., Sect. B* **1976**, *32*, 2248.
32. Neidle, S. *Biochim. Biophys. Acta* **1976**, *454*, 207.
33. Longo, J.; Richardson, M. F. *Acta Crystallogr. Sect. B* **1982**, **38**, 2724.
34. Chu, S. S. C. *Acta Crystallogr. Sect. B* **1976**, *32*, 1583.
35. Wei, C. H.; Einstein, J. R. *Acta Crystallogr. Sect. B* **1978**, *34*, 205.
36. Ternay, A. L., Jr.; Cushman, J. A.; Harwood, J. S.; Yu, C. P. *J. Heterocycl. Chem.* **1987**, *24*, 1067.
37. Johnson, C. E.; Bovey, F. A. *J. Chem. Phys.* **1958**, *29*, 1012.
38. See, for example, Curtin, D. Y.; Carlson, C. G.; McCarty, C. G. *Can. J. Chem.* **1964**, *42*, 565.

39. Casper, M. L.; Casper, J. N. L.; Seiber, J. N.; Matsumoto, K. Abstr. 146th National Meeting of the American Chemical Society, Denver, CO, 1964, p. 30C.

40. Anteunis, M.; Tavernier, D.; Borremans, F. *Bull. Soc. Chim. Belg.* **1966**, *75*, 396.

41. (a) Ternay, A L., Jr.; Ens, L.; Herrmann, J.; Evans, S. *J. Org. Chem.* **1969**, *34*, 940. (b) Lansbury, P. T.; Bieron, J. F.; Lacher, A. J. *J. Am. Chem. Soc.* **1968**, *88*, 1482. (c) Colson, J. G.; Lansbury, P. T.; Saeva, F. D. *J. Am. Chem. Soc.* **1967**, *89*, 4987.

42. Nagata, W.; Terasawa, T.; Tori, K. *J. Am. Chem. Soc.* **1964**, *86*, 3746.

43. Ternay, A. L.. Jr.; Brinkmann, A.; Evans, S.; Hermann, *J. Chem. Commun.* **1969**, 654.

44. (a) Chu, S. S. C. *Acta Crystallogr., Sect. B* **1973**, *29*, 1690. (b) Chu, S. S. C.; Chung, B. *Acta Crystallogr. Sect. B* **1973**, *29*, 2253.

45. Ternay, A. L., Jr.; Chasar, D. W. *J. Org. Chem.* **1967**, *33*, 3814.

46. Chasar, D. W. Ph.D. Dissertation, Case Western Reserve University, Cleveland, OH, 1968.

47. Ivanov, G. E.; Turov, A. V.; Kornilov, M. Yu. *Ukr. Khim. Zh. (Russ. Ed.)* **1987**, *53*, 743; *Chem. Abstr.* **1988**, *108*, 130727b.

48. Ens, L. A. Ph.D. Dissertation, Case Western Reserve University, Cleveland, OH, 1969; *Diss. Abstr. Int. B.* **1970**, *30*, 4041.

49. Sharpless, N. E.; Bradley, R. B.; Ferretti, J. A. *Org. Magn. Reson.* **1974**, *6*, 115.

50. Williams, D. H.; Fleming, I. *Spectroscopic Methods in Organic Chemistry*, 4th ed.; McGraw-Hill: New York, 1987; pp. 114–120.

51. Chu, S. S. C.; Chung, B. *Acta. Crystallogr., Sect. B* **1974**, *30*, 235.

52. Chen, J. S.; Watson, W. H.; Austin, D.; Ternay, A. L., Jr. *J. Org. Chem.* **1979**, *44*, 1989.

53. Vazquez, S.; Castrillon, J. *Spectrochim. Acta Part A* **1974**, *30*, 2021.

54. Ternay, A. L , Jr.; Herrmann, J.; Hayes, B. A.; Joseph-Nathan, P. *J. Org. Chem.* **1980**, *45*, 189.

55. Ternay, A. L., Jr.; Herrmann, J.; Harris, M.; Hayes, R. R. *J. Org. Chem.* **1977**, *42*, 2010.

56. Ternay, A. L., Jr.; Cushman, J. A.; Harwood, J. S.; Yu, C. P. *J. Heterocycl. Chem.* **1987**, *24*, 1062.

57. Evans, S. A., Jr. Ph.D. Dissertation, Case Western Reserve University, Cleveland, OH, 1970; *Diss. Abstr. Int. B* **1970**, *31*, 3257.

58. Schindler, O.; Lehner, H.; Michaelis, W.; Schmutz, J. *Helv. Chim. Acta* **1963**, *46*, 1097.

59. (a) Michaelis, W.; Schindler, O.; Signer, R. *Helv. Chim Acta* **1966**, *49*, 42. (b) Schindler, O.; Michaelis, W.; Gauch, R. *Helv. Chim Acta* **1966**, *49*, 1483.

60. Ternay, A. L., Jr.; Chasar, D. W., Sax, M. *J. Org. Chem.* **1967**, *32*, 2465.

61. Chu, S. S. C.; Rosenstein, R. D., Ternay, A. L., Jr. *Acta. Crystallogr., Sect. B* **1979**, *35*, 2430.

62. Chu, S. S. C.; Grant, W. K.; Napolene, V.; Ternay, A. L., Jr.; Massah, F. *Acta Crystallogr, Sect. B* **1981**, *37*, 1948.

63. Jackobs, J.; Sundaralingam, M. L. *Acta Crystallogr., Sect. B.* **1969**, *25*, 2487.

64. Chu, S. S. C.; Book, L. *Acta Crystallogr., Sect. C* **1983**, *39*, 643.

65. Chu, S. S. C.; Napoleone, V.; Massah, F.; Ternay, A. L., Jr. *Acta. Crystallogr., Sect. B* **1981**, *37*, 775.

66. Chu, S. S. C.; Napoleone, V. *Acta Crystallogr., Sect. C* **1983**, *39*, 646.

67. Chu, S. S. C.; Napoleone, V. *Acta. Crystallogr., Sect. B* **1982**, *38*, 2288.

68. Chu, S. S. C.; Rosenstein, R. D. *Acta Crystallogr., Sect. B* **1980**, *36*, 989.

69. Chu, S. S. C.; *Acta Crystallogr., Sect. B* **1975**, *31*, 1082.

70. Book, L.; Chu, S. S. C.; Rosenstein, R. D. *Acta Crystallogr., Sect. C* **1983**, *39*, 648.

71. Evans, S. A.; Ternay, A. L., Jr. *J. Org. Chem.* **1975**, *40*, 2993.

72. Hori, M.; Kataoka, T.; Shimizu, H. *Chem. Lett.* **1974**, 1073.

73. Ohno, S.; Shimizu, H.; Kataoka, T.; Hori, M. *Chem. Pharm. Bull.* **1984**, *32*, 3471.

74. Chu, S. S. C.; Yang, H. T. *Acta Crystallogr., Sect. B* **1976**, *32*, 2248.
75. Kostitsyn, A. B.; Bychkov, N. N.; Negrebetskii, V. V.; Stepanov, B. I. *Zh. Obshch. Khim.* **1986**, *56*, 1282; *Chem. Abstr.* **1987**, *106*, 213709g.
76. Chu, S. S. C.; Napoleone, V.; Chu, T. L. *J. Heterocycl. Chem.* **1987**, *24*, 143.
77. Chu, S. S. C.; Kow, W. W. van der Helm, D. *Acta Crystallogr., Sect. B* **1978**, *34*, 308.
78. (a) Ohno, S.; Shimizu, H.; Kataoka, T.; Hori, M. *Chem. Pharm. Bull.* **1984**, *32*, 3471. (b) Hori, M.; Kataoka, T. Shimizu, H.; Ohno, S. *Heterocycles* **1979**, *12*, 1417.
79. Lee, C. C.; Chowdhury, R. L.; Piorko, A.; Sutherland, R. G. *J. Organometal. Chem.* **1986**, *310*, 391.
80. Ohno, S.; Shimizu, H.; Kataoka, T.; Hori, M.; Kido, M. *J. Org. Chem.* **1984**, *49*, 3151.
81. Hori, M.; Kataoka, T.; Shimizu, H.; Ohno, S. *Heterocycles* **1979**, *12*, 1555.
82. Price, C. C.; Hori, M.; Parasaran, T.; Polk, M. *J. Am. Chem. Soc.* **1963**, *85*, 2278.
83. Muren, J. F. *J. Med. Chem.* **1970**, *13*, 140.
84. Hori, M.; Kataoka, T.; Shimizu, H.; Ohno, S.; Narita, K. *Tetrahedron Lett.* **1978**, 251.
85. Maryanoff, C. A.; Hayes, K. S.; Mislow, K. *J. Am. Chem. Soc.* **1977**, *99*, 4412.
86. (a) Tamura, Y.; Nishikawa, Y.; Sumoto, K.; Ikeda, M.; Murase, M.; Kise, M. *J. Org. Chem.* **1977**, *42*, 3226. (b) Tamura, Y.; Mukai, C.; Nishikawa, V.; Ikeda, M. *J. Org. Chem.* **1979**, *44*, 3296. (c) Tamura, Y.; Mukai, C.; Nakajima, N.; Ikada, M. *J. Org. Chem.* **1980**, *45*, 2972.
87. Tamura, Y.; Mukai, C.; Ikeda, M. *Chem. Pharm. Bull.* **1982**, *30*, 4069.
88. Tamura, Y.; Matsushima, H.; Minamikawa, J.; Ikeda, M.; Sumoto, K. *Tetrahedron* **1975**, *31*, 303.
89. Tamura, Y.; Nishikawa, Y.; Mukai, C.; Sumoto, K.; Ikeda, M.; Kise, M. *J. Org. Chem.* **1979**, *44*, 1684.
90. Hori, M.; Kataoka, T.; Shimizu, H. *Chem. Lett.* **1974**, 1073.
91. Mukai, C.; Ikeda, M.; Kido, M. *Chem. Lett.* **1981**, 619.
92. Ando, W.; Yagihara, T.; Tozune, S.; Imai, I.; Suzuki, J.; Toyama, T.; Nakaido, S.; Migita, T. *J. Org. Chem.* **1972**, *37*, 1721.
93. Ternay, A. L., Jr.; Craig, D.; O'Neal, H. R. *J. Org. Chem.* **1980**, *45*, 1520.
94. (a) Tamura, Y., Mukai, C.; Nakajima, N.; Ikeda, M.; Kido, M. *J. Chem. Soc., Perkin Trans. 1* **1982**, 212. (b) Tamura, Y.; Mukai, C.; Ikeda, M. *Heterocycles* **1979**, *12*, 1179.
95. Galloy, J.; Watson, W. H.; Craig, D.; Guidry, C.; Morgan, M.; McKellar, R.; Ternay, A. L., Jr.; Martin, G. *J. Heterocycl. Chem.* **1983**, 399.
96. Ternay, A. L., Jr., Baack, J. C.; Chu, S. S. C.; Napoleone, V.; Martin, G.; Alfaro, C. *J. Heterocycl. Chem.* **1982**, *19*, 833.
97. Abbady, A.; Askari, S.; Morgan, M.; Ternay, A. L., Jr.; Galloy, J.; Watson, W. H. *J. Heterocycl. Chem.* **1982**, *19*, 1473.
98. Ternay, A. L., Jr.; Cushman, J. A.; Harwood, J. S.; Yu, C. P. *J. Heterocycl. Chem.* **1987**, *24*, 1067.
99. Kluba, M.; Harwood, J.; Casey, P. K.; Ternay, Jr., A. L. *Heterocycl. Chem.* **1985**, *22*, 1261.
100. Stothers, J. B. *Carbon-13 NMR Spectroscopy*; Academic Press: New York, 1972; Chapter 11.

7

Application of Empirical Force-Field Calculations to the Conformational Analysis of Cyclohexenes, Cyclohexadienes, and Hydroaromatics

Kenny B. Lipkowitz

7.1 INTRODUCTION

Molecular mechanics is a nonquantum mechanical method of computing structures, energies, vibrational spectra, and some physicochemical properties of molecules. It is a fast, reliable computational method when properly used and has been extensively documented. In the Appendix of this book I describe the method, its underlying philosophy, and why it is especially applicable to conformational analysis. I highlight many of the assumptions, approximations, and pitfalls that may be encountered when using this procedure. In this chapter, however, I focus on the conformational analysis of cyclohexenes, cyclohexadienes, and related hydroaromatics in keeping with the subject in this book.

Before I begin, allow me to digress somewhat and mention that as a lecturer at the summer workshops in applied quantum chemistry sponsored by the Quantum Chemistry Program Exchange, I had the opportunity to institute and teach the topic of molecular mechanics. Early into the lectures I would hold a Dreiding stereo model in one hand, a magnetic tape representative of the QCPE programs in the other, and make the attendees aware that both are models, each with inherent strengths and weaknesses, and each of which can be abused.

In molecular mechanics one chooses a model of the structure of matter, selects a mathematical expression that describes this, calculates a numerical value that can be compared with experiment, and improves upon the model to obtain a better fit. The mechanical model I carried to the lecture hall was 1,4-cyclohexadiene. The workshop participants and I would delight in placing the model in its planar form on the lecture podium and with a smack of the lectern watch the model pop into a boat form. This for me has become a classic example of how mechanical models are sometimes unreliable representatives of molecular structure and can be terribly misleading. Needless to say, if mechanical models can be misleading, so can computer models.

Clearly, the advantages of computer modeling with molecular mechanics has been realized especially in the field of conformational analysis, where the method is regularly used. It nicely complements vibrational spectroscopy, microwave, electron diffraction, and X-ray crystallographic studies, and molecular mechanics has profoundly enhanced our understanding of molecular structure and dynamics in the area of NMR spectroscopy. The method is now being used by synthetic organic, inorganic, physical, and of course biological chemists. Molecular mechanics applications are numerous and the future looks bright for this fast developing field. It is reasonable then to know something about the philosophy of the method along with what it can and cannot do. If you already know about molecular mechanics keep reading; if not, see the Appendix of this book.

7.2 CONFORMATIONAL ANALYSIS OF PARTIALLY SATURATED SIX-MEMBERED RINGS

Even with the approximations, pitfalls, and deficiencies described in the Appendix molecular mechanics does quite well; especially in conformational analysis where relative energies are being compared and deficiencies in the methodology tend to cancel. In keeping with the theme of this book we now consider partially unsaturated six-membered rings.

7.2.1 Cyclohexene

Cyclohexene exists as a degenerate set of half-chair forms rapidly interconverting at room temperature. Empirical force-field studies of the interconversion pathway were somewhat controversial with regard to the boat form being a

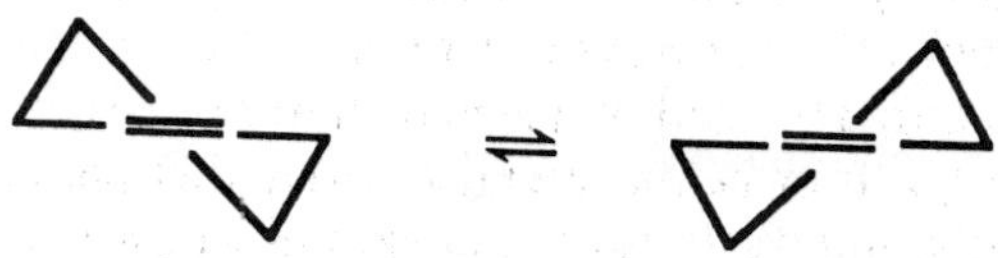

1 1

minimum or a maximum during ring inversion. The earliest comprehensive examination of cyclohexenes by force-field methods was published by Bucourt and Hainaut.[1] They considered the progressive deformation of cyclohexene from the half-chair up to boat conformations, where the + and − refer to the sign of the

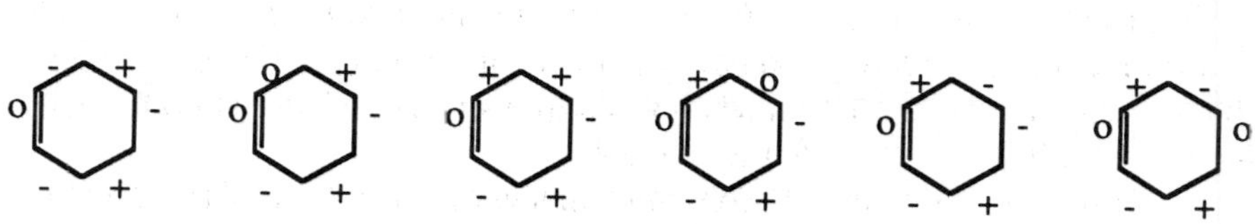

1a 1b 1c 1d

Table 7.1

Form	E_b	$E\omega$	(kcal/mol)
1a	0.15	1.42	
1b	0.31	2.48	
1c	1.01	5.64	
1d	1.05	7.43	

corresponding dihedral angle. **1a** is the monoplanar or half-chair form; **1b** is the 1,2-diplanar or sofa form; **1c** is the 1,3-diplanar form; and **1d** is the 1,4-diplanar or boat form.

The force field employed considered angle bending, torsional, and nonbonded terms only. For all conformations considered, the van der Waals terms do not exceed 0.1 kcal/mol. The values for bending and torsional contributions are given in Table 7.1. Hence, based on this fairly incomplete force field, there appears to be a smooth progression from half-chair to boat conformation with the boat form corresponding to a transition state for the interconversion. The boat form is about 7 kcal/mol less stable than the half-chair. Using a more sophisticated force field and implementing an energy minimization scheme, Favini's results[2] were not much different from those of Bucourt and verified earlier suggestions that a structure flatter than the half-chair is probably most appropriate for cyclohexene.

In contrast, Allinger et al.[3] assessed the conformational states of cyclohexene and concluded that the pseudorotation pathway of cyclohexene involved a boat form (C_s symmetry) that is a high-energy intermediate rather than a transition state. The transition structure was 1.60 kcal/mol higher in energy than the boat form and 5.93 kcal/mol above the half-chair form (Figure 7.1). Later using a

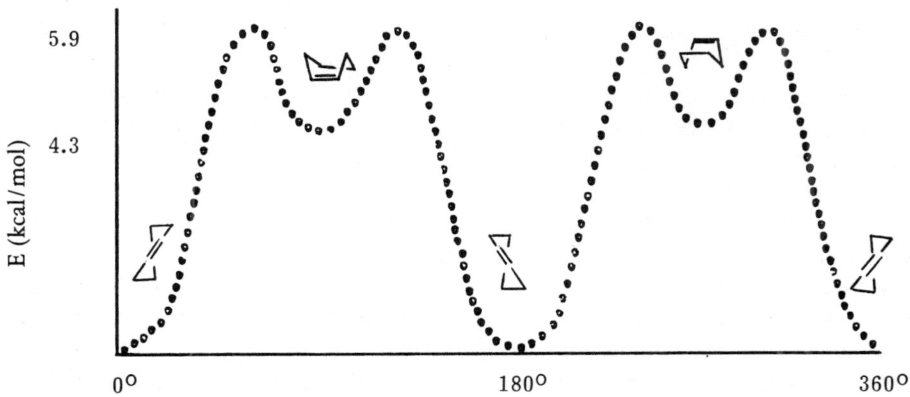

Figure 7.1 Cyclohexene interconversion.

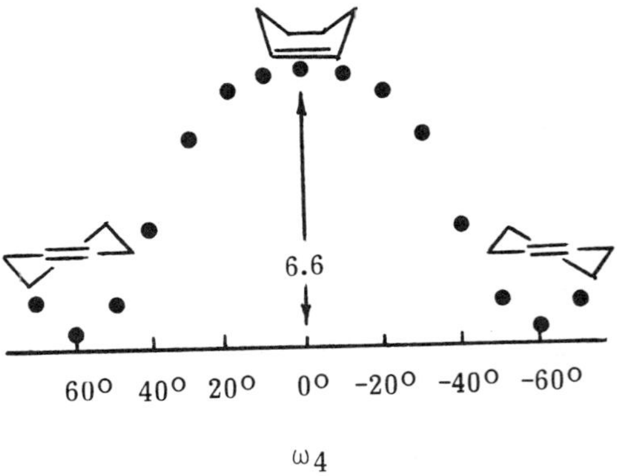

Figure 7.2 Calculated strain-energy profile for ring inversion in cyclohexene.

modified version of that force field which better treated stretch–bend cross-terms together with a cubic correction for bond stretching, Allinger and Sprague[4] found the boat form to be the transition state for cyclohexene ring flipping with $\Delta H^{\ddagger}$ = 6.39 kcal/mol.

Using Boyd's consistent force field (which has far less bilinear cross-terms than the Ermer–Lifson field[5]) suitable parameters were developed by Anet and Yavari[6] to study cycloalkenes. They found that the strain energy for the half-chair conformation of cyclohexene is a minimum and the boat form is a transition state. The transition structure was located with the criterion that a single negative eigenvector exists in the Hessian matrix. This was not done in the other studies of cyclohexene. The difference in strain energies between half-chair and boat forms of cyclohexene is 6.3 kcal/mol. A plot of the strain energy as a function of ω (defined as the torsion angle between atoms 3, 4, 5, and 6) is presented in Figure 7.2. Table 7.2 lists the computed strain energies in kcal/mol for the various geometries of cyclohexene. Again, realize that the total energies are more meaningful values to consider but, within the framework of Boyd's force field, it is instructive to see how the individual components contribute to the total steric energy. For most terms there is a smooth gradation when going from $\omega = 60°$ to $\omega = 0°$. However, the angle bending and nonbonded interactions change nonuniformly. Interestingly, the nonbonded terms reach a minimum at $\omega = 40°$ and begin increasing again as the half-chair form is approached; the angle bending deformations are maximized at $\omega = 30°$. Such trends, albeit informative, may be meaningless until several different force-field studies give the same trends.

Empirical force-field studies of cyclohexene by other groups have been reported[7,8] but only the half-chair conformations were considered. The general consensus based on molecular mechanics (see Appendix, Ref. 22) is that two half-chair conformations are interconverted in a pseudorotation-like motion through a symmetrical boat conformation transition state. The calculated energies of the transition states range from 6.09 to 7.00 kcal/mol, which is only slightly higher than the experimental $\Delta H^{\ddagger}$ of 5.31 kcal/mol. A study of cyclohexene and *cis*-$\Delta^{6,7}$-

Table 7.2 Calculated Strain Energies (kcal/mol) in Various Geometries of Cyclohexene

Strain-Energy Contributions	Half-Chair $\omega_4 = 60°$	$\omega_4 = 50°$	$\omega_4 = 40°$	$\omega_4 = 30°$	$\omega_4 = 20°$	$\omega_4 = 10°$	Boat $\omega_4 = 0°$
Bond stretching	0.14	0.14	0.17	0.21	0.23	0.23	0.23
Bond Angle bending	0.21	0.53	1.25	1.60	1.18	1.10	1.08
Out-of-plane bonding	0.00	0.00	0.00	0.00	0.00	0.00	0.00
Torsional strain	1.24	1.94	3.04	4.77	6.10	6.46	6.54
Nonbonded interactions	3.29	3.03	3.01	3.22	3.51	3.63	3.67
Total strain energy	4.88	5.64	7.47	9.80	11.02	11.42	11.52

octalin interconversion pathways has been analyzed in detail by Vanhee et al.[8] Since the nomenclature in this paper is somewhat nonconventional and to conserve space, we omit it from this review but suggest it be read.

7.2.2 Monosubstituted Cyclohexenes

7.2.2.1 Methylcyclohexene

The most comprehensive examination of a series of related alkyl substituted cyclohexenes is an early study by Allinger et al.[3] of methyl cyclohexenes. The enthalpy of each conformer was computed as described earlier. To the force field's strain energy was added the appropriate bond energy. The enthalpy of each isomer was then obtained by mixing the conformers according to their Boltzmann distribution. The relative entropy was calculated for each isomer by considering the symmetry number (i.e., if the ring is meso or *dl*) and then adding the entropy of mixing if there were two conformers for that isomer. With $\Delta H°$ and $\Delta S°$ it was possible to obtain $\Delta G°$. Table 7.3 contains the computed thermodynamic properties of the methylcyclohexene isomers.

The most stable isomer is, as anticipated, the trisubstituted alkene. Of the remaining disubstituted olefins, the endocyclic arrangement is preferred over the exocyclic arrangement. Interestingly, this preference has its origin in the entropic rather than the enthalpic portion of $\Delta G°$. The 3-methyl and 4-methylcyclohexenes each consist of two *dl* pairs with axial and equatorial substituents for a total of four half-chair conformations. The additional entropy of mixing along with symmetry differences between the endocyclic and exocyclic isomers makes the $\Delta G°$ of the endocyclic isomers more stable than the exocyclic isomer.

In 3-methylcyclohexene the axial form is approximately 0.8 kcal/mol less stable than the equatorial form. In the 4-methylcyclohexene the axial form is approximately 1 kcal/mol less stable. Furthermore, the 4-substituted ring system is more stable than the 3-substituted isomer. An explanation for this was not given but the effect probably has its origins in allylic strain.

7.2.2.2 Other Monosubstituted Cyclohexenes

Molecular mechanics studies of cyclohexenes are rare in contrast to their saturated counterparts, cyclohexanes. Most of my literature files contain calculations on substituted cyclohexenes as ancillary material only; the focus is not on cyclohexene per se. Nonetheless, several examples of monosubstituted cyclohexene force-field studies directed toward understanding the shape of the ring with substituents do exist.

In one of these studies carried out by Todeschini and his group[9] in Italy, the conformation of 1-vinylcyclohexene was studied by molecular mechanics. Most of the previous studies showed the most stable form to be the planar s-trans conformation but considerable disagreement existed as to whether the second most stable form has a coplanar s-cis structure or a nonplanar s-cis structure. The force field used was a form of Schleyer's that was modified with a quantum

Table 7.3 The Thermodynamic Properties (kcal/mol and eu) of the Methylcyclohexene Isomers

	Compound	Σ Bond Energies	Steric Energy	ΔH_i	ΔS_i	$\Delta H°$	$\Delta S°$	$\Delta G°_{298}$
2	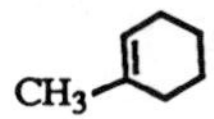	−12.49	2.41	0.00	1.38	0.00	0.00	0.00
3	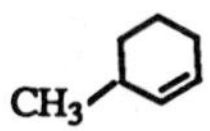	−10.11	ax 3.26	3.23	1.38	2.66	1.10	2.33
			eq 2.50	2.47	1.38			
4	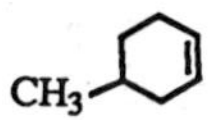	−10.11	ax 3.06	3.03	1.38	2.14	0.83	2.14
			eq 2.02	1.99	1.38			
5	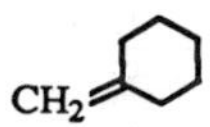	− 9.61	1.67	2.14	0	2.14	−1.38	2.55

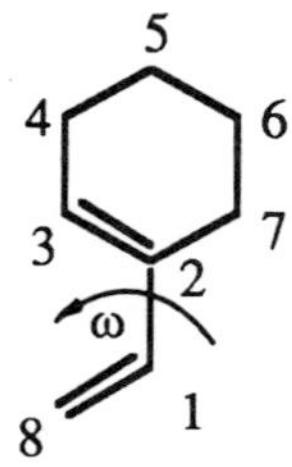

6

mechanical procedure to account for delocalized π systems. The strain energies computed for the s-cis and s-trans conformations are 8.7 and 7.9 kcal/mol, respectively. Full relaxation of all internal degrees of freedom for the s-cis form skewed ω to 176.6° with a final strain energy of 7.8 kcal/mol. Unfortunately, these authors did not do a comparable full relaxation for the s-trans conformer nor were they adequately able to compute the s-cis/s-trans rotational barrier. For all orientations of the vinyl group, however, the half-chair form of cyclohexene is 4–5 kcal/mol more stable than the boat form.

A second example of monosubstituted cyclohexenes studied by force-field methods is the conformational analysis of 3-cyclohexene-1-carboxaldehyde, 7.[10] The impetus for this study was the observation that both aldehyde and alkene functionality demonstrate reduced reactivity for a series of standard chemical reactions.[11] This reduced reactivity was attributed to an interaction between alkene and aldehyde such that electron density of the former group is transferred to the latter group, that is, 7a–7c.

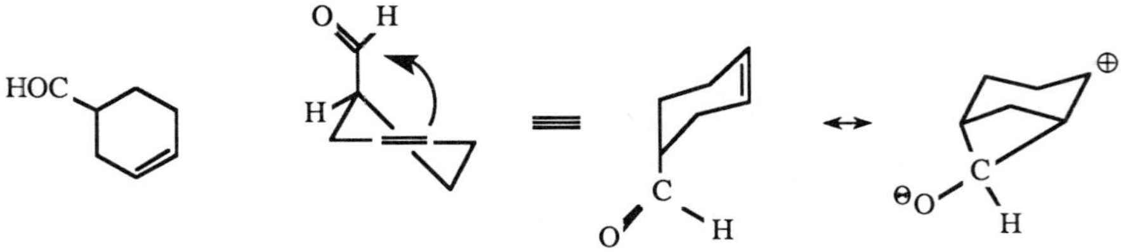

7 7a 7b 7c

To effect such a transannular interaction, termed the "supra-annular effect," the carbonyl must adopt an axial position. Indeed, it was suggested that electron-withdrawing groups in general would exist in the axial form.

3-Cyclohexene-1-carboxaldehyde was examined with the MMP2 force field, which allows for calculation of conjugated π systems by implementing a quantum mechanical calculation. The approach taken was to evaluate the energies of axial and equatorial aldehyde (three unique orientations of CHO were considered) first with the quantum mechanical interaction between aldehyde and alkene neglected and second with the overlap between carbonyl and double bond carbons explicitly considered in the MMP2 SCF routine. The energy difference between these sets of calculations is the stabilization energy due to nonbonded overlap. The stabilization due to this supra-annular effect was found to be zero (~1 cal/mol). Note that orbital overlap falls off sharply with distance and the interatomic separation between carbonyl and alkene carbons, even in the axial form is large (in excess of 3.2 Å). Single-point *ab initio* molecular orbital calculations confirm that no charge is transferred from alkene to olefin as suggested by resonance structure **7c** but these calculations employ a restrictive minimal basis set. In summary then, the deactivation of alkene and aldehyde is due to something other than a supra-annular effect.

As a matter of fact, the conformation with the equatorial aldehyde was found to be more stable than the axial form with $\Delta G = 0.83$ kcal/mol. This implies approximately 80% equatorial form at room temperature and is in agreement with NMR studies of 3-cyclohexene-1-carboxaldehyde and related systems which indicate $\Delta G = 0.4$–0.5 kcal/mol.

7.2.3 Disubstituted Cyclohexenes

7.2.3.1 Dimethylcyclohexenes

Disubstituted cyclohexenes have also been investigated by Allinger with molecular mechanics.[3] This work was initiated to explore an earlier claim in the literature that although a methyl group on the 3 position of cyclohexene prefers the equatorial form by ~0.8 kcal/mol, this preference would be diminished or even inverted if the group was a substituent on the adjacent carbon. Allinger found, by calculating enthalpies for conformational equilibria between axial and equatorial forms of 1,6-dimethylcyclohexene, that very little difference between the equilibria is observed (see Table 7.4). The reason for this is that the dihedral angle which the axial hydrogen at C-3 in cyclohexene makes with the C-2 hydrogen is not much greater than that between the equatorial hydrogen at C-3 and the hydrogen at C-2. When the hydrogen at C-2 is replaced by methyl, this methyl interacts almost equally with the hydrogens (or methyl groups) at C-3. Other dimethyl substituted cyclohexenes evaluated by Allinger are listed in Table 7.5. The energy difference between the 3,4-dimethylcyclohexenes is quite small and the 4,4-dimethylcyclohexene is more stable than the 3,3-isomer.

Table 7.4 Calculated Enthalpies (kcal/mol) for Conformational Equilibria: Equatorial ⇌ Axial Methyl

	Compound	$\Delta H°$
8	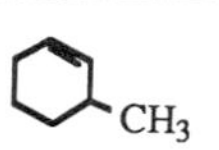	+0.76
9	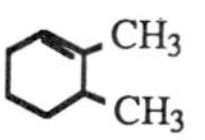	+0.84

Table 7.5 Calculated Enthalpy Data (kcal/mol) for Dimethylcyclohexenes

			ΔH° (kcal/mol)
10	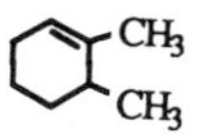	trans-Diequatorial	0.00
		trans-Diaxial	0.71
		cis-3-Axial,4-equatorial	0.25
		cis-3-Equatorial,4-axial	0.46
11	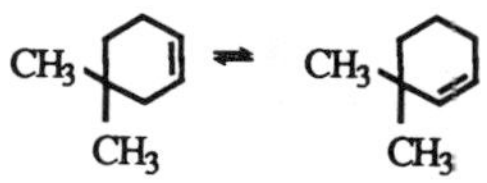		0.84

7.2.3.2 Monoterpenes

A detailed consistent force-field analysis of monoterpenes has been carried out by Singh and Keiderling.[12] This work represents an interesting application of molecular mechanics for the calculation of vibrational circular dichroism (VCD) spectra. Among other terpenes studied, the VCD of (+)-menthene, **12**, (+)-limonene, **13**, and (+)-menthenol, **14**, were determined. Two basic ring

12

13

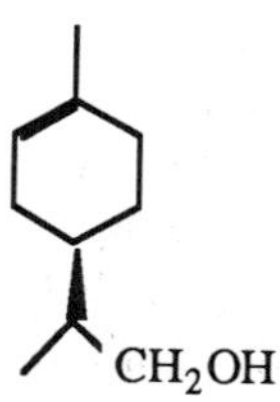

14

conformers were considered and energies for all rotomeric states of the isopropyl group were determined. In the half-chair conformations, the isopropyl group can be either axial (AX) or equatorial (EQ) with the EQ structures always lower in energy (Table 7.6). The ring structures of **12–14** are all very similar. Presented in Figure 7.3 are minimum energy conformations of various forms of menthene to illustrate the general shapes of these disubstituted cyclohexenes. The perspective is chosen to emphasize the planarity of the carbons at the 6, 1, 2, and 3 positions (numbering from the methyl through the double bond toward the isopropyl) and the displacement of C-4 above and C-5 below the plane. It should be noted that the AX conformer is shown in the opposite configuration to emphasize structural similarity. The rotamers can be characterized as having the methyl C—H bonds trans, EQ(T), or gauche, EQ(G) and EQ(G′). While EQ(G) and EQ(G′) are very close in energy, they are significantly lower than EQ(T) and the three AX conformers.

Using these conformations and point charges derived from CNDO/S-CI,

Table 7.6 CFF Relative Total Energies of Menthene, Limonene, and Menthenol

Menthene		Limonene		Menthenol[a]	
Conformer	Energy	Conformer	Energy	Conformer	Energy
EQ(G′)	−1.748	EQ(C)	−3.282	EQ(G′)	−2.970
EQ(G)	−1.536	EQ(T)	−1.988	EQ(G)	−1.870
EQ(T)	−0.748	AX(T)	0.000	EQ(T)	−0.998
AX(G′)	−0.125			AX(T)	0.000
AX(T)	−0.055				
AX(G)	0.000				

[a] Energies in kcal/mol referenced arbitrarily to the highest-energy conformer in the set. T, G, G′, and C stand for trans, gauche 1, gauche 2, and cis rotamers, respectively, while EQ and AX stand for equatorial and axial orientations of the substituent with respect to the ring.

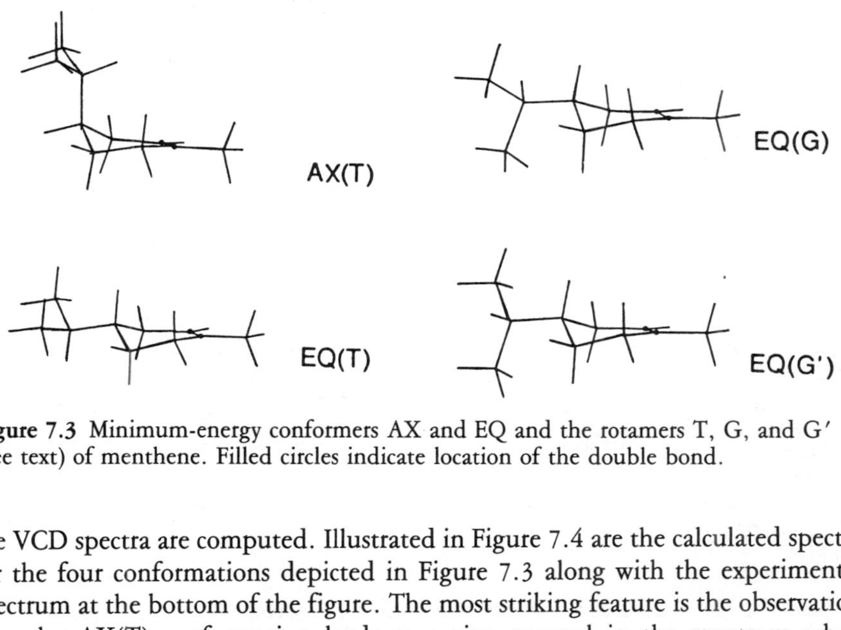

Figure 7.3 Minimum-energy conformers AX and EQ and the rotamers T, G, and G′ (see text) of menthene. Filled circles indicate location of the double bond.

the VCD spectra are computed. Illustrated in Figure 7.4 are the calculated spectra for the four conformations depicted in Figure 7.3 along with the experimental spectrum at the bottom of the figure. The most striking feature is the observation that the AX(T) conformation leads to a sign reversal in the spectrum when compared to the EQ forms, which is inconsistent with the experimental curves. This pattern was found for calculated VCD spectra of molecules **12–14**.

The calculated CH stretching VCD is dominated by the CH_2 groups, which in turn are conformationally dependent. The axial versus equatorial nature of the isopropyl groups affects the phase of the quasidegenerate methylene stretching vibrations resulting in a sign change for the resultant CH_2 VCD. Thus, the patterns observed are conformationally sensitive. As shown in Figure 7.4, however, the rotation of the isopropyl group has only a minor effect on the VCD spectrum. A more detailed analysis is described in the original paper but this

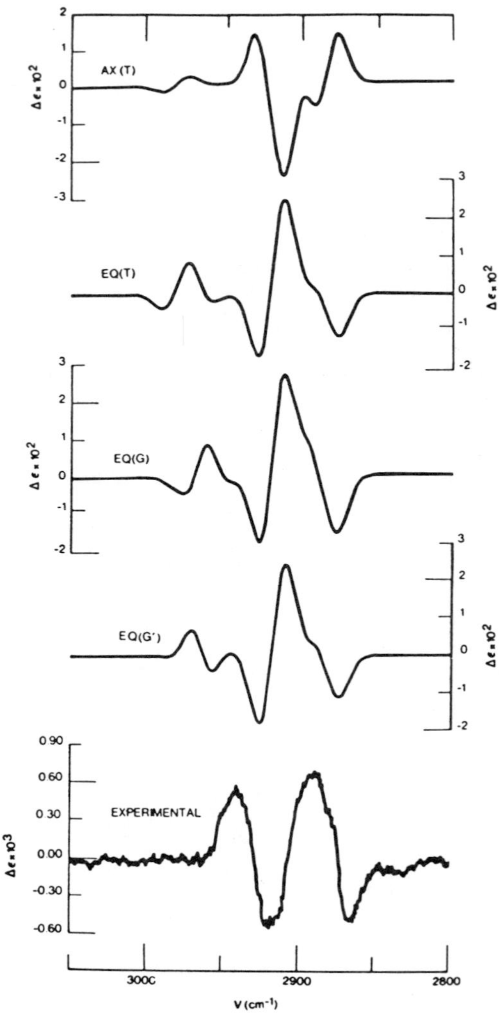

Figure 7.4 Experimentally observed (0.2 M in CCl_4, resolution 10.6 cm^{-1}, bottom) and theoretically calculated VCD using the CFF parameters and SCNDO charges for (+)-menthene for the conformers of Figure 7.3.

emphasizes the conformational aspects of disubstituted cyclohexenes and shows how empirical force fields can be used to solve a complex problem.

7.2.3.3 Bicyclo[4.1.0] Derivatives

Carenes and Norcarenes: The conformations of bicyclo[*n*.1.0] derivatives **15–19** have been investigated by molecular mechanics.[13] The structure of these molecules has been somewhat controversial. An electron diffraction study of 2-norcarene concluded two conformers were present in a 7:3 mixture in the gas phase with the pseudo-chair conformation predominating.[14] This was confirmed by photoelectron spectroscopy.[15] In contrast, NMR analysis of 2-norcarene derivatives[16] indicates a pseudo-boat form to be favored unless an endo-7-methyl group is present which sterically interferes with an axial hydrogen on C-7 in the

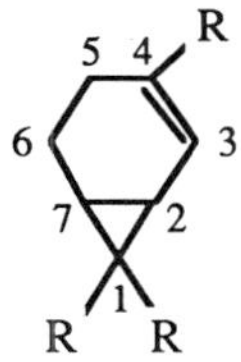

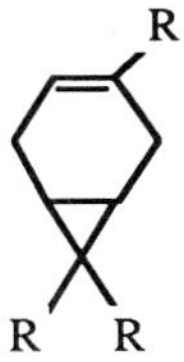

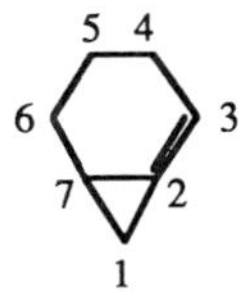

19

15, R = H (2-norcarene)
16, R = Me (2-carene)

17, R = H (3-norcarene)
18, R = Me (3-carene)

pseudo-boat form. This is the case for 2-carene, **16**. Conformational studies of **17** have been reported with the syn-boat form expected to be preferred,[17] but a flat form is also consistent with NMR spectra.[18] Three conformations, syn-boat, planar, and anti-boat, have been proposed for 3-carene [citations 7, 10–13 in Ref. 13].

By using a modified version of the Schleyer–Andose–Mislow force field, the stable conformations and energies of **15–19** were obtained. The results are summarized in Table 7.7. For 2-norcarene the pseudo-chair conformation depicted in Figure 7.5 is preferred over the pseudo-boat conformation by less than 1 kcal/mol and is in agreement with the electron diffraction results. The energy barrier to interconversion between these two forms is computed to be less than 3 kcal/mol. The pseudo-chair conformation is also found in 2-carene, **16**, and the interconversion energy barrier is calculated to be less than 2.5 kcal/mol. The 2.1 kcal/mol preference for the pseudo-chair conformation over the pseudo-boat form amplifies the argument that a steric interference between an endo-7-methyl group with an axial H on C-7 destabilizes the pseudo-boat form.

For 3-norcarene, **17**, little difference in energy between the syn-boat and

Table 7.7 Heats of Formation (kcal/mol) for Compounds **15–19**

Compound	Form	ΔH_f^{298}
15	Pseudo-chair	31.37
	Pseudo-boat	31.79
	Biplanar	34.01
16	Pseudo-chair	10.46
	Pseudo-boat	12.57
	Biplanar	12.81
17	Syn-boat	30.15
	Anti-boat	31.27
	Biplanar	30.58
18	Syn-boat	10.74
	Anti-boat	10.16
	Biplanar	9.64
19	Biplanar	51.58

2-Norcarene pseudo-chair 3-Norcarene syn-boat

Figure 7.5 Most stable conformations of **15** and **17** based on empirical force-field calculations.

biplanar (flat) conformations confirms the experimental arguments in Refs. 18 and 19. The syn-boat form is represented in Figure 7.5. For 3-carene a planar, six-membered ring (biplanar) is calculated to be slightly more stable than either of the boat forms. Finally, the only minimum energy structure located for $\Delta^{1,2}$-bicyclo[4.1.0]heptene, **19**, is designated as biplanar but the ring is not actually flat. The torsion angle ω_{1234} which is 0° in biplanar forms of **15–18** is −47° in **19**. Likewise, ω_{2345} which is 0° in biplanar **15–18** is 41° in **19**. Also note that the validity of an empirical force-field calculation for **19** may not be upheld because of the unusual bonding in that molecule.

Norcaradienes: The valence tautomerism between 1,3,5-cycloheptatriene, **20**, and bicyclo[4.1.0]hepta-2,4-diene, **21** (norcaradiene), can be shifted toward the norcaradiene side of the equilibrium by various factors (usually electronic effects, i.e., X = π electron acceptor). These equilibria, in addition to being

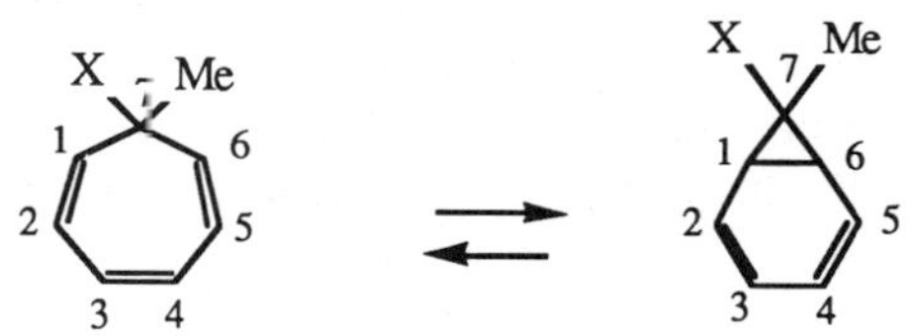

20 (CHT) **21** (NCD)

sensitive to electronic effects of substituents, also appear sensitive to steric effects. Thus, the equilibrium is shifted further to the norcaradiene for 7-methyl-7-methoxycycloheptatriene than in 7-methoxycycloheptatriene[19] presumably because of transannular repulsion between the axial methyl group at C-7 and the C-3—C-4 π bond in the triene. The norcaradiene population was also found to increase with size of alkyl groups in 7-alkyl-7-cyanocycloheptatrienes but it was argued[20] that the triene was destabilized due to repulsions of the equatorial 7-alkyl groups with H-1 and H-6 in the triene.

The structural effects of *t*-butyl groups on the cycloheptatriene–norcaradiene valence tautomerism were investigated experimentally and computationally by Takeuchi et al.[21] The following equilibria were explored but calculations (MMPI) were done with a hydrogen replacing the CN groups since cyano parameters were not available. Several interesting experimental observations were

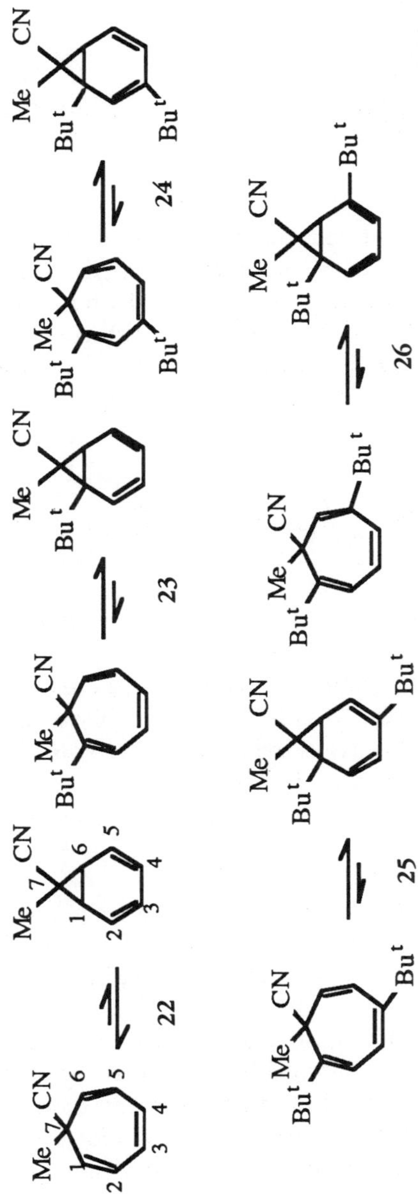

well reproduced by the empirical force field. First, the 7-methyl group in **23–26** takes on an endo geometry. The steric energy of the endo isomer, **27**, is approximately 3 kcal/mol more stable than the exo isomer, **28**, in agreement with

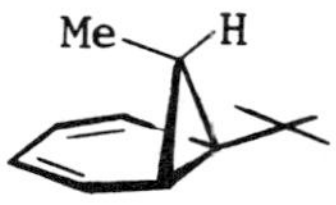

27

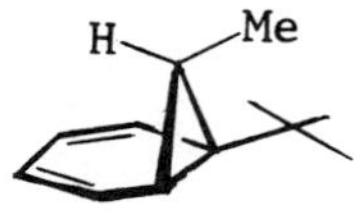

28

experiment. Second, the introduction of a *t*-butyl group at the 3, 4, or 5 position of 20 favors the NCD form in the order 4-*t*-butyl < 3-*t*-butyl < 5-*t*-butyl. Computationally, it is found that the 3-*t*-butyl, 4-*t*-butyl, and 5-*t*-butyl favor the NCD form by 1.1, 0.7, and 2.4 kcal/mol, which compare favorably with the experimental values of 1.5, 1.0, and 2.4 kcal/mol, respectively.

These steric effects are exemplified in the angles α and β:

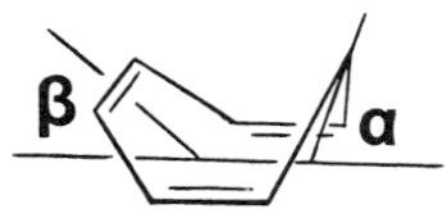

CHT

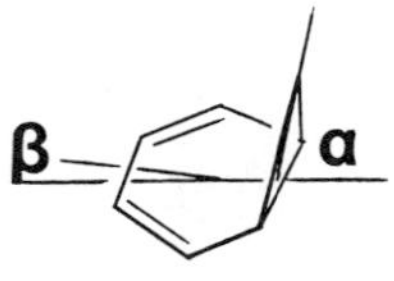

NCD

It is found that the *t*-butyl group on positions 3, 4, 5 causes further bending of the CHT ring as compared with NCD by 3–5° in β while α remains unchanged. In contrast, the structural changes in the NCDs are very small. Thus, the empirical force-field analysis indicates that the *t*-butyl group sterically destabilizes the CHT form more than the NCD form, shifting the equilibrium toward the norcaradiene.

7.2.3.4 Octalins

Octalins can be thought of as disubstituted cyclohexenes. An assessment of *cis*- and *trans*-octalins was provided by Bucourt and Hainaut[1] but a more complete evaluation of structures and energies has been compiled by Allinger et al.[3] using a more robust force field.

Of the six octalin isomers, three of these have two diastereomeric structures resulting in a study of nine conformations. Of these, only one is meso while the others are *dl* pairs. Allinger calculated the concentrations of the six isomers in equilibrium for several temperatures. The results for a single temperature are compiled in Table 7.8. The results for 298 K and higher temperatures agree fairly well with experiment.

Trans ring fusion is seen to be preferred over cis fusion. Comparing *trans*-1-octalin with *cis*-1-octalin, the enthalpy difference is 1.4 kcal/mol while the enthalpy difference for the *trans*- versus *cis*-2-octalins is 2.3 kcal/mol. A large

Table 7.8 Enthalpies and Entropies for the Octalin System

Isomer	Calculated $H°$ (kcal/mol)	Symmetry Number	*dl* or meso	Calculated $\Delta H°_{298}$ (kcal/mol)	Calculated $\Delta S°_{298}$ (eu)	Percentage at 298 K: Calculated	Percentage at 298 K: Experimental
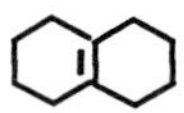 29	0.31	2	meso	0.00	0.00	86.2	90.7
	0.00	4	*dl*				
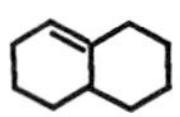 30	1.84 (eq)	1	*dl*	1.88	1.99	11.4	7.7
	. . . (ax)	1	*dl*				
trans 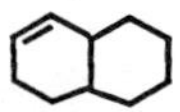 31	3.10	1	*dl*	3.14	1.45	0.86	0.6
trans 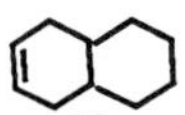 32	2.57	2	*dl*	2.46	0.07	1.4	1.0
cis 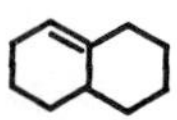 33	(eq)	1	*dl*	4.50	2.79	0.2	
	4.74 (ax)	1	*dl*				
cis 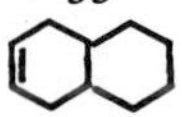 34			*dl*	5.28	1.45	0.0	

contribution to the 2.3 kcal/mol difference for *trans*- and *cis*-2-octalins originates from torsional energies. These torsional deformations are less extreme in the 1-octalins. Although such observations are insightful, keep in mind that a single component of the total steric energy from a single force field may be meaningless; if several force fields indicate the torsional deviation as the origin of the above thermodynamic stabilities, one may feel more confident in such an interpretation. Also observed in Table 7.8 is that 4,5-ring fusions of both substituents are equatorial. The 2-octalin is lower in energy than the 1-octalin when the ring juncture is trans by approximately 0.5 kcal/mol. For cis ring fusion, in contrast, the 1-octalin is preferred over the 2-octalin by approximately 0.8 kcal/mol. Other octalin calculations have been reported.[22]

7.2.3.5 Other Disubstituted Cyclohexenes

Very few examples of molecular mechanics calculations directed toward studying the preferred conformations of disubstituted cyclohexenes exit. The free energies of methyl substituted 3-cyclohexene-1-carboxaldehydes, however, have been computed with the MMP2 force field. Both cis and trans forms of 2-methyl-3-cyclohexene-1-carboxaldehyde and 6-methyl-3-cyclohexene-1-carboxaldehyde have been explored considering the three unique orientations of the aldehyde group.[10]

For 2-methyl-3-cyclohexene-1-carboxaldehyde, **35**, the most stable cis conformation, **35a**, has the aldehyde group equatorial and the methyl group axial.

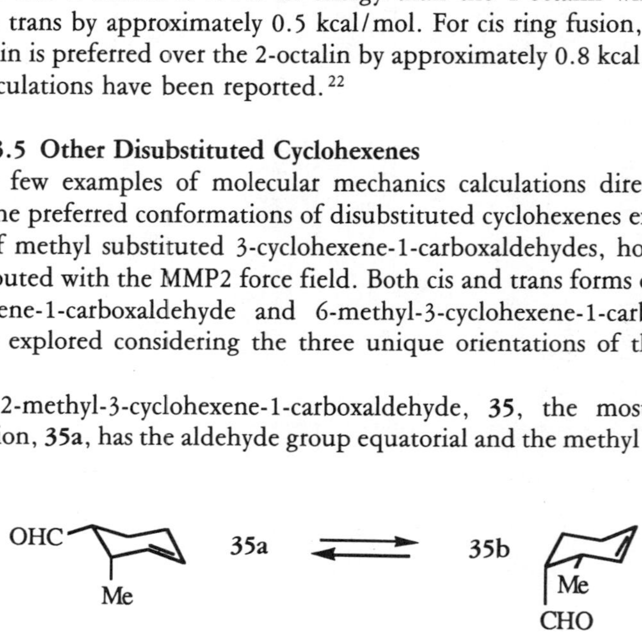

The MMP2 strain energy favoring **35a** is approximately 0.85 kcal/mol. This energy difference is attributed to bending and van der Waals effects. Presumably, the allylic strain of the equatorial methyl group in **35b** contributes to this destabilization.

For the trans isomer of 2-methyl-3-cyclohexene-1-carboxaldehyde the most stable conformation is **36a**, which has both substituents equatorial. The strain energy between these two forms is approximately 0.74 kcal/mol. The principal source of destabilization in **36b** compared to **36a** originates from torsional strain attributed to unfavorable gauche interactions.

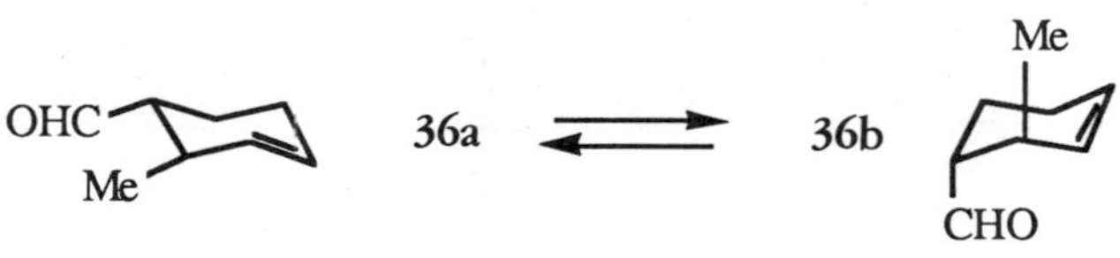

For *cis*-6-methyl-3-cyclohexene-1-carboxaldehyde the MMP2 final steric energies favor an axial aldehyde, **37b**, by only 0.16 kcal/mol over the equatorial form **37a**. Converting these raw numbers into free energies by computing the

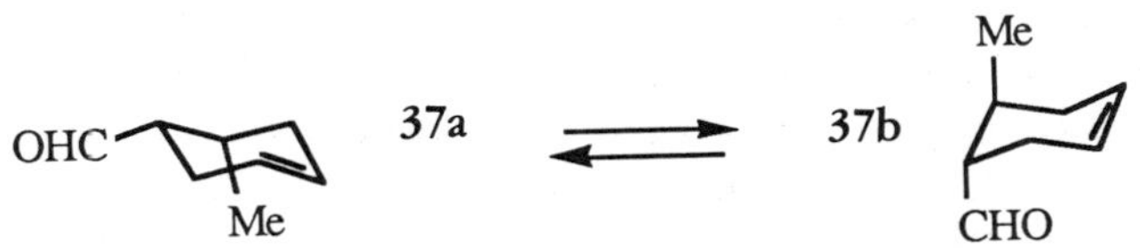

Boltzmann distribution of all conformers present, summing to determine the average heat content ΔH, and subtracting the entropies of mixing give a free-energy difference of 0.24 kcal/mol favoring **37b**. As in the parent cycloalkene there is no supra-annular effect observed here.

For *trans*-6-methyl-3-cyclohexene-1-carboxaldehyde the diequatorial conformer, **38a**, is more stable than the diaxial form **38b** by approximately 1.0 kcal/

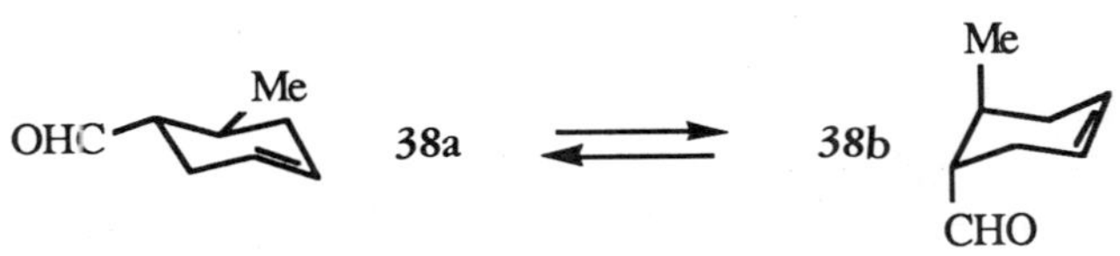

mol. The free-energy difference between these two forms favors **38a** by 0.24 kcal/mol, however. Based on these results the A value of a CH_3 and CHO group are expected to be similar. As of 1986, no A value for CHO has been reported.

7.2.4 Tri- and Tetrasubstituted Cyclohexenes

Again very few examples of tri- or tetrasubstituted cyclohexene conformations computed by molecular mechanics exist. One example from the literature, however, does assess the structures of polysubstituted cyclohexenes in some detail. White and Bovill,[23a] using their own force field, developed a molecular mechanics-based procedure to explore the outcome of multicenter stereoselective reactions. This was based on early work of Wipke and Gund.[23b]

Wipke and Gund initially devised the method to estimate the steric congestion of nucleophilic attack at sterically hindered ketones. The measure of each atom's ability to hinder the incoming reagent is governed by its van der Waals radius. For each participating atom, a cone of preferred approach perpendicular to the carbonyl plane and tangent to the sphere of the van der Waals radius around that atom is defined. This cone intersects a sphere of unit radius on the carbonyl carbon to generate a solid angle A that is equated with the accessibility of the site of attack for each hindering atom. The reciprocal of the accessibility is the congestion. The total congestion is the sum of congestion for all hindering atoms above or below the plane of the carbonyl and the preferred direction of attack is on the side with the lowest congestion.

Rather than use this atom-centered type calculation derived for nucleophilic addition to carbonyls, White and Bovill considered an alternative method that allows for studies of electrophilic additions to carbon–carbon double bonds. In their method they computed the congestion on each side of the midpoint of the double bond. An alternative to this is to find the total congestion in terms of the sums of congestions for each sp^2 carbon. This is called the summed atom-centered congestion factor, which was then used to understand the stereoselectivity of hydroboration–oxidation of hindered alkenes.

The alkenes considered are alkylcyclohexenes, **39–46**, which can adapt one (or both) of two possible half-chair conformation, A and B. The minimum energy conformations were calculated with the WBFF rather than the more recent WBFF2 because the latter is especially bad at predicting the inversion barrier in cyclohexane. In addition to the half-chair conformers considered, the populations

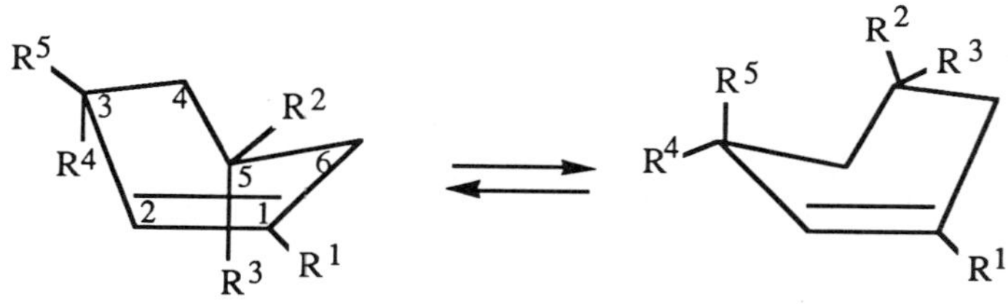

A **B**

of nonequivalent isopropyl rotamers were computed. Before computing the preferred direction of attack, the preferred conformation (A or B) and the percentage of each rotamer present (based on a Boltzmann weighting of steric energies) were obtained. The results are presented in Table 7.9.

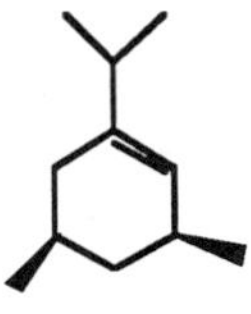

39

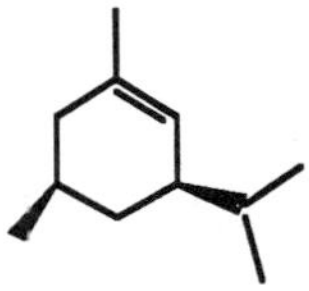

40

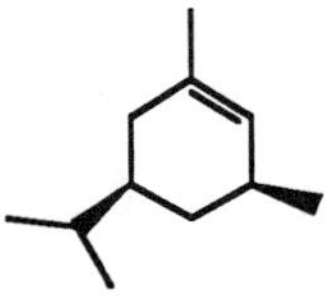

41

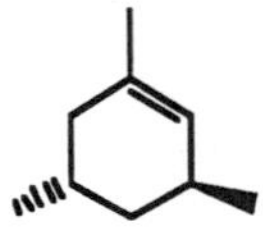

42

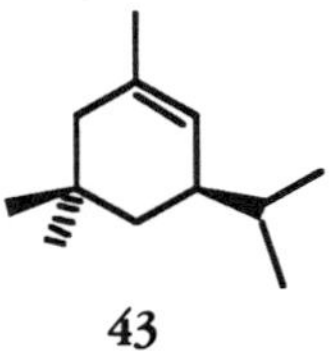

43

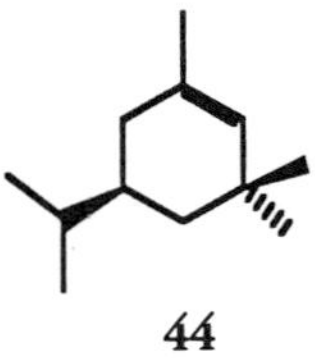

44

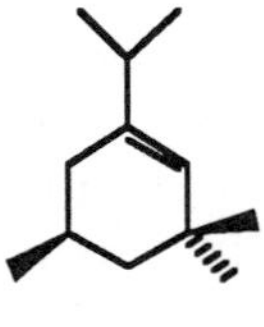

45

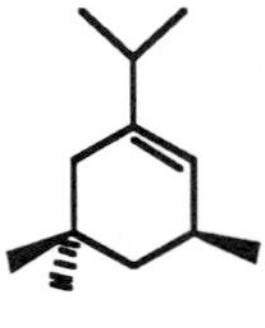

46

Table 7.9 WBFF Geometries of Cyclohexenes **39–46**

	39			**40**			**41**			**42**		
	E_{sj}	x_j	P_{bj}	E_{sj}	x_j	P_{bj}	E_{sj}	x_j	P_{bj}	E_{sj}	x_j	P_{bj}
Conformation A												
$j = 1$	4.09	45	46	4.30	57	95	5.35	37	55	4.95	31	91
$j = 2$	4.36	29	66	4.74	28	93	5.37	36	58	5.40	15	91
$j = 3$	4.42	26	49	5.15	15	61	5.54	27	60	5.73	9	32
Conformation B												
$j = 1$	6.00			6.78			7.19			5.19	21	3
$j = 2$										5.36	16	94
$j = 3$										5.78	8	75

	43			**44**			**45**			**46**		
	E_{sj}	x_j	P_{bj}	E_{sj}	x_j	P_{bj}	E_{sj}	x_j	P_{bj}	E_{sj}	x_j	P_{bj}
Conformation A												
$j = 1$	5.35	56	93	6.02	38	58	4.81	43	57	5.17	42	28
$j = 2$	5.79	28	83	6.12	32	41	5.05	29	40	5.30	34	32
$j = 3$	6.12	16	30	6.19	30	38	5.06	28	52	5.51	24	45
Conformation B												
$j = 1$	7.26			7.70			6.49			6.64		
$j = 2$												
$j = 3$												

In this table, E_{sj} denotes the steric energy (kcal/mol) of the alkylcyclohexene with the isopropyl group in each of its j rotameric states, x_j denotes the percent population of the alkylcyclohexene containing the jth rotamer in the equilibrium mixture, and p_{bj} represents the percentage of product resulting from reagent attack below the plane of the double bond of the alkylcyclohexene for the isopropyl group in its jth rotameric state.

All cyclohexanes except **42** prefer the half-chair form A by 1.3–2.5 kcal/mol. For cyclohexene, **42**, the two half chair forms A and B are within 0.2 kcal/mol of each other. Additionally, each of the rotomeric states accessible by the isopropyl group are appreciably populated. Based on these empirical force-field results the aforementioned scheme for calculating steric congestion of both faces of the cyclohexene was invoked and the results were found, qualitatively, to agree with experiment.

The suggestion that high stereoselectivity of hydroboration in **39–41** is due to R^5 hindering attack from the top face of the alkenes was only partly supported by the calculations because the pseudoequatorial isopropyl group in **40** strongly hinders top side attack while the 3-methyl groups in **39** and **41** have little effect on the calculated stereoselectivities.

7.2.5 Methylenecyclohexanes

Methylenecyclohexanes represent an interesting class of partially unsaturated six-membered rings that have been extensively studied experimentally and which constitute a portion of this book. Methylenecyclohexanes have also been investigated with molecular mechanics; especially the parent hydrocarbon which adopts a chair conformation that interconverts through a high-energy boat form. The computed energy of this ring flipping is 8.1 kcal/mol, in agreement with experiment. This barrier is somewhat higher than that of cyclohexanone but is substantially lower than that of cyclohexane itself.

In a detailed investigation by Anet and Yavari,[6] seven unique ways of converting the chair form to a boat form were explored. The conformational pathways were approximated by driving around a single torsional angle or two adjacent torsional angles. The minimum energy reaction pathway found with

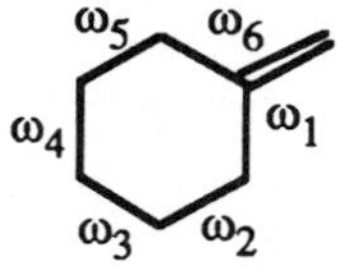

Boyd's consistent force field corresponds to rotation around ω_1 (calculated strain-energy barriers = 8.1 kcal/mol). Driving around ω_2 or ω_1 and ω_6 together is almost as favorable a process, however. The various boat and twist-boat conformations of methylenecyclohexane are close in energy and are generally 5–6 kcal/mol above the global minimum. Table 7.10 lists the total strain energies along with the contributing components of the chair form, two boat forms, and two twist-boat forms of methylenecyclohexane. In all cases it is seen that the barrier to ring flipping has its origins in torsional strain and to a slightly lesser extent, nonbonded repulsions. Using a different force field, Allinger also studied the conformational interconversion of this molecule.[3] The chair form was constrained to have C_3 symmetry but otherwise allowed to fully relax. Three extreme boat forms were considered which are connected by a pseudorotation cycle. One has C_s symmetry where the methylene is at the prow of the boat, another has the methylene on the side of the boat (C_1 symmetry), while the third maintains C_2 symmetry as a twist structure. The most favorable arrangement of the boat is the twist-boat form which is computed to be 4.44 kcal/mol above the chair.

Table 7.10 Calculated Strain Energies (kcal/mol) in Various Conformations of Methylenecyclohexane

Strain-Energy Contributions	Chair (C_s)	Boat-1 (C_s)	Boat-2 (C_1)	Twist-Boat-1 (C_2)	Twist-Boat-2 (C_1)
Bond stretching	0.18	0.25	0.24	0.24	0.25
Bond-angle bending	0.44	0.53	0.56	0.46	0.50
Torsional strain	0.23	4.80	4.65	4.35	4.38
Out-of-plane bending	0.01	0.05	0.02	0.00	0.04
Nonbonded interactions	2.87	4.31	3.95	3.86	4.05
Total strain energy	3.73	9.42	9.42	8.91	9.22

Few examples of substituted methylenecyclohexenes have been studied by molecular mechanics. Two examples that have been assessed by molecular mechanics are **47** and **48**, where the calculated enthalpies of axial and equatorial

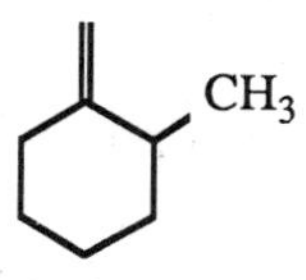

47

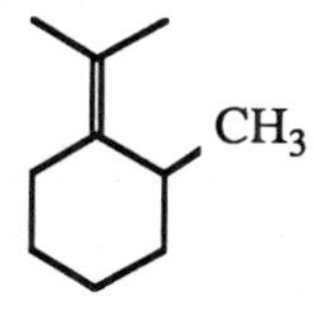

48

methyl groups were computed. The impetus for this particular study was to address the allylic strain concepts of Johnson who suggested that an increase in the size of substituents on the exo carbon of methylenecyclohexene would shift the normal equatorial preference of a C-2 substituent toward the axial position. The equatorial ⇌ axial methyl calculated enthalpies are $\Delta H^\circ = +0.50$ kcal/mol for **47** and $\Delta H^\circ = -2.62$ kcal/mol for **48**, confirming Johnson's early ideas about allylic strain.

Another interesting example of a methylenecyclohexane is **49**, which was originally suggested by one of the authors in this book as existing in a boat or twist form based on NMR coupling constants.[24] The pseudorotation cycle interconnecting the various forms of 1,4-dimethylenecyclohexane, qualitatively similar to methylenecyclohexane itself, is presented in Figure 7.6. The twist-boat form is seen to be more favorable than the boat form but less stable than the chair form by 3.53 kcal/mol. These are calculated enthalpies only. The boat forms of 1,4-dimethylenecyclohexane are relatively flexible and are expected to have higher

49

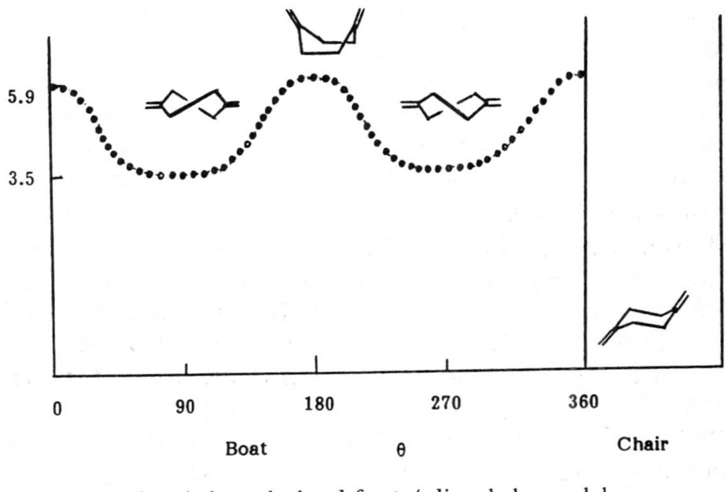

Figure 7.6 The conformational enthalpy calculated for 1,4-dimethylenecyclohexane.

entropies than the chair forms. It would be interesting to compute the free energies of this system but to my knowledge this has not yet been done.

7.2.6 Cyclohexenes as Part of Other Ring Systems

Cyclohexenes incorporated into bi-, tri-, and polycyclic ring systems have been studied by the empirical force-field method. Most of these studies have focused on olefin pyramidalization, especially in sesquinorbornenes, and olefin twisting, especially with regard to stability to bridgehead olefins.

7.2.6.1 Sesquinorbornenes

Norbornene, **50**, contains a cyclohexene, albeit as part of a bicyclic ring system. *Syn-* and *anti*-sesquinorbornenes, **51** and **52**, can be considered as two fused norbornenes sharing a common double bond. These molecules display

50

51

52

unusual chemical properties as well as perplexing and controversial structural features that have recently been summarized in a book.[25]

One of the more compelling reasons for studying these systems with molecular mechanics is that X-ray crystallographic analysis of derivatives of the anti-isomer, **52**, reveals both planar[26] and bent[27] double bonds. This bending is actually a pyramidalization of the olefin which implies a hingelike motion for diastereoisomerism is possible. Out-of-plane bending in norbornene, norbornadiene, and related bicyclic hydrocarbons due to olefin pyramidalization had been studied earlier by Wipff and Morokuma and by Houk and his group employing *ab initio* molecular orbital methods.[28] The first empirical force-field study of the out-of-plane bending of hydrogen in norbornadiene preceded these studies by six years where, using a consistent field, Ermer calculated a hydrogen movement in the endo direction of 3.9–4.1°.[29] This bending was also assessed with the MM2 force field where a 1.8° endo out-of-plane bending was computed.[30]

The oop distortions in these and related ring systems are modest when compared to the large bendings (12–18°) found in derivatives of **51** and **52**. In 1983 a series of papers from independent laboratories emerged that attempted to explain most of the peculiar structural, chemical, and spectroscopic properties of these molecules.[27,28,31,32]

Derivatives of *syn*-sesquinorbornene, **51**, have been studied by X-ray crystallography and all are characterized by a nonplanar π system with endo bending of 12–18°. With MM2 it was found that the endo conformation (Structure A in Figure 7.7) deviates from planarity by 18° and is more stable than the exo conformation which is computed to deviate from planarity by 35°.[31,32] Both the 18° bending and preference for the exo conformation are consonant with experiment. Using his consistent force field, Ermer reported the exo conformation to be more stable with substantially exaggerated out-of-plane bending.[27] These nonplanar double-bond distortions arise from large angle openings of the exocyclic sp^3-sp^2-sp^3 carbons when keeping the olefin planar, but they have been attributed to deficiencies in the force field.

For the anti-sesquinorbornene the preferred conformation is calculated by MM2 to be puckered 26°[31,32] (structure D in Figure 7.7). The planar conformation represents a transition state between the two puckered forms which is less than 2 kcal/mol destabilized with respect to the minimum. Interestingly, two X-ray structures of *anti*-sesquinorbornene derivatives have been reported; one has a planar double bond[26] and the other is puckered approximately 13°.[27] If the force field results are at all meaningful, crystal packing effects alone can account for these contradicting results.

The MM2 energies for the different conformations of **51** and **52** as a function of the pucker angle are summarized in Figure 7.8. The numbers in the figure are heats of formation computed by Jorgensen[31] with the block diagonal Newton–Raphson optimizer implemented by Allinger in MM2. Relative steric energies are reported by Johnson[32] who also used MM2 but with a full Newton–Raphson optimization to locate more precisely minima and maxima on the potential surface. Both studies effectively give the same results.

Two as yet unknown compounds, **53** and **54**, were also investigated by Johnson.[32] The ground-state structure of **53** has exact C_{2v} symmetry and is bent

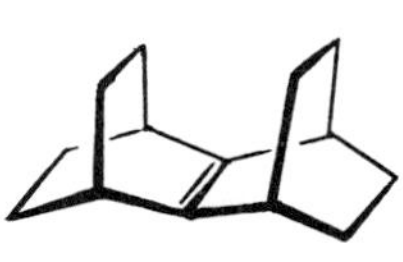

53

54

7°. A planar transition state of D_{2h} is calculated to lie only 100 cal above the interconverting minima. The most stable structure of **54** also has C_{2v} symmetry but in contrast to **53** is puckered 35°. Since MM2 overestimated the puckering of the *anti*-sesquinorbornene (26° MM2 versus 13.2° X-ray of a derivative), we might expect this value also to be overestimated. Nonetheless, a substantial pyramidalization of this olefin is expected. A planar transition state with D_{2h} symmetry was computed to be 6.2 kcal/mol above the ground state.

Several comments about these results are needed. First, the inversion process

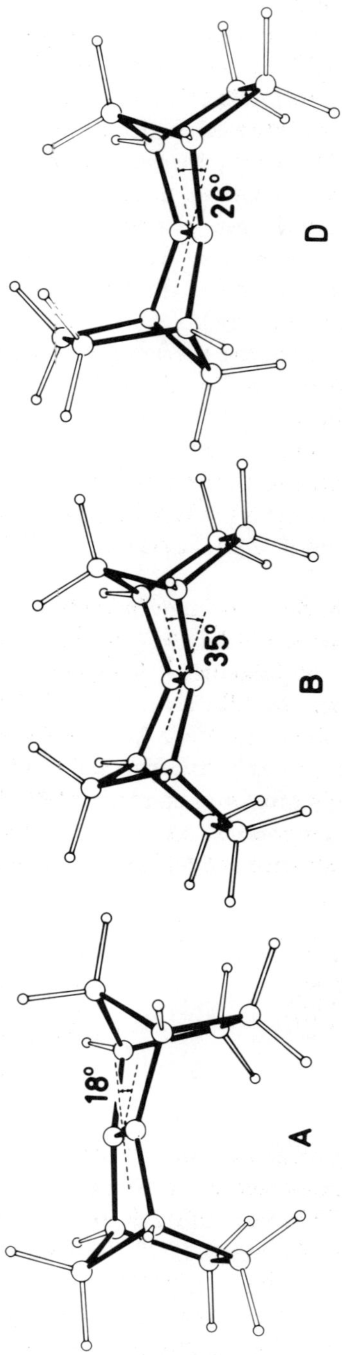

Figure 7.7 Computer-generated three-dimensional drawings of the minimum-energy conformations of the sesquinorbornenes **51** (A and B) and **52** (D).

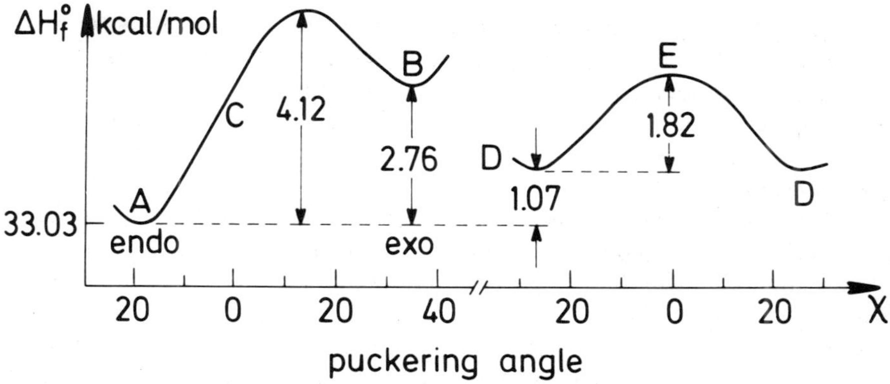

Figure 7.8 MM2 calculated energies for the different conformations of the sesquinorborenes **51** and **52** as function of the puckering angle.

in **51** represents a diastereoisomerization that is unlikely to be observed because less than 1% of the minor diastereomer is expected to exist at room temperature. Second, the inversion processes in **52–54** represent topomerizations that might be observable on the NMR time scale for **54** only. Finally, a comment about the MM2 force field. My opinion is that MM2 oop bending force constants are inadequate; they are too low for alkenes and alkynes. Thus, we expect and find in these studies that MM2 will overestimate the oop bendings.

Two very different explanations for alkene pyramidalizations in these ring systems have been presented.[25] One, based on quantum chemical arguments, concludes that distortion occurs to minimize hyperconjugative mixing between olefin π orbitals with high-lying σ orbitals especially in the methylene bridges, while the other concludes distortions originate from torsional effects in which the unsymmetrical arrangement of the three allylic bonds forces the alkene to pyramidalize. This was first proposed by Burkert[30] and later verified by Houk et al.[28] based on molecular mechanics calculations.

7.2.6.2 Bridgehead Olefins

Another application of molecular mechanics involves the analysis of olefinic strain (OS) in bridgehead alkenes. Many of these alkenes, as a consequence of being incorporated in a bi- or polycyclic substructure, may be construed as cycloalkenes larger than six carbons. Nonetheless, many of these complex ring systems do contain a cyclohexene moiety and are therefore considered here.

The evaluation and prediction of the stability of bridgehead olefins was placed on a firm basis by Maier and Schleyer who, in a seminal paper, formalized the concept of olefinic strain (OS) which has supplanted many of the previous empirical rules and computational methods for estimating alkene stability.[33] OS is calculated by subtracting the total strain energy of the most stable conformer of the parent hydrocarbon from the total strain energy of the olefin in its most stable conformation. Since OS values are obtained by subtraction, many of the errors inherent to molecular mechanics are presumed to cancel. For small systems this definition of OS does not create practical problems in performing the analysis because relatively few minima on the multidimensional potential energy surface exist and the global minimum for alkane and alkene can be located with some confidence. For larger systems more work and care is needed to ensure the global minima have been located.

The dominating influence on the bridgehead olefin stability is the strain associated with the double bond; the smaller of the two olefinic rings always incorporates a cis cycloalkene whereas the larger ring maintains a trans configuration. When the double bond is incorporated in the larger bridge of the bicycloalkene, the conformational influence decreases in the order: *cis*-cyclopentene > *cis*-cyclohexene > *cis*-cycloheptene > *cis*-cyclooctene.

An empirical set of rules derived from comparison of OS values computed with molecular mechanics with experimental data allow classification of bridgehead alkenes into three categories:

1. Isolable: OS $\leq$ 17 kcal/mol.
2. Observable: 17 $\leq$ OS $\leq$ 21 kcal/mol.
3. Unstable: OS $\geq$ 21 kcal/mol.

Definitions of these categories are found in the original paper.[33] These rules do not apply to all types of strained alkenes. Alkenes lacking twisted geometries like cyclopropenes or alkenes located between two bridgeheads that can bend out of plane as in the above section are not considered to be Bredt olefins and do not obey the OS rules. Note that the dynamic properties of the bicyclic molecules related to sesquinorbornene mentioned above involve a hinge like bending about the double bond and that the transition states for diastereoisomerization or topomerization do not result form twisting about the π bond, but rather from bending only.[32] In the OS calculations most of the strain energy originates from twisting rather than bending. Interestingly, an earlier force-field analysis of bridgehead olefin stability by Ermer[34] suggested a quantitative reactivity criterion could be based on the out-of-plane bending energy of the double bond.

Many bridgehead alkenes have been studied with molecular mechanics, only a few of which will be mentioned here.[33] Bicyclo[2.2.1]hept-1-ene, **55**, a norbornene isomer, contains a boatlike cyclohexene that has an OS value of 34.9 kcal/mol and has not been observed directly. The homolog, bicyclo[2.2.2]oct-1-ene, **56**, has an OS in excess of 40 kcal/mol and has been presumed to exist only

55

56

57

from trapping experiments. The bicyclo[3.2.2]nonene, **57**, has an OS calculated to be 20.6 kcal/mol. This "cyclohexene" has been investigated at -80 °C but dimerizes at higher temperatures as do many of these strained alkenes. The bicyclo[3.3.1]non-1-enes **58a** and **58b** have been studied with the former having

58a

58b

an OS value of 44.2 kcal/mol and the latter only 15.2 kcal/mol. Other examples of cyclohexenes incorporated in bicyclic rings have also been studied this way.

An interesting extension of the above work was the realization that in medium-size polycyclic ring systems, some bridgehead double bonds may be more stable than the corresponding alkane, resulting in negative olefin strain energies. These compounds are considered "hyperstable olefins" and have been thoroughly investigated by McEwen and Schleyer.[35] Here the strain energy of the olefin is less than that of its parent hydrocarbon.

An interesting series of cyclohexenes investigated for olefinic strain with molecular mechanics is the bicyclo[n.2.2] bridgehead olefins **59**, where $n = 1–10$. All molecules in this series have a *cis*-cyclohexene and a trans double bond in the second ring. The olefinic strain values for the series $n = 5–10$ is quite different

59a

n = 1 - 10

59b

than in the series $n = 1–4$, the former series being hyperstable in contrast to the latter. These results are summarized in Figure 7.9. Hyperstability of bridgehead olefins is related to the stability of the cyclic moieties. Qualitatively, as the systems increase in size, hyperstability of bridgehead olefins begins when both a *trans*-cyclononaene and a *cis*-cyclohexene ring are present.

The ramifications of negative OS manifest themselves in, among other effects, reduced chemical reactivity. For example, some cyclophanes that undergo only partial hydrogenation resist further reduction because of hyperstability. Thus, [2.2.2]-(1,2,4)cyclophane and [2.2.2]-(1,2,4)-(1,2,5)cyclophane incompletely hydrogenate to cyclohexenes **60** and **61**, respectively.

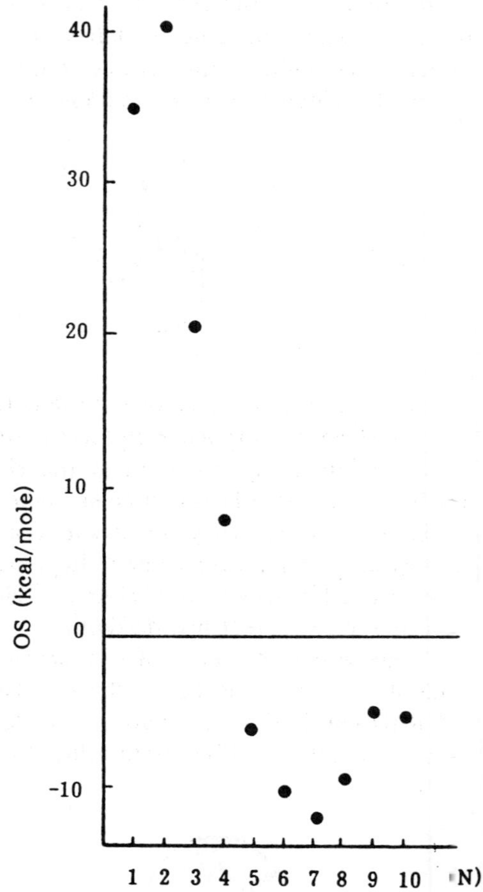

Figure 7.9 OS value versus chain size (n) for bicyclo[n.2.2]-1(n + 40)-enes.

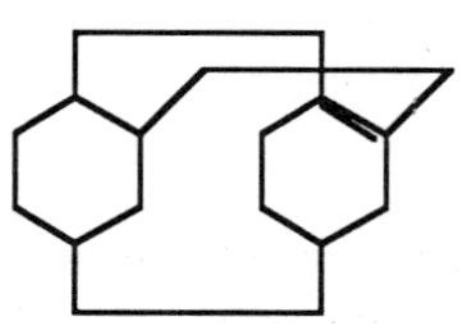

60

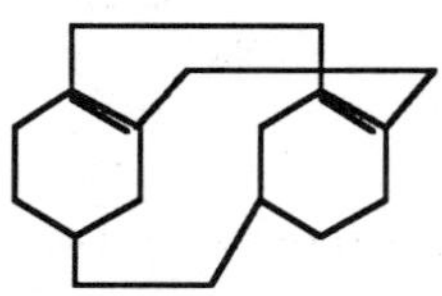

61

Many other examples of cyclohexenes incorporated into bicyclic or polycyclic rings have been studied for olefinic strain by molecular mechanics. One last example of OS induced by π bond twisting in cyclohexenes will be presented here. It provides an example of how molecular mechanics can be used to study an as yet unknown molecule by providing insight about its stability and whether or not an attempt should be made to synthesize it. Tetracyclo[8,2,2,$0^{2,7}$,$0^{3,10}$]tetradecene-2(3), **62**, referred to as orthogonene and its parent hydrocarbon, **63**, have been studied with MM2.[36] Orthogonene has a twisted olefin with an internal strain

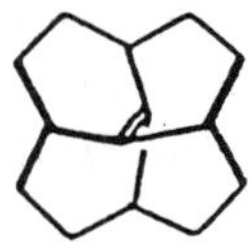

62

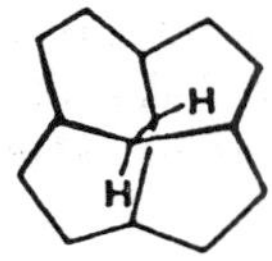

63

energy of 118 kcal/mol while its parent is calculated to have a strain energy of 99.6 kcal/mol. The OS of this interesting molecule is 18.4 kcal/mol, which classifies it as borderline between an isolable and observable bridgehead olefin. The parent hydrocarbon is itself highly strained mainly due to transannular interactions between internal hydrogen atoms which may be an important factor to enhance the kinetic stability of **62**. Additionally, the polycyclic structure forces the vinyl bond angles to 107° rather than the normal values in alkenes. This, in turn, enhances the coupling between the orthogonal *p* atomic orbitals with the appropriate σ bonds in the cyclohexenes which hyperconjugatively help stabilize the twisted form of the π bond. Keep in mind that even though the OS indicates orthogonene as a marginally isolable molecule this may be an artifact of the force-field calculation; certainly this molecule is not one that is especially well suited to molecular mechanics calculations because of the severity of the olefin distortion and the unusual electronic effects not properly accounted for in molecular mechanics.

7.2.7 Dihydroaromatics

1,2-Cyclohexadiene is not amendable to study by empirical force fields; 1,3- and 1,4-cyclohexadienes are and have been explored with molecular mechanics. The 1,4-isomer in particular has been thoroughly studied by quantum and molecular mechanics. The conformational properties of this molecule, along with its benzannulated homologs dihydronaphthalene and dihydroanthracene, are well characterized and provide a focal point of this book.

7.2.7.1 1,3-Cyclohexadiene

This conjugated system has a twisted C_2 geometry with the diene moiety twisted out of plane 12–18° depending on the experimental determination. Thermodynamically, it has almost the same free energy as the isomeric 1,4-diene. Favini and co-workers[37] examined cycloalkadienes with molecular mechanics by making the C^2_{sp}—C^2_{sp} bond length a function of the dihedral angle, while simultaneously making the dihedral angle a function of the *p-p* π resonance integral. Allinger, using a more generalized approach to calculating structure and energies of conjugated π systems,[38] calculated the torsion angle $\omega_{1,2,3,4}$ to be 12°. The nonplanarity of the ring is attributed to (1) the nonbonded interactions of the methylene protons at C-5 with those of C-6 (they are eclipsed when the ring is planar) and (2) the angle strain that would exist in the sp^3 carbons if the ring were planar. Factors opposing nonplanarity are the nonbonded repulsions of the vinyl protons on C_1 and C_4 with the methylenes on C-5 and C-6, and the disruption of conjugation in the π system. These opposing forces tend to make the potential

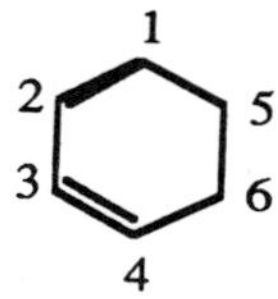

64

energy surface very shallow. The barrier to conformation inversion in 1,3-cyclohexadiene appears not to have been investigated.

7.2.7.2 Dihydrobenzene, Dihydronaphthalene, and Dihydroanthracene

The preferred geometries of 1,4-dihydrobenzene (DHB), **65**, 1,4-dihydronaphthalene (DHN), **66**, and 9,10-dihydroanthracene (DHA), **67**, has been the source of many investigations and the center of considerable controversy.[39] At

65

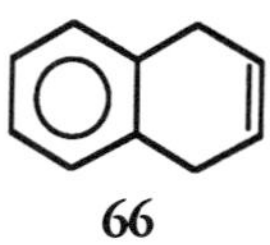

66

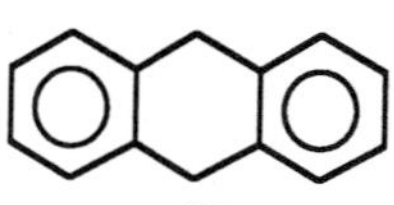

67

issue was whether the preferred conformation of dihydroaromatics correspond to planar structures (**65a**) or two pairs of equilibrating boat conformations (**65b**). A historical account of this controversy, especially with regard to misinterpretation of

65b C_{2v} 65a D_{2h} 65b C_{2v}

spectra, has been provided by Rabideau in Chapter 4 and is not reconstructed here. Early conformational analysis, as in most studies, relied heavily on information gleaned from mechanical models; unfortunately, mechanical models completely misrepresent reality in these ring systems. Furthermore, a very early force calculation by Herbstein[40] on 1,4-cyclohexadiene did not adequately address the questions of planarity because the influence of torsional strain was not considered. High-quality molecular mechanics calculations by Allinger and Sprague[4] and Ermer and Lifson[5] on **65**, however, clearly depicted the molecule with a planar geometry. This was later confirmed with yet another force field developed by White and Bovill.[23a]

Some of the controversy regarding the conformation of dihydro aromatic compounds can be traced to the choice of the model on which expectations are based. Consider first the question of angle strain. For a planar six-membered ring the sum of the interior angles must be 720° (i.e., the average bond angle in the ring must be 120°). Using 120° and 109.5° as the optimum angles at sp^2 and sp^3 centers, respectively, substantial angle strain would be expected for planar conformations of **65–67**. This situation is reflected by the behavior of Dreiding models, which show a clear preference for nonplanarity. However, neither 109.5° nor 120° bond angles may be optimal. While toluene may be a good model for the aromatic derivative **67**, a simple alkene such as propene with a C=C—CH_3 bond angle of 124° is a much more appropriate model for the dihydrobenzene, **65**. Propane, with a C—C—C bond angle of 112.4°, can be used to model the C—CH_2—C angle of **65–67**.

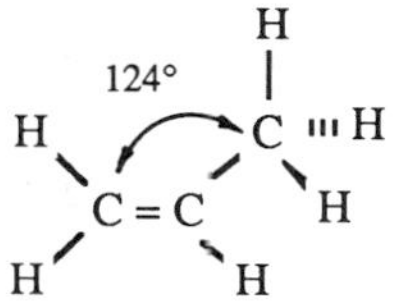

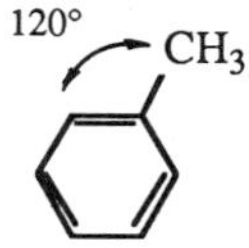

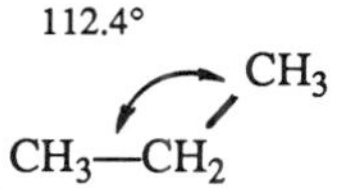

Using the appropriate bond angles for toluene, propene, and propane as optimal, the sum of the interior angles in a model planar structure for **67** would be 705° [i.e., (4 × 120°) + (2 × 112.4°)], whereas the interior angles in the corresponding model planar structure for dihydrobenzene, **65**, would add up to 721° [i.e., (4 × 124°) + (2 × 112.4°)]. Consequently, planarity could be achieved in dihydroanthracene **67** only with some angle distortion (705° versus 720°). In contrast, planar 1,4-dihydrobenzene should have no angle deformation. The angle strain for planar dihydronaphthalene **66** would be intermediate, and the trend for angle strain would favor nonplanar structures in the order **67** > **66** > **65**.

Torsional effects also are important. In the planar conformations of **65–67**, the two hydrogens of each CH_2 group are symmetrically staggered with respect to the adjacent vinyl hydrogens or aryl ring. Propene and toluene again model the effects to be expected in **65–67**. The preferred conformation of propene, with the vinyl hydrogen staggered between two hydrogens of the adjacent methyl group, would favor a planar conformation for **65**. Any distortion from planarity will therefore lead to unfavorable torsional (eclipsing) interactions. (Note that this effect is not incorporated into Dreiding models which only reflect angle strain.)

On the other hand, toluene exhibits little rotational preference; hence little torsional effect is predicted for **67**. Once again the dihydronaphthalene **66** should be intermediate. Torsional effects therefore should favor planar structures in the order **65** > **66** > **67**. The nonplanar forms of **65–67** should also be favored by their lower symmetry. This would contribute modestly (0.4 kcal/mol at 298 K) to the entropy term ($T\,\Delta S$).

The subtle interplay between torsional and angular distortions in dihydroaromatics noted by Lifson and by Allinger prompted Rabideau and Lipkowitz to carefully assess the conformations of **65–67** and their substituted derivatives by quantum and molecular mechanics.[41,43,44] (The quantum chemical results are discussed in detail in Chapter 6.) The puckering of rings **65–67** is defined by the

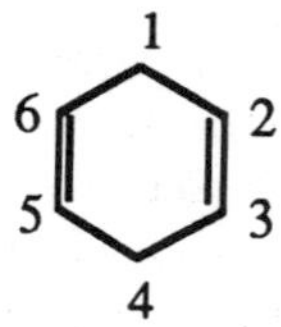

improper torsion angle α = [(5)-(4)-(1)-(3) + (6)-(4)-(1)-(2)]/2. Using the MMI force field, the energy profiles in Figures 7.10–7.12 were obtained; similar results are found with the MM2 force field.

All computational methods have planar minimum energy conformations. Dihydroanthracene, **67**, has an additional minimum energy conformation predicted by MM2 to be in a boat form. This result contrasts somewhat with the CFF calculation of Sygu/a and Holak[42] who find only boat conformations for **67**. The potential energy curves for **65** and **67** as a function of ring pucker computed with the QCFF/PI + MCA method is depicted in Figure 7.13. Whether the triple minima surface in Figure 7.12 or double minima surface in Figure 7.13 is correct is

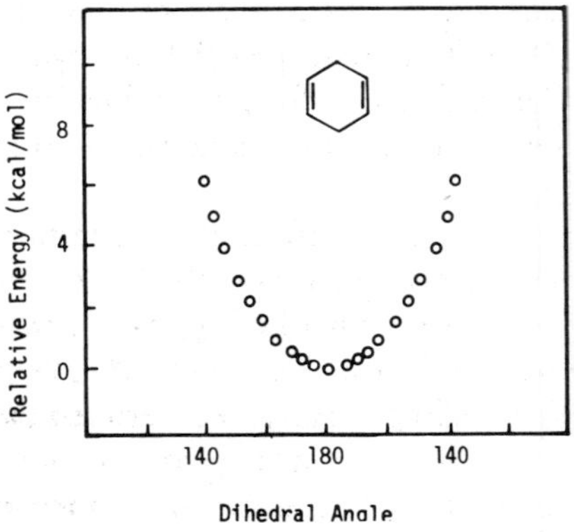

Figure 7.10 MMI energy profile of **65**.

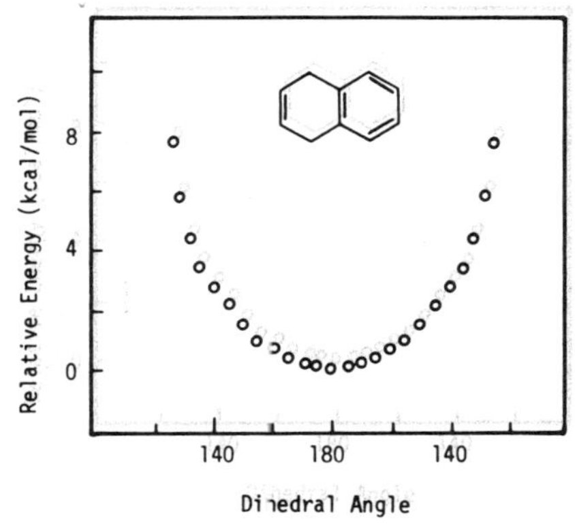

Figure 7.11 MMI energy profile of **66**.

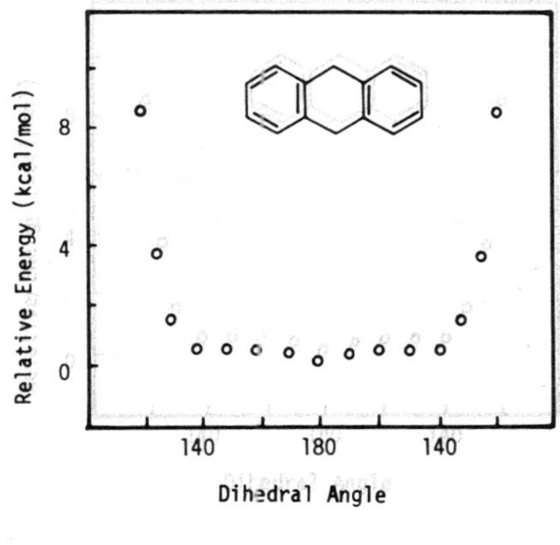

Figure 7.12 MMI energy profile of **67**.

a moot point; both force fields indicate an effectively planar structure with minima separated by extremely low barriers.

The shape of these slices through the multidimensional energy surfaces for molecules **65–67** may be summarized as follows. Although the minimum energy structures for **65–67** in each case were indicated to be planar, the calculated energy wells (Figures 7.10–7.13) are quite flat. The energies required for 20° distortions from planarity for **65–67** are about 1 kcal/mol or less and decrease in the order **65** > **66** > **67** due to the interplay or angle strain and torsional effects. The particularly small energy required for distortion of **67** (Figure 7.12) permits explanation of the otherwise contradictory experimental result for this compound. The extremely flat energy surface for **67** suggests that the nonplanar X-ray structure may not represent the preferred geometry in solution or in the gas phase. The thermal energy available at room temperature (H_{RT} = 0.6 kcal/mol) and the entropy contributions will presumably result in large puckering amplitudes, particularly in the case of **67**, and this will influence significantly any experimental

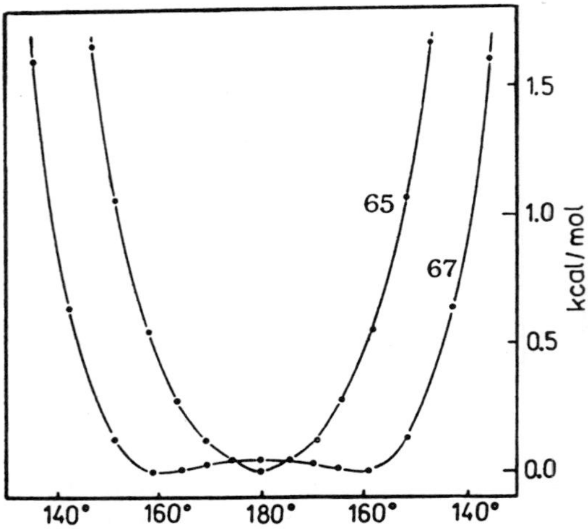

Figure 7.13 QCFF potential energy curves for 1,4-dihydrobenzene, **65**, and 9,10-dihydroanthracene, **67**.

measurements made on these compounds. Such perturbation of experimental observations has already been suggested for **65**, and a recent study showing a temperature dependence of 1,4-proton coupling constants in partially deuterated **65** provides additional support for this conclusion. While the parent compounds **65–67** have minimum energy structures that are effectively planar, this may not be true for their derivatives. The actual conformations of substituted dihydroaromatic compounds are determined by the conformational preferences of whatever substituents are present (see below).

7.2.8 Monosubstituted Dihydroaromatics

The introduction of a single substituent on the saturated carbon of molecules **65–67** reduces the symmetry and a planar conformation with $\alpha = 180°$ is no longer expected to be an energy minimum. By placing substituents on these rings, we introduce yet another subtle effect: steric repulsions. A force-field study of the effect of alkyl groups on DHB, DHN, and DHA has been reported using the MMI and MMPI force fields.[43,44] Molecules **68–70** were explored in detail.

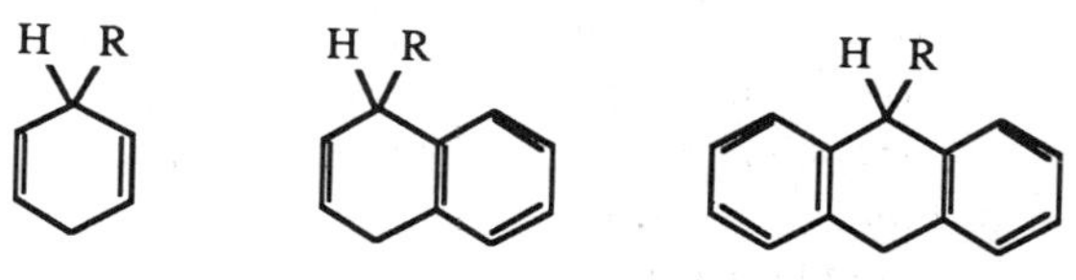

68 **69** **70**

a, R = H
b, R = CH_3
c, R = C_2H_5
d, R = i-C_3H_7
e, R = t-C_4H_9

Table 7.11 Optimized Structures of 1-Substituted 1,4-Dihydrobenzenes[a,b]

Compound	R	Folding Angle, α (degrees)	Strain Energy (kcal/mol)
68a	H	180	−1.6
68b	CH_3	174	−0.7
68c	C_2H_2	173	−0.7
68d	*i*-C_3H_7	169	0.1
68e	*t*-C_4H_9	160	1.8
68f	C_6H_5	176	*c*

[a] Complete geometry optimization with MMI.
[b] The preferred orientation of the substituent is axial in each case.
[c] Not calculated; MMI does not contain the necessary heat of formation parameters for aromatic compounds.

7.2.8.1 Monosubstituted DHBs

The geometry optimized structures of 1-substituted DHBs are presented in Table 7.11. As expected $\alpha = 180°$ is no longer the minimum energy structure. It is seen that the distortion from planarity increases with the steric bulk of the substituent. While the optimum geometry for **68b-e** is nonplanar in each case, we are reluctant to describe these structures as "boats" because the distortions are small. Even the *tert*-butyl derivative ($\alpha = 160°$) is bent by only 20° whereas Dreiding models of the dihydroaromatics exhibit distortions about twice as large ($\alpha = 145°$). Moreover **68a-e** each exhibit a single energy minimum. Thus, the boat–boat equilibrium does not describe the behavior of this system. The energy profile for strain energy versus folding angle of 1-methyl-1,4-dihydrobenzene is typical and is shown in Figure 7.14 together with the energy profile for the parent compound **68a**.

Figure 7.14 clearly shows that the major effect of substitution at the 1 position is to skew the energy profile relative to that of **68a**. The energy minimum for **68b** is at 174° rather than 180°, but the curves are very similar in other

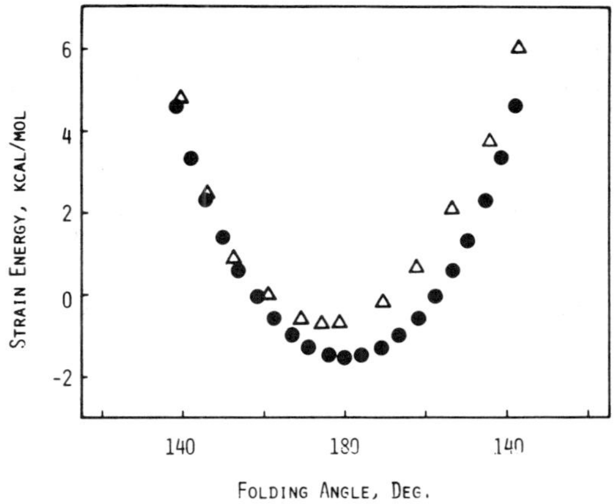

Figure 7.14 Energy versus folding angle for 1,4-dihydrobenzene (**68a**, ●) and 1-methyl-1,4-dihydrobenzene (**68b**, △).

respects. One major consequence of these calculations is the demonstration that a bulky substituent in a nonplanar conformation preferentially occupies the pseudoaxial position. There is no energy minimum for conformations in which the substituent is pseudoequatorial even though it has long been assumed that the equatorial position would be the favored location for substituents. This original suggestion was made on the basis of space-filling models and appears to have been accepted without serious question for almost 10 years. This erroneous interpretation was almost certainly aided by the behavior of Dreiding models (which exhibit a clear preference for nonplanar geometries for **65–67**) and by analogy with cyclohexane derivatives. Unfortunately, such comparisons are invalid and misleading. Dreiding models reflect only angle strain, whereas the conformations of the dihydroaromatics are determined by the interplay of angle strain, torsional effects, and nonbonded interactions. The preferred equatorial conformations of substituted cyclohexane reflect a decrease in the unfavorable 1,3-diaxial interactions of the axial conformer. In contrast, nonbonded interactions in the dihydroaromatic series are minimized in the axial conformation and maximized in the equatorial conformation, where the substituent is eclipsed with the adjacent vinyl hydrogen.

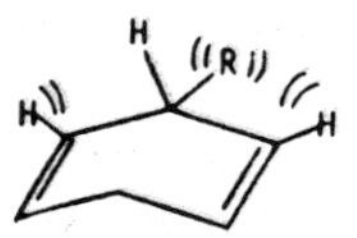

The calculations of **68a-f** fully support this interpretation. In each instance the preference for a nonplanar conformation can be viewed as a destabilization of the planar form resulting from steric interactions between the substituent and the adjacent vinyl groups. These nonbonded interactions are alleviated by distortion from planarity, and this in turn causes an increase in angle strain. Increasingly steric bulk of the substituent results in a monotonic change in folding angle along the series **68a-e** and a corresponding change in strain energy. Introduction of an alkyl substituent at the 1 position of 1,4-dihydrobenzene therefore leads to several changes in the energy profile (Figure 7.14). The optimum structure is no longer planar, and the strain energy of this structure is greater than for the unsubstituted

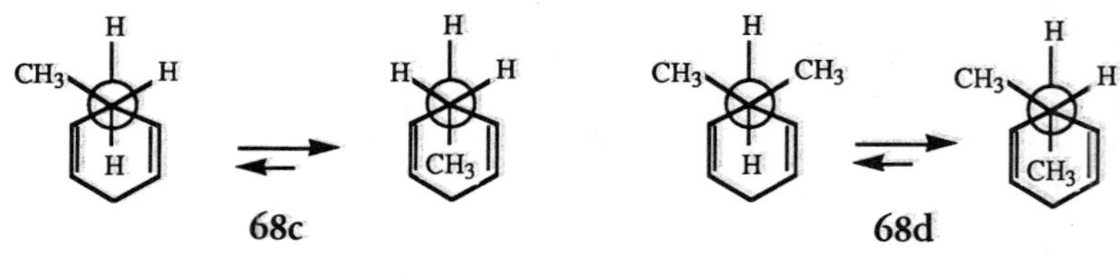

68c **68d**

Strain energy (kcal/mol)	0.2	-0.7	1.7	0.1
α	170°	173°	160°	169°

compound (**68a**). An even greater increase in strain energy is seen for those geometries in which the methyl group has an equatorial orientation. For geometries that are distorted from the optimum geometry but with an axial methyl group, the curve for **68b** becomes nearly superimposable on that for **68a**.

The ethyl and isopropyl derivatives **68c** and **68d** provide further evidence that the major structural feature governing the nonplanar distortion is the interaction of the alkyl substituent with the adjacent vinyl groups (allylic strain). Both of these compounds have two nonequivalent rotamers of the alkyl substituent, each of which is an energy minimum. The more stable rotamer for each compound is that in which interactions with the vinyl CH are minimized. Both the strain energy and the folding angle appear to be determined almost entirely by the groups that are gauche to vinyl CH. For example, the less stable (C_s symmetry) rotamer of **68d** is folded to the same extent as the *tert*-butyl derivative, **68e**, and also has nearly the same strain energy. The same comparison can be made between the more stable rotamer of **68c** and the methyl derivative **68a**, and between the two unsymmetrical rotamers of **68c** and **68d**. Finally, the small distortion from planarity of the phenyl derivative reinforces these arguments. Although the phenyl group has a greater steric bulk than an isopropyl group in cyclohexyl systems, the phenyl derivative **68f** exhibits a smaller distortion than any of the alkyl derivatives **68b-e**. By adopting a perpendicular geometry it is possible for **68f** to minimize the unfavorable interactions between the phenyl ring and vinyl CH groups.

7.2.8.2 Monosubstituted DHNs

The unfavorable nonbonded interactions of alkyl substituents found in monosubstituted DHBs are expected to be even more severe here. When an alkyl substituent is in a pseudoequatorial arrangement, the nonbonded interactions

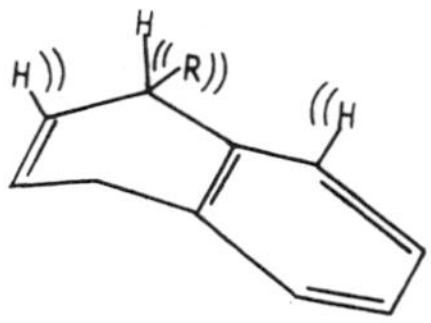

involve the vinyl hydrogen on one side and the aromatic ring with its peri hydrogen on the other side. The calculations for **69a**, **69b**, and **69e** (Table 7.12) substantiate these expectations. The change in strain energy along the series **69a-e** is greater than for **68a-e**, and the distortions from planarity are also greater for derivatives of **69**.

The energy profile of the methyl derivative, **69b**, is shown in Figure 7.15 together with that of the parent compound **69a**. As in the case of the dihydrobenzenes, the major effect of methyl substitution is to increase the energy of those geometries where the substituent is pseudoequatorial. Thus, the portion

Table 7.12 Optimized Structures of 1-Substituted 1,4-Dihydronaphthalenes[a,b]

Compound	R	Folding Angle, α (degrees)	Relative Strain Energy[c] (kcal/mol)
69a	H	180	0
69b	CH_3	164	1.3
69e	t-C_4H_9	149	4.3
		132[d]	13.6[d]

[a] Complete geometry optimizations with MMI.
[b] Preferred orientation of the substituent is axial in each case.
[c] Relative to the planar form of **69a**. The absolute strain energy is not calculated because MMI does not contain the necessary heat of form parameters for aromatic compounds.
[d] The substituent is equatorial

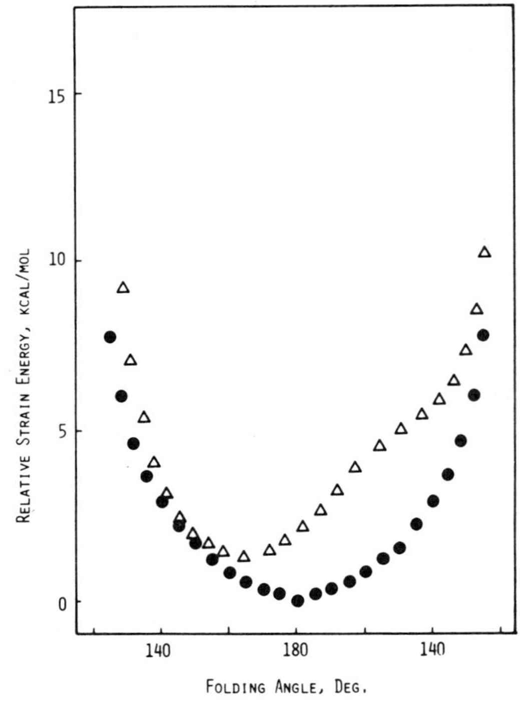

Figure 7.15 Energy versus folding angle for 1,4-dihydronaphthalene (**69a**, ●) and 1-methyl-1,4-dihydronaphthalene (**69b**, △).

of the curve for **69b** where the methyl group is axial is nearly superimposed on the curve for **69a**. Only when the substituent becomes equatorial does the relative strain energy differ substantially from that of **69a**.

The *tert*-butyl derivative, **69e**, exhibits a new feature in its energy profile (Figure 7.16). Unlike the dihydrobenzenes, **68a-f**, and the other dihydronaphthalenes, **69a** and **69b**, this compound has two distinct energy minima. From a theoretical point of view this is a boat-to-boat equilibrium, but in practice it is a

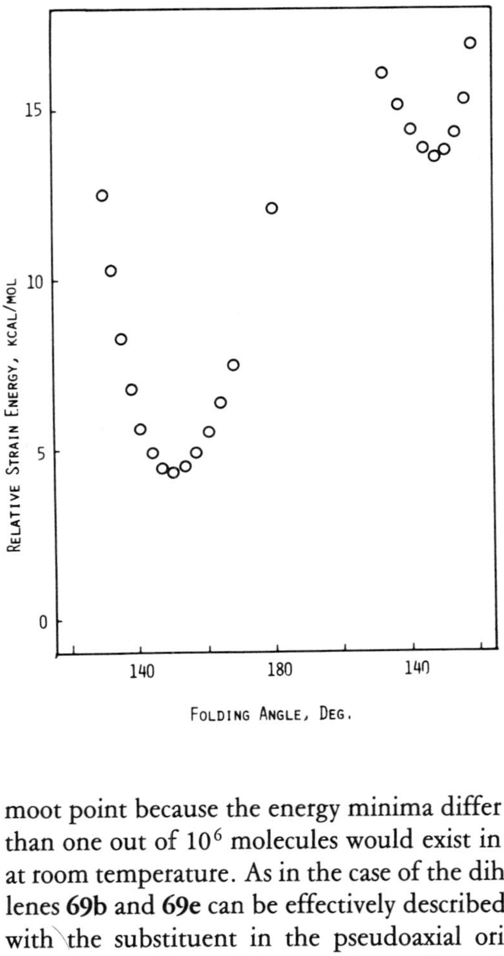

Figure 7.16 Energy versus folding angle for 1-*tert*-butyl-1,4-dihydronaphthalene (**69e**).

moot point because the energy minima differ by more than 9 kcal/mol and fewer than one out of 10^6 molecules would exist in the pseudoequatorial conformation at room temperature. As in the case of the dihydrobenzenes, the dihydronaphthalenes **69b** and **69e** can be effectively described as existing in a single conformation with the substituent in the pseudoaxial orientation. These results are also in agreement with previous experimental conclusions that many derivatives of **69** exist as a "flattened boat" and that greater distortions from planarity are found for dihydronaphthalenes with a single large substituent.

The energy of the parent dihydroanthracene, **70a**, is insensitive to the folding angle. The calculated energy minimum with MMI is a planar structure, although a folding angle of 145° is found in the solid state. It can be anticipated that alkyl substitution would substantially alter the situation because nonbonded interactions involve two aromatic rings. Indeed, these nonbonded effects strongly destabilize the planar form, resulting in folded structures that are preferred for both **70b** and **70e** (see Table 7.13).

As was observed with **69e**, both of these substituted dihydroanthracenes show two energy minima (Figure 7.17). Once again the energy difference between the two forms is quite large; for example, the minima for **70b** differ by 3.7 kcal/mol. At room temperature the pseudoequatorial form would not be experimen-

Table 7.13 Optimized Structures of 9-Substituted 9,10-Dihydroanthracenes[a,b]

Compound	R	Folding Angle, α (degrees)	Relative Strain Energy[c] (kcal/mol)
70a	H	180	0
70b	CH_3	148	0.9
		131[d]	4.6[d]
70c	*t*-C_4H_9	143	5.2
		126[d]	15.0[d]

[a] Complete geometry optimization with MMI.
[b] Preferred orientation of the substituent is axial in each case.
[c] Relative to the planar form of **70a**. The absolute strain energy is not calculated because MMI does not contain the necessary heat of formation parameters for aromatic compounds.
[d] The substituent is equatorial

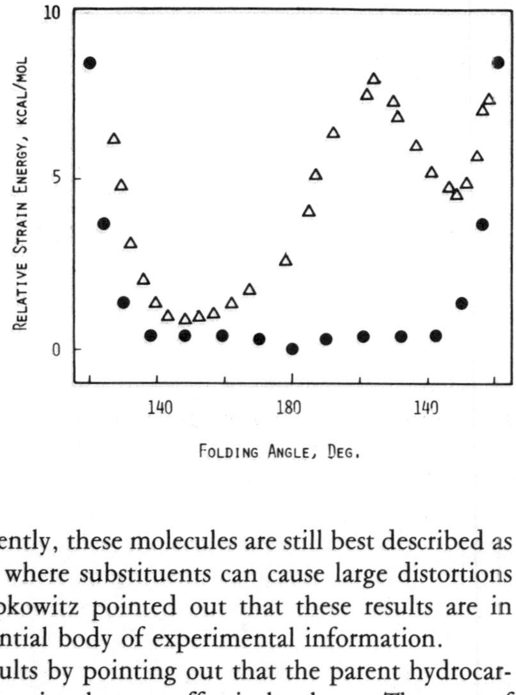

Figure 7.17 Energy versus folding angle for 9,10-dihydroanthracene (**70a**, ●) and 9-methyl-9,10-dihydroanthracene (**70e**, △).

tally detectable by NMR. Consequently, these molecules are still best described as existing in a single conformation, where substituents can cause large distortions from planarity. Rabideau and Lipkowitz pointed out that these results are in complete agreement with a substantial body of experimental information.

We can summarize these results by pointing out that the parent hydrocarbons, **68–70**, have optimum geometries that are effectively planar. The ease of distortion is not greatly affected by 1-alkyl substitution for dihydrobenzenes, and the cost of a 15° folding distortion from the optimum structure is about 0.6 kcal/mol for both **68a** and **68b**. Alkyl substitution has an increasingly large effect on the benzannulated derivatives **69** and **70**, however. A 15° distortion requires only 0.5 kcal/mol for **69a**, and comparable folding for the methyl derivative requires 0.5 and 0.7 kcal/mol depending on the direction. The energy for 15° distortion of the parent dihydroanthracene is further decreased to less than 0.4 kcal/mol, but the methyl derivative requires 0.9 or 2.5 kcal/mol depending on the direction of the distortion. These effects can be seen graphically by comparison of the steepness of the various curves in the figures. Clearly, alkyl substitution in the dihydroanthracenes, and to a lesser extent in the dihydronaphthalenes and dihydrobenzenes, results in a much steeper energy well. This in turn should greatly reduce the vibrational amplitudes and result in experimental behavior which is much more consistent with a single, well-defined molecular geometry.

In all three of the dihydroaromatic systems, substitution at one of the reduced positions distorts the optimum geometry from a planar conformation to one in which the substituent occupies an axial position. The preferred axial orientation results largely from minimizing nonbonded interactions between the substituent and the adjacent vinyl or peri hydrogens. These steric interactions are greater in the derivatives of **69** and **70**. Thus, the energy profiles are much steeper for derivatives of **69** and **70** than for **68**. Similarly, the nonplanar distortion of the optimum geometry increases from **68** to **69** to **70** with any particular alkyl substituent. A comment concerning the validity of the empirical force-field results should also be made here. The force field employed for these studies was MMI, which is now known to invoke hydrogens that are too "large" and too "hard." This ultimately overestimates the H---H repulsions. Since the steric effects that exist in these dihydroaromatics are predominantly H---H repulsions, we point out that in some instances the shapes of the potential energy curves are not very accurate. In particular, the double minima found in Figures 7.16 and 7.17 may not exist. The origin of the double minimum is the nonbonded repulsion of the alkyl group hydrogens with the peri hydrogens of the aromatic ring. As the ring is puckered the axial substituent encounters the peri hydrogens and eventually slides past them. It could be expected that with smaller and softer hydrogens, as found in most other force fields as well as in the MM2 force field, these repulsions are removed and the higher energy minima revert to uneventful inflections that look similar to the methylnaphthalene energy surface in Figure 7.15.

7.2.9 Disubstituted Dihydroaromatics

Traditional views on the stereochemical preferences of disubstituted dihydroaromatics would include planar forms (particularly for DHBs) and boat-to-boat interconversions (particularly for DHAs). The results of molecular mechanics calculations by Rabideau and co-workers,[44] however, show that the conformational properties of DHBs, DHNs, and DHAs may be far more complex than previously believed (illustrated in Figure 7.18). The origin of this greater complexity is due chiefly to nonbonded interactions between the alkyl substituents and the peri hydrogens. This interaction increases, of course, in the series DHB $<$ DHN $<$ DHA as well as Me $<$ *t*-Bu. Two significant consequences of these peri interactions are illustrated with **71** and **72**. We may view **71** as inherently planar but where nonbonded interactions produce a distortion into a chairlike conformation. Interestingly, although considerable controversy has existed concerning the stereochemistry of these dihydrobenzenes, chair structures

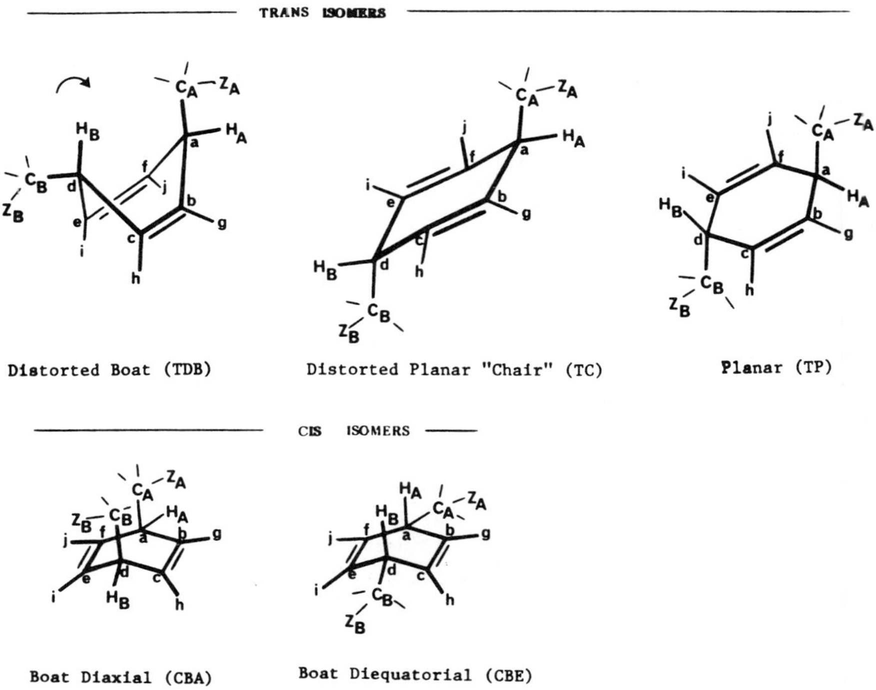

Figure 7.18 Conformations of *cis*- and *trans*-dimethyl- and di-*tert*-butyl-dihydrobenzenes (DHB), -dihydronaphthalenes (DHN), and -dihydroanthracenes (DHA).

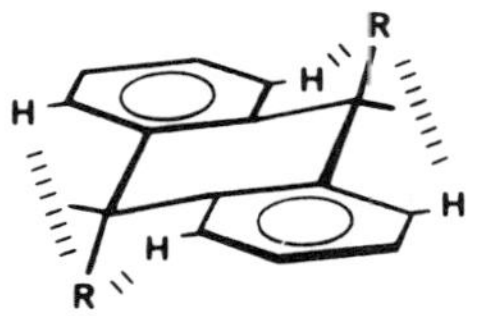

71

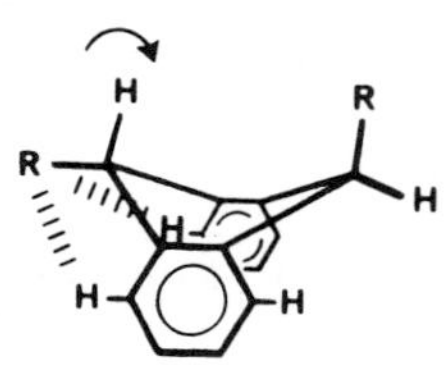

72

have never been seriously proposed. The only experimental evidence for such a deformation is an X-ray structure of *trans*-bis(trimethylsilyl)-9,10-dihydroanthracene. Structure **72** illustrates another way in which these peri interactions can be avoided. An unusually high degree of ring folding will result from the R group sterically interacting with (and "passing by") the ortho hydrogens. Such a distorted boat was calculated as a minimum in several cases with the trans isomers (vida infra).

Another point of interest in disubstituted dihydroaromatics in the orientation of *tert*-butyl substituents, which is depicted in Figure 7.19. The ss rotation

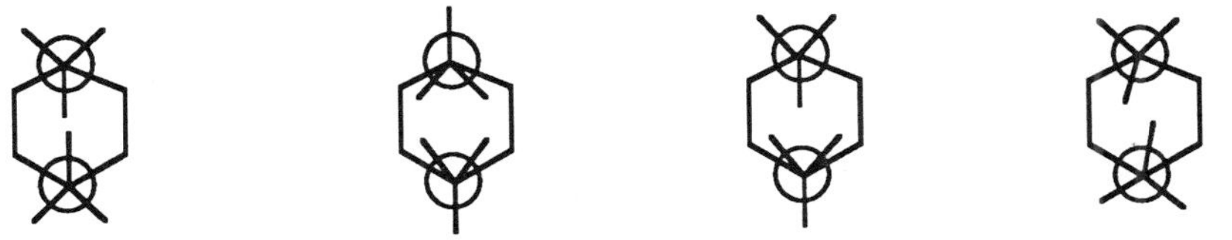

ss ee se geared

Figure 7.19 Orientations of the *tert*-butyl groups in the DHA system. The nomenclature we adopt refers to the Newman projection of the *tert*-butyl groups on the six-membered ring; ss = staggered–staggered, ee = eclipsed–eclipsed, se = staggered–eclipsed.

appears to minimize peri interactions and is usually adopted in the absence of any transannular interference. However, incorporation of a second *tert*-butyl group on the same side of the ring will cause a "gearing" effect of the two groups, whereas a hydrogen across the ring can "split" the methyls of a *tert*-butyl causing an ee geometry. Three of these four possibilities actually result as minima.

7.2.9.1 cis-Disubstituted DHBs

The calculated folding angle for *cis*-di-MeDHB indicates a flattened boat with only a 4° increase in pucker as compared to the monomethyl derivative (174°). The methyl groups are both in the pseudoaxial position. Interestingly, *cis*-di-*t*-BuDHB shows a slight pseudoequatorial preference for the *tert*-butyl groups indicating that transannular steric interactions may be important. It should be noted that the preferred orientation of large substituents in DHBs has been controversial, but recent experimental results indicate that a single large substituent forces a modest deviation from planarity with the substituent pseudoaxial (NMR results with a dimethylcarbinol group), whereas two large groups in a cis isomer produce a nearly planar ring with the groups very slightly pseudoequatorial (X-ray results with *cis*-phenyl and *cis*-trityl substituents). Thus, the MMPI calculations are in agreement with available experimental results. Rotational preferences also become important with tert-butyl groups and in this case have the ee arrangement.

7.2.9.2 cis-Disubstituted DHNs

As expected, *cis*-di-MeDHN and *cis*-di-*t*-BuDHN show more ring folding, 160° and 157°, respectively, due to increased peri interactions from the ortho aryl hydrogens. This folding produces an additional steric problem with *cis*-di-*t*-BuDHN, in that the *tert*-butyl groups are brought relatively close together. This results in a gearing of these substituents.

7.2.9.3 cis-Disubstituted DHAs

cis-di-MeDHA presents an interesting case in that MMPI shows two minima: a diaxial boat and a diequatorial boat. With methyl substituents, continued folding of the diequatorial boat results in the methyls eventually passing the peri hydrogens, leading to a decrease in energy. However, the calculations suggest a 7 kcal/mol difference in total steric energy between the two possible boat forms and so the diequatorial conformation is not expected to be observed. As was noted above, the MMI force field overestimates nonbonded repulsions which are important here. Not unexpectedly, *cis*-di-*t*-BuDHA is highly folded (153°), and the *tert*-butyl groups are once again in a geared arrangement.

7.2.9.4 trans-Disubstituted DHBs

Force-field results agree with experimental evidence and confirm that a planar geometry for *trans*-di-*t*-BuDHB is beginning to show a chairlike distortion, which becomes important for the DHNs and DHAs discussed below. The *tert*-butyl groups are oriented in the ee arrangement, which is the preferred rotational orientation in the absence of serious, transannular steric effects (cf. *cis*-di-MeDHB).

7.2.9.5 trans-Disubstituted DHNs

trans-di-*t*-BuDHN is especially interesting since the chair and distorted boat conformations are energetically different by less than 1.5 kcal/mol, and both conformations represent somewhat nonconventional structures for this ring system (see Figure 7.20 for stereoviews). In both cases, serious distortions that relieve steric interactions between the *tert*-butyl group and the peri hydrogens are evident. In the boat form the DHN is considerably folded (142°) with one end turned up allowing the pseudoequatorial *tert*-butyl to "pass" the orthohydrogens. In the chair form, similar steric interactions lead to twisting about the olefinic (17°) and aromatic (9°) double bonds. Furthermore, the orientation of the *tert*-butyl groups is different for these conformational isomers. In the flattened state, the groups are ee, minimizing the interaction with the peri positions. In the distorted boat, one of the transannular methine hydrogens is forced into the pseudoaxial *tert*-butyl group across the ring and so "splits" the methyl producing an se relationship. Analogous geometries arising from similar transannular interactions have been observed experimentally with isopropyl groups in the DHA series.

7.2.9.6 trans-Disubstituted DHAs

Both MMPI and MM2 calculations suggest two conformations of nearly equal energy for *trans*-di-MeDHA. The planar structure shows only a slight chairlike distortion by MMPI, and even less so by MM2. This is expected since the origin of the distortion is van der Waals repulsions which MMPI overestimates. The boat conformation shows a considerable distortion and the phenomenon of a methyl group "passing by" the peri interactions appears significant for both MMPI and MM2.

As expected, the *trans*-di-*t*-BuDHA provides the most distorted structures in both the chair and boat forms (see Figure 7.21). In the boat form, two stable conformations were found. The lowest energy structure possessed staggered

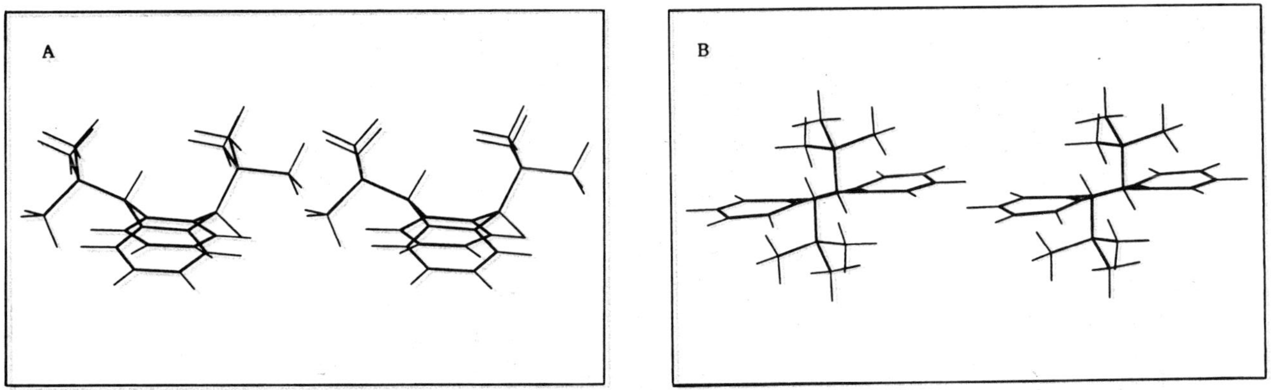

Figure 7.20 Stereoviews of *trans*-1,4-di-*tert*-butyl-1,4-dihydroanthracene. A is the boat form and B is the chair form.

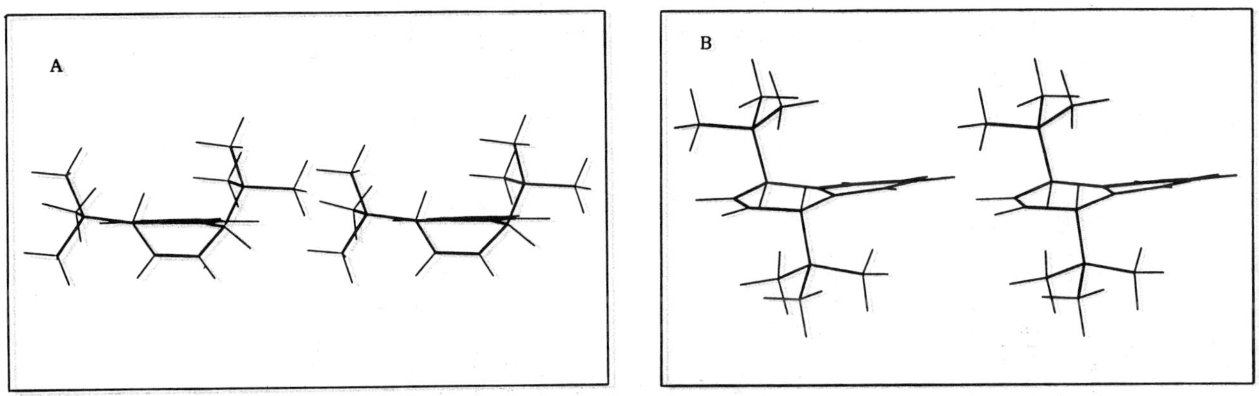

Figure 7.21 Stereoviews of *trans*-9,10-di-*tert*-butyl-9,10-dihydronaphthalene. A is the boat form and B is the chair form.

equatorial and eclipsed axial *tert*-butyl groups while the 0.3 kcal/mol less stable boat conformer has both *tert*-butyl groups staggered. The transition state for axial *tert*-butyl rotation between these boat forms is 1.1 kcal/mol above the most stable boat conformation.

The chair conformation was found to be 7.1 kcal/mol (MM2) and 4.0 kcal/mol (MMPI) higher in energy than the se boat conformation. Although the symmetry of the chair form (C_{2h}) requires the mean planes of the aryl rings to be coplanar, there is considerable twisting of each ring (endocyclic torsion angles up to 7.7° by MM2 and 13° by MMPI). The central ring is a flattened chair with torsion angles near ±15°. The *tert*-butyl groups in the chair conformer are both staggered. The staggered–eclipsed chair conformation represents the transition state for *tert*-butyl rotation among chair topomers; the barrier is only 0.5 kcal/mol.

7.2.9.7 Interconversion Pathways

The geometric constraints of trans-disubstituted DHNs and DHAs are such that two boat forms and two chair forms exist. The dynamics of boat-to-boat, chair-to-chair, and boat-to-chair interconversions have been considered.[44] An interesting example is the interconversion of topomeric boat forms of 9,10-di-*t*-Bu-DHA. Dreiding models suggest that all DHAs undergo boat-to-boat inversion through a planar transition state. The boat-to-boat interconversion pathway based on computer models is more complex (see Figure 7.22). It was found that the most stable boat conformation (a, 23.40 kcal/mol) must first rotate the axial *tert*-butyl group to the eclipsed conformation (b, 23.68 kcal/mol, E_a = 1.09 kcal/mol). From this conformation, the tricyclic system begins to flatten. The asymmetric transition state (c, E_a = 11.16 kcal/mol) is reached just before the C_{2h} chair conformation is attained (d, 30.46 kcal/mol). Continuation of the interconversion process is isometric to that just described. The dominant structural distortion involves passing the equatorially substitution ring carbon through the mean plane of the ring system. There are also small orthogonal distortions involving *tert*-butyl rotation.

This boat–chair–boat interconversion is reminiscent of the cyclohexane

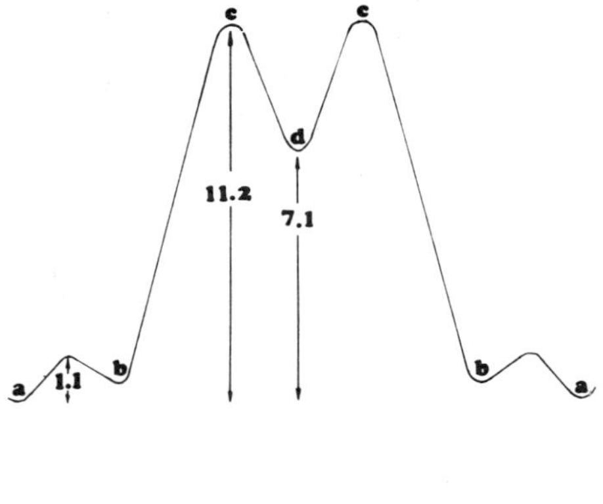

Figure 7.22 The MM2 reaction coordinate for boat-to-boat isomerization of *trans*-9,10-di-*tert*-butyl-dihydroanthracenes. Computed structures a–d are shown in Figure 7.23. Energies are in kcal/mol.

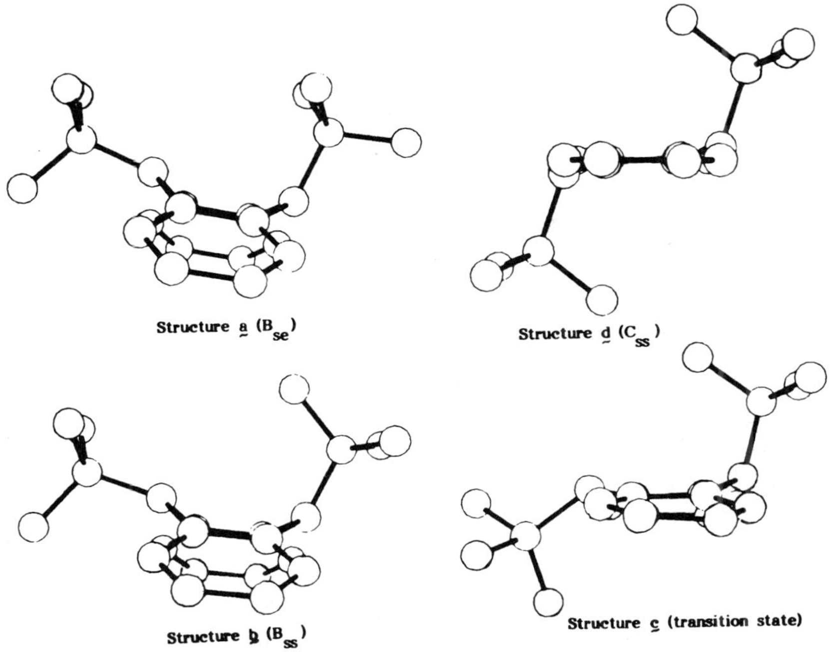

Figure 7.23 MM2 computed conformations of structures a–d along the reaction coordinate in Figure 7.22. The computed symmetries are the following: a (C_s); b (C_s); c (C_1); d (C_{2h}).

chair–boat–chair isomerization (both barriers near 11 kcal/mol). On the basis of these computed isomerization barriers, it is anticipated that low-temperature DNMR studies are feasible.

7.2.10 Beltenes

The lateral fusion of 1,4-cyclohexadienes and linkage into a macrocyclic beltlike assembly provide an interesting set of hydrocarbons designated by Alder and Sessions[45] as [*n*]beltenes, **73**. Mechanical models indicate beltenes with $n > 5$ are relatively strain-free, are circular, and, for $n > 6$, possesses a substantial cavity with which other molecules can bind. An assessment of the beltene architecture along with the structural changes induced by selective hydrogenation of olefins was undertaken with the MM2 force field.

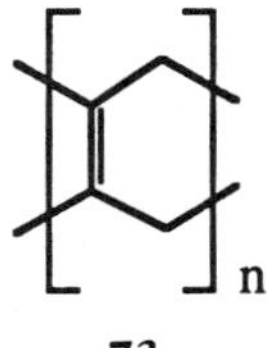

73

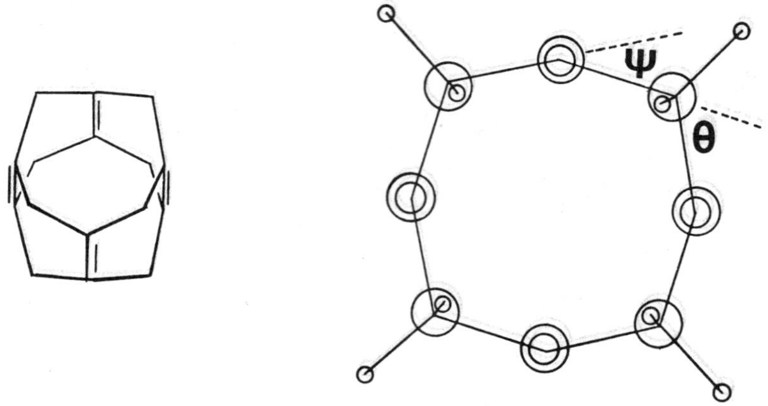

Figure 7.24 [4] Beltene: (a) perspective drawing, (b) top view as calculated by MM2; double circle, double bonded bridgehead carbon; circle, methylene carbon; small circle, hydrogen.

The [*n*]beltenes have minimum energy structures with D_{nh} symmetry for $n = 3–13$ and no other conformational minima exist. This is a consequence of the shape of the energy curve in Figure 7.10; any tendency to flatten one ring is offset by the increased strain energy of the other rings. A perspective drawing and a top view of [4]beltene computed with the MM2 force field is presented in Figure 7.24. The geometries and energies for the [*n*]beltenes are presented in Table 7.14. The monotonic trends found in the table speak for themselves. No transannular interactions are important and for the larger beltenes with $n > 6$ the strain energy originates predominantly from puckering of the cyclohexadiene moiety (the strain energy per ring unit is approximately equal to that of a similarly bent 1,4-cyclohexadiene).

For comparative purposes the linearly fused 1,4-cyclohexadienes, **74**, were

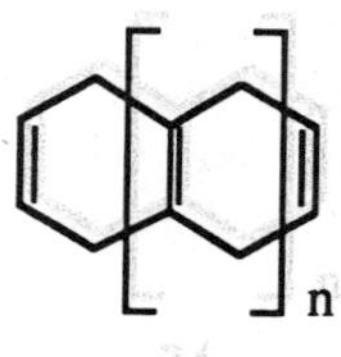

74

also computed with MM2. No bent structures could be found. All minimum energy structures were observed to be planar and a constant increment of 12.8 kcal/mol was found for the strain energy of an inner C_4H_4 unit. This may be considered as the strain energy per unit of an infinite beltene.

Hydrogenated beltenes were also studied with molecular mechanics. Two, three, or four pairs of hydrogen atoms were added from outside the belt. Addition

Table 7.14 Geometry and Energy (kJ/mol) for the [*n*]Beltenes

n	3	4	5	6	7	8	9	10	11	12
θ	68.4	59.7	54.6	50.6	46.9	43.4	39.7	36	32.7	30
see Figure 7.24b										
Ψ	51.6	30.3	17.4	9.4	4.5	1.6	0.3	0	0	0
Cavity diameter (Å)	0	0	0	0.6	1.5	2.4	3.3	4.1	5.0	5.8
Heat of formation	524	456	452	465	484	508	535	564	597	630
Strain energy (per C_4H_4 unit)	141	81.0	57.9	45.4	37.3	31.7	27.8	25.0	22.9	21.2
Compression	7.4	2.7	1.5	1.2	1.0	0.9	0.8	0.8	0.7	0.7
Bending	21.7	12.0	8.5	5.5	3.5	2.3	1.6	1.3	1.2	1.2
van der Waals, 1,4-	26.9	17.9	15.2	13.7	12.6	11.8	11.2	10.8	10.4	10.1
van der Waals, other	5.5	−2.3	−5.3	−5.4	−4.9	−4.4	−4.1	−4.0	−3.9	−3.9
Torsion	79.5	47.5	34.1	26.2	20.8	16.9	14.1	11.9	10.2	8.9
Olefinic torsion	76.8	31.8	10.9	3.2	0.8	0	0	0	0	0
Strain energy of cyclohexa-1,4-diene at same θ	84.9	59.4	49.4	42.7	38.1	33.9	29.7	26.4	23.8	22.2

of 2 × H_2 to the opposite ring junctions of [4]beltene leads to a decrease in strain but perhydrogenation then increases the strain. In the perhydrogenated molecule the cyclohexanes are flattened boats unable to twist. This is contrasted with [6]beltene where the strain energy begins to rise with the addition of 3 × H_2. The larger partially saturated beltenes have several minimum energy structures. Furthermore, they appear to be quite floppy. Many of the structures are oval in shape and some are flexible enough so as to not have a central cavity.

7.3 CONCLUDING REMARKS

The conformational features of partially unsaturated six-membered rings originate from the interplay of angular and torsional deformations. The addition of alkyl substituents to the parent ring system further enriches the conformational complexity of these simple rings by introducing subtle steric effects. Relying on mechanical models to account for the forces responsible for molecular shape can often be misleading. This was stressed throughout the chapter. Molecular mechanics, albeit a model with its own limitations and deficiencies, can, when properly used, be much more informative.

The application of molecular mechanics to the conformational analysis of partially saturated six-membered rings has profoundly enhanced our understanding of the preferred shape of these molecules along with some dynamical information that is otherwise impossible to obtain. Future applications of empirical force fields as employed in Monte Carlo and molecular dynamics simulations will invariably enhance our knowledge of conformational dynamics and will address questions of how solvents and ions perturb the molecular conformations.

REFERENCES

1. Bucourt, R.; Hainaut, D. *Bull. Soc. Chim. France* **1965**, 1366.
2. Favini, G.; Buemi, G.; Raimondi, M. *J. Mol. Struct.* **1968**, *2*, 137.
3. Allinger, N. L.; Hirsch, J. A.; Miller, M. A.; Tyminski, I. J. *J. Am. Chem. Soc.* **1968**, *90*, 5773.
4. Allinger, N. L.; Sprague, J. T., *J. Am. Chem. Soc.* **1972**, *94*, 5734.
5. Ermer, O.; Lifson, S., *J. Am. Chem. Soc.* **1973**, *95*, 4121.
6. Anet, F. A. L.; Yavari, I. *Tetrahedron* **1978**, *34*, 2879.
7. See, e.g., White, D. N. J.; Bovill, M. J. *J. Chem. Soc., Perkin Trans. 2* **1983**, 225.
8. Vanhee, P.; Tavernier, D.; Baas, J. M. A.; Van de Graff, B. *Bull. Soc. Chim. Belg.* **1981**, *90*, 697.
9. Pitea, D.; Moro, G.; Tantardini, G. F.; Todeschini, R. *J. Mol. Struct. THEOCHEM.* **1983**, *105*, 291.
10. Bowen, P.; Allinger, N. L. *J. Org. Chem.* **1986**, *51*, 1513.

11. Kugatova-Shemyakina, G. P.; Nikolaev, G. M.; Andreev, V. M. *Tetrahedron* **1967**, *23*, 2721.
12. Singh, R. D.; Keiderling, T. A. *J. Am. Chem. Soc.* **1981**, *103*, 2387.
13. Favini, G.; Sottocornola, M.; Todeschini, R. *J. Mol. Struct. THEOCHEM.* **1982**, *90*, 165.
14. Hagen, K.; Traetteberg, M. *Acta Chem. Scand.* **1972**, *26*, 3636.
15. Heilbronner, E.; Gleiter, R.; Hoshi, T.; DeMeijere, A. *Helv. Chim. Acta.* **1973**, *56*, 1594.
16. Paquette, L. A.; Wilson, S. E. *J. Org. Chem.* **1972**, *37*, 3849.
17. Plemenkov, V. V.; Bredikhin, A. A.; Aminova, *Dokl. Akad. Nauk. SSR* **1976**, *231*, 893.
18. Paquette, L. A.; Liao, C. C.; Liotta, D. C.; Fristad, W. E. *J. Am. Chem. Soc.* **1976**, *98*, 6412.
19. Tsuruta, H.; Mori, S.; Mukai, T. *Chem. Lett.* **1974**, 1127.
20. Takahashi, K.; Takase, K.; Toda, H. *Chem. Lett.* **1981**, 979.
21. Takeuchi, K.; Kitagawa, T.; Ueda, A.; Senzaki, Y.; Okamoto, K. *Tetrahedron* **1985**, *41*, 5463.
22. Vanhee, P.; Tavernier, D. *Bull. Chim. Soc. Belg.* **1983**, *92*, 767.
23. (a) White, D. N. J.; Bovill, M. *J. Chem. Soc. Perkin Trans 2* **1983**, 225. (b) Wipke, W. T.; Gund, P. *J. Am. Chem. Soc.* **1974**, *96*, 299; **1976**, *98*, 8107.
24. Lambert, J. B.; *J. Am. Chem. Soc.* **1967**, *89*, 1836.
25. Watson, W. H. "Stereochemistry and Reactivity of Systems Containing π Electrons," *Methods in Stereochemical Analysis,* Vol. 3; Verlag Chemie International: Deerfield Beach, 1983.
26. Watson, W. H.; Galloy, J.; Bartlett, P. D.; Roof, A. A. M. *J. Am. Chem. Soc.* **1981**, *103*, 2022.
27. Ermer, O.; Bödecker, C.-D. *Helv. Chim. Acta.* **1983**, *66*, 943.
28. Houk, K. N.; Rondan, N. G.; Brown, F. K.; Jorgensen, W. L.; Madura, J. D.; Spellmeyer, D. C. *J. Am. Chem. Soc.* **1983**, *105*, 5980.
29. Ermer, O. *Tetrahedron* **1974**, *30*, 3103.
30. Burkert, U. *Agnew. Chem., Int. Ed. Engl.* **1981**, *20*, 572.
31. Jorgensen, F. S. *Tetrahedron Lett.* **1983**, *24*, 5289.
32. Johnson, C. A. *J. Chem. Soc. Chem. Commun.* **1983**, 1135.
33. Maier, W. F.; Schleyer, P. v. R. *J. Am. Chem. Soc.* **1981**, *103*, 1891.
34. Ermer, O. *Z. Naturforsch. B: Anorg. Chem., Org. Chem.* **1977**, *32B*, 837.
35. McEwen, A. B.; Schleyer, P. v. R. *J. Am. Chem. Soc.* **1986**, *108*, 3951.
36. Jeffrey, D. A.; Maier, W. F. *Tetrahedron* **1984**, *40*, 2799.
37. Favini, G.; Zuccarello, F.; Buemi, G. *J. Mol. Struct.* **1969**, *3*, 385.
38. Allinger, N. L.; Sprague, J. T. *J. Am. Chem. Soc.* **1973**, *95*, 3893.
39. For a comprehensive review see Rabideau, P. W. *Acc. Chem. Res.* **1978**, *11*, 141.
40. Herbstein, F. A. *J. Chem. Soc.* **1959**, 2292.
41. Lipkowitz, K. B.; Rabideau, P. W.; Raber, D. J.; Hardee, L. E.; Schleyer, P. v. R.; Kos, A. J.; Kahn, R. A. *J. Org. Chem.* **1982**, *47*, 1002.
42. Syguła, A.; Holak, T. A. *Tetrahedron Lett.* **1983**, *24*, 2893.
43. Raber, D. J.; Hardee, L. E.; Rabideau, P. W.; Lipkowitz, K. B. *J. Am. Chem. Soc.* **1982**, *104*, 2843.
44. Rabideau, P. W.; Lipkowitz, K. B.; Nachbar, Jr., R. B., *J. Am. Chem. Soc.* **1984**, *106*, 3119.
45. Alder, R. W.; Sessions, R. B. *J. Chem. Soc., Perkins Trans. 2* **1985**, 1849.

8

Conformational Analysis of Hydroaromatic Metabolites of Carcinogenic Hydrocarbons and the Relation of Conformation to Biological Activity

Ronald G. Harvey

While the carcinogenic properties of certain polycyclic aromatic hydrocarbons (PAHs) have been known for many years, only recently has it been recognized that PAHs require metabolic activation to express their bioactivity.[1] Metabolism of PAHs in mammalian tissues takes place principally on the microsomes of the endoplasmic reticulum and affords a variety of oxidized products, including phenols, dihydrodiols, tetraols, quinones, and their conjugates, most of which are biologically inert.[2,3] However, certain hydrocarbon metabolites are highly tumorigenic and/or mutagenic. These include arene oxides, dihydrodiols, diol epoxides, and structurally related epoxide derivatives. There is now substantial evidence that the mechanism of PAH carcinogenesis entails covalent interaction of certain of these intermediate species with nucleic acids, leading initially to mutation and ultimately to tumor induction.[1-4] Stereochemistry and conformation are believed to play important roles in both the enzymatic activation of PAHs and the covalent binding of their active metabolites with nucleic acids.

This chapter reviews current evidence concerning the stereochemistry and conformational properties of the active PAH metabolites, including dihydrodiols, diol epoxides, and arene oxides, as well as related PAH epoxide derivatives. These findings are related to the DNA binding properties of these molecules, and the relation between adduct conformation and biological activity is discussed.

8.1 METABOLIC ACTIVATION AND DNA BINDING

Metabolism of PAHs takes place primarily on the microsomes and is catalyzed by the family of enzymes known as the cytochrome P-450 mixed-function oxidases, or monooxygenases.[5] The primary metabolites are arene oxide intermediates formed by the addition of an atom of oxygen to an aromatic bond (Figure 8.1). The less stable arene oxides rearrange spontaneously to phenols, and the direction of isomerization is predictable from the β-delocalization energies

Metabolism of PAHs by Microsomal Enzymes

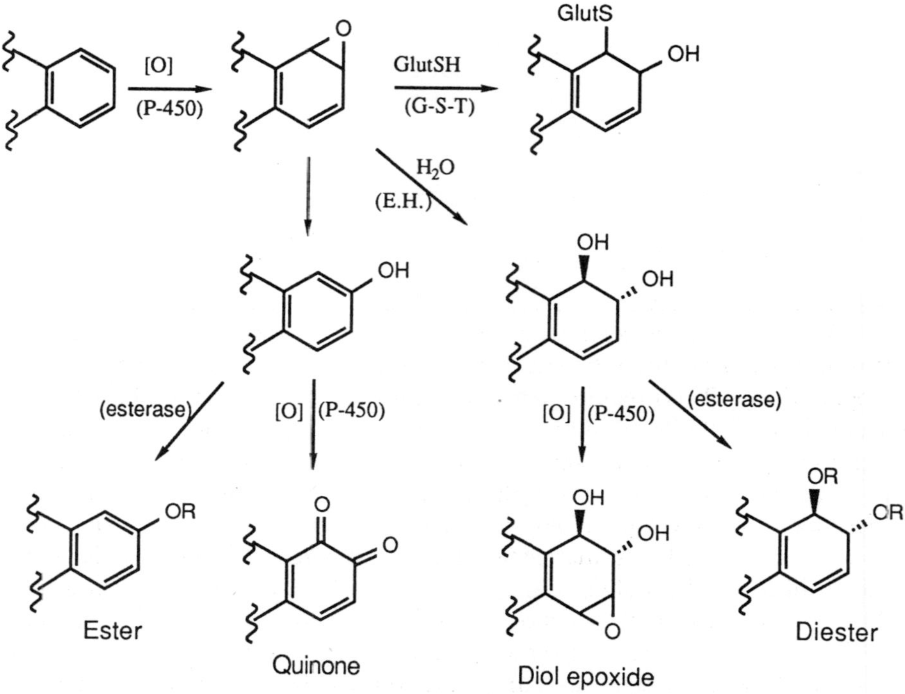

Figure 8.1 Pathways of metabolism of polycyclic aromatic hydrocarbons (PAHs). E.H. = epoxide hydrase; G-S-T = glutathione-*S*-transferase.

calculated by theoretical MO methods.[6] The more stable arene oxide intermediates survive sufficiently long to undergo hydration catalyzed by epoxide hydrase to yield *trans*-dihydrodiols or add glutathione catalyzed by glutathione-S-transferase. The phenols and dihydrodiols are conjugated and excreted as their water-soluble glucuronic and sulfate esters. Further oxidative metabolism yields a variety of products, including quinones, diol epoxides, and tetraols. The diol epoxides are generally not isolated as such due to the facility of their reactions with water and other cellular nucleophiles. These reactive intermediates were initially detected and identified as the principal DNA-bound adducts arising from metabolism of PAHs in mammalian tissues.[1] Diol epoxide metabolites are now believed to be the principal active forms of carcinogenic alternant PAHs in animal tissues.

The structures of the diastereomeric *anti*- and *syn*-diol epoxide derivatives of benzo[*a*]pyrene are shown in Figure 8.2. The *anti*-isomer (*anti*-BPDE) has the benzylic hydroxyl group and the epoxide oxygen atom on the opposite faces of the

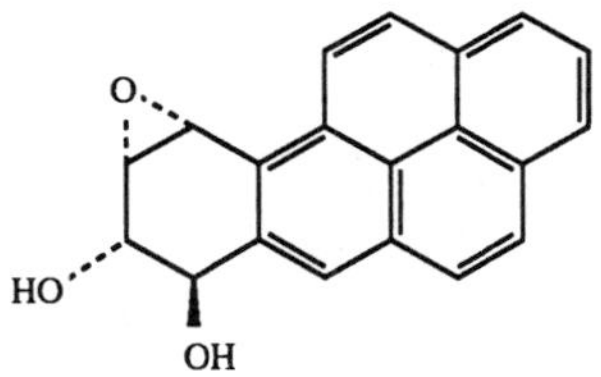

(+)-*anti*-BPDE

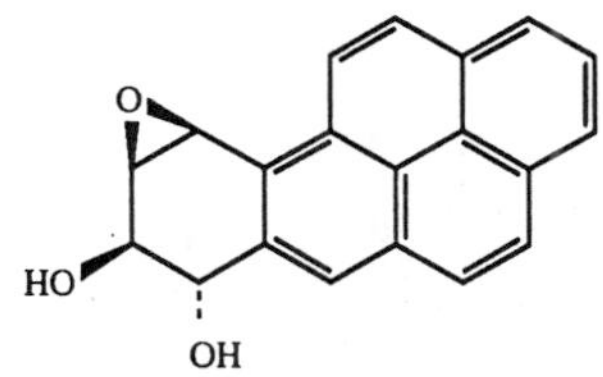

(−)-*anti*-BPDE

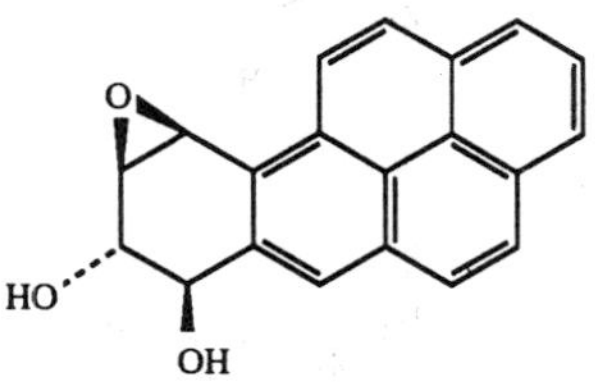

(−)-*syn*-BPDE

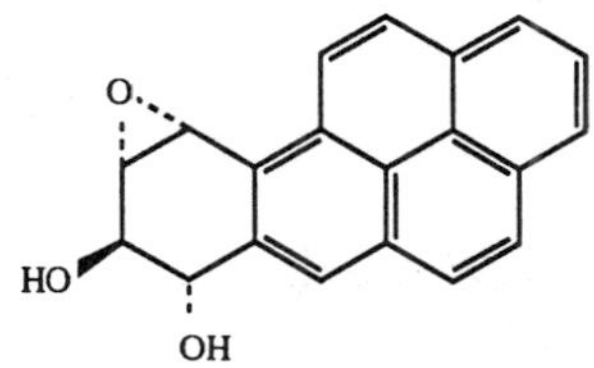

(+)-*syn*-BPDE

Figure 8.2 The structures and absolute configurations of *trans*-7,8-dihydroxy-*anti*-9,10-epoxy-7,8,9,10-tetrahydrobenzo[*a*]pyrene, more commonly known as *anti*-benzo[*a*]pyrene diol epoxide (*anti*-BPDE) and its diastereomer *syn*-BPDE.

molecule, whereas the *syn*-isomer (*syn*-BPDE) has these groups on the same face. Both diastereomers exist as pairs of optically active enantiomers. The absolute configuration of the principal DNA-bound metabolite of benzo[*a*]pyrene in mammalian cells, including human lung, was assigned by Nakanishi et al.[7] as (+)-*anti*-BPDE (Figure 8.2) using the exciton chirality dichroism method.

This adduct was shown by Fourier transform NMR and high-resolution mass spectral analysis to be a guanosine derivative covalently linked to the 2-NH_2 group (Figure 8.3).[8] Minor adducts arising from reaction of *anti*-BPDE on the 7-N and 6-O positions of guanosine and the 6-NH_2 group of adenosine as well as from reaction of *syn*-BPDE on various base sites have also been detected.[9] Analogous nucleic-acid-bound adducts have subsequently been characterized as the principal products formed in the reactions of the active metabolites of other carcinogenic PAHs in animal tissues.[3,10,11] The role of diol epoxide metabolites in carcinogenesis is also supported by mutagenicity and tumorigenicity assays. *Anti*- and *syn*-BPDE and *trans*-7,8-dihydroxy-7,8-dihydrobenzo[*a*]pyrene, which is their metabolic precursor, are all strong mutagens in bacterial and mammalian cells; *anti*-BPDE generally exhibits greater activity than *syn*-BPDE in most tests.[12] *Anti*-BPDE also shows higher activity than its diastereomer as a carcinogen on mouse skin and in the newborn mouse lung.[13] Structurally analogous diol epoxide derivatives of other PAHs containing an epoxide ring in a molecular bay region are implicated by similar biological evidence as the active forms of other carcinogenic PAHs.[3]

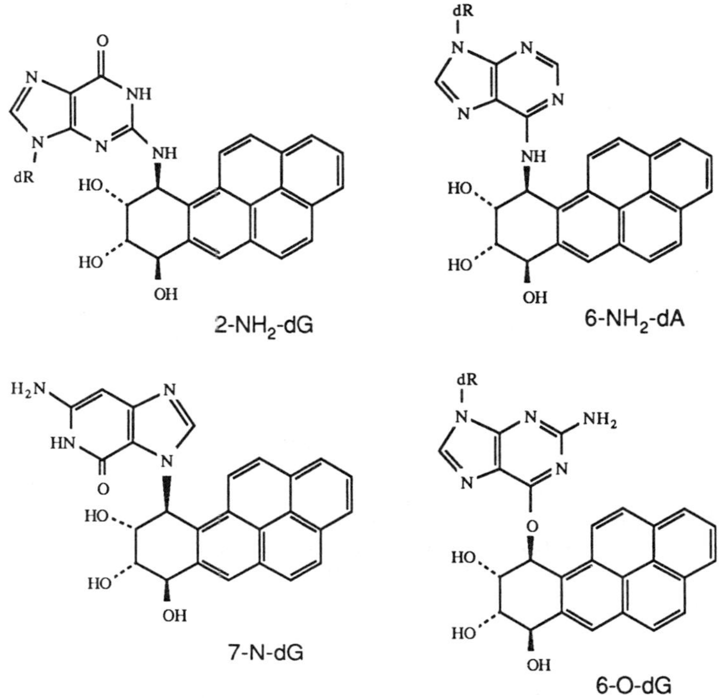

Figure 8.3 The structures of the major (2-NH$_2$-dG) and minor nucleoside adducts formed by covalent binding of *anti*-BPDE to nucleic acids.

8.2 STEREOCHEMISTRY OF DIHYDRODIOLS

Since the stereochemical and biological properties of PAH dihydrodiols are dependent on the molecular region of substitution, it is convenient to separate the discussion of the various types of dihydrodiols into the three subcategories: K region, bay region, and other dihydrodiols. The term "K-region" derives from the early quantum mechanical theories of PAH carcinogenesis proposed by Pullman and Pullman.[14] A K-region is an aromatic bond, such as the 5,6-bond of benz[*a*]anthracene, which is located between two fused aromatic rings; excision of a K-region leaves an intact aromatic ring system (Figure 8.4). The term "bay region" comes from NMR spectroscopy and refers to a sterically crowded region, such as the region between the 1 and 12 positions of benz[*a*]anthracene, which is formed by the angular fusion of additional aromatic rings on naphthalene or higher aromatic ring systems. Methods for the synthesis of PAH dihydrodiols have recently been reviewed.[15]

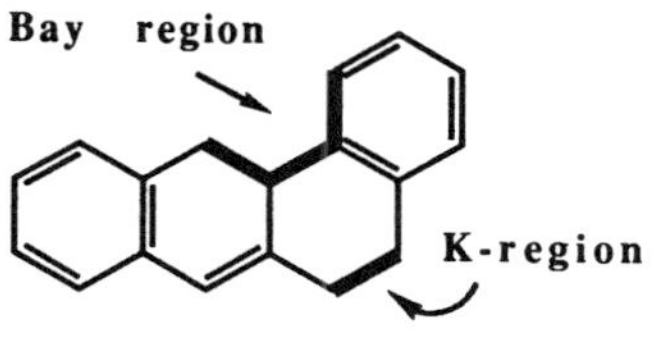

Figure 8.4 The K-region and bay region of benz[*a*]anthracene.

8.2.1 K-Region Dihydrodiols

Dihydrodiols of this type may be either cis or trans dependent on whether the vicinal hydroxyl groups are on the same or opposite faces of the molecule. Each of these diastereomers may exist in two possible conformations in relatively dynamic equilibrium (Figure 8.5). For *trans*-dihydrodiols, such as *trans*-5,6-dihydroxy-5,6-dihydrobenz[*a*]anthracene, the hydroxyl groups may be oriented diaxially or diequatorially. For the corresponding *cis*-diastereomers, the equilibrium is between axial–equatorial and equatorial–axial conformers. Since most K-region dihydrodiols are unsymmetrical, the *cis*- and *trans*-dihydrodiols generally also occur as pairs of optically active enantiomers.

X-ray crystallographic analysis of the structure of the *cis*-5,6-dihydrodiol of 7,12-dimethylbenz[*a*]anthracene (DMBA) shows that the ring system is severely buckled and the 6-hydroxyl group is axial, while the 5-hydroxyl group is equatorial.[16] The distortion of the ring system from planarity is largely a consequence of the steric interaction of the bay-region 12-methyl group with the benzo ring. The axial conformation of the 6-hydroxyl group is attributed to the

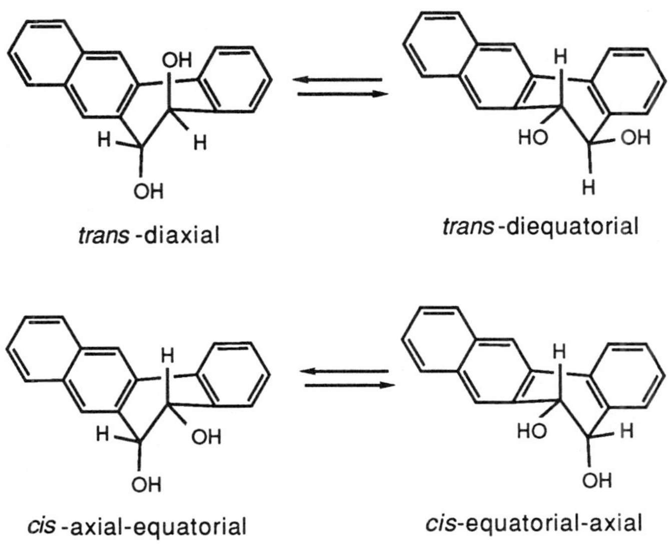

Figure 8.5 Equilibrium between the diaxial and the diequatorial conformers of the K-region dihydrodiols of benz[*a*]anthracene.

steric effect of the methyl group in the adjacent 7 position. Since similar steric interaction is also present in the corresponding *trans*-stereoisomer, it may be anticipated that the *trans*-5,6-dihydrodiol of DMBA exists in a preferred *trans*-diaxial conformation. This conformational assignment is supported by the relatively low value of the ^{1}H-NMR coupling constant of the K-region protons of the *trans*-5,6-dihydrodiol of DMBA ($J_{5,6}$ = 3.2 ± 0.5 Hz),[17] consistent with the diequatorial orientation of the benzylic protons. However, the value of $J_{5,6}$ for the *cis*-5,6-dihydrodiol of DMBA was closely similar,[17] indicating that stereoisomeric assignment of sterically hindered dihydrodiols of this type on the basis of ^{1}H-NMR data alone is unsafe.

K-region dihydrodiols unsubstituted in the aromatic ring positions peri to the benzylic positions of the saturated ring are not subject to the same steric constraint, and they should be relatively free to adopt conformations having one or both hydroxyl groups equatorially oriented. Analysis of ^{1}H-NMR coupling constant data (Table 8.1) for a series of K-region *trans*-dihydrodiol derivatives of benz[*a*]anthracene (**1**), benzo[*a*]pyrene (**2**), and dibenz[*a,h*]anthracene (**3**) (Figure 8.6)[18] reveals that in the absence of steric constraints these dihydrodiols exhibit a relatively large coupling constant between the carbinol protons (9.9–10.3 Hz), indicative of a preferred diequatorial conformation.[19] On the other hand, structurally related dihydrodiols, which contain a substituent adjacent to one of the hydroxyl groups, generally show a smaller coupling constant (2.6–3.5 Hz), consistent with the diaxial conformation. Esterification of the *trans*-dihydrodiols also generally shifts the conformational equilibrium in favor of the diaxial conformers as a consequence of the greater steric requirements of the ester

Table 8.1 NMR Coupling Constants for the Carbinol Protons of PAH K-Region *trans*-Dihydrodiols and Diesters[a]

Dihydrodiol	*J* (Hz)	Dihydrodiol	*J* (Hz)
1	10.3	8-CH_3-**1**	9.9
1 diacetate	5[c]	8-CH_2OH-**1**	9.9
1 BMA[b]	4.6[d]	12-CH_3-**1**	10.3
1-CH_3-**1**	10.2	7,12-di-CH_3-**1**	3.2[e]
4-CH_3-**1**	3.1	7-F-**1**	2.6
7-CH_3-**1**	3.2	7-Br-**1**	2.9
2		6-CH_3-**2**	3.0
2 BMA[b]	4[d]	7-CH_3-**2**	10.3
3-CH_3-**2**	2.9	6-Br-**2**	2.9
3	10.2		
3 diacetate	4.5		

[a] The data are from Ref. 18 except as noted.
[b] BMA = bismenthoxyacetate.
[c] Reference 20.
[d] Reference 21.
[e] Reference 17.

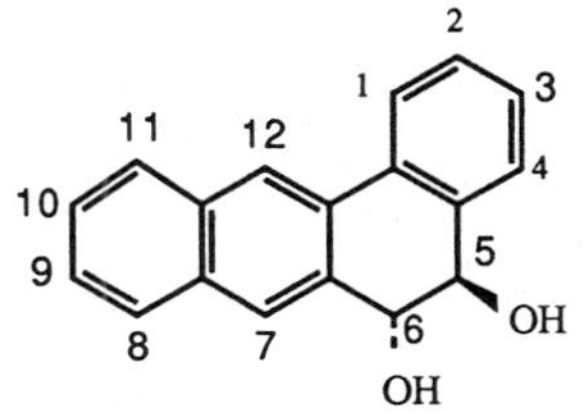

1

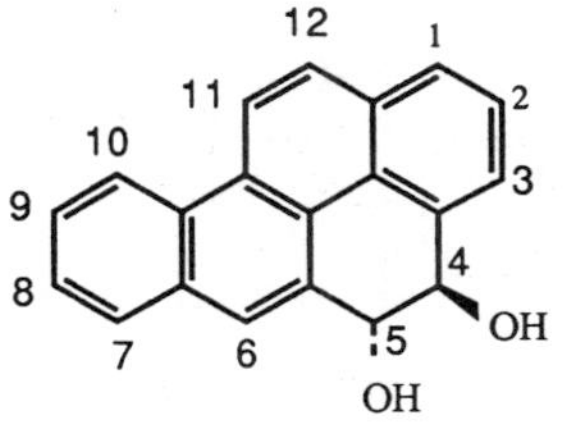

2

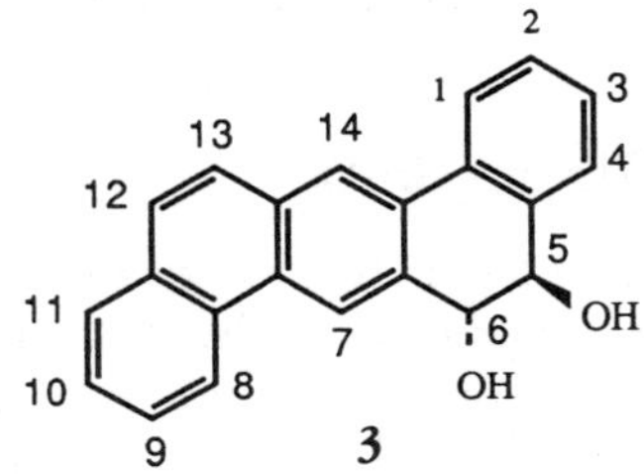

3

Figure 8.6 K-region *trans*-dihydrodiol derivatives of benz[*a*]anthracene (**1**), benzo[*a*]pyrene (**2**), and dibenz[*a,h*]anthracene (**3**).

functions. This effect is evident in the lower values of the coupling constants of the diesters in comparison with the free dihydrodiols (Table 8.1).

The position of the conformational equilibrium for these dihydrodiols may be calculated from the relationship

$$J_{obs} = xJ_{ee} + (1 - x)J_{aa} \qquad [8.1]$$

where J_{obs} is the observed coupling for the benzylic protons and J_{ee} and J_{aa} are the coupling constants for these protons in the conformations in which the hydroxyl groups are diaxial and diequatorial, respectively. Using the values of J_{aa} = 16 and J_{ee} = 2 Hz, the fraction (x) of the conformers with the hydroxyl groups in the diaxial orientation are calculated to be 40–43% for the unsubstituted K-region dihydrodiols and 89–95% for the related peri-substituted dihydrodiols. The position of the conformational equilibrium may also be expected to be influenced by solvents. In aprotic solvents, such as chloroform and acetone, hydrogen bonding between the hydroxyl groups in the diequatorial conformation contributes to the stability of this conformer, whereas in protic solvents this effect is expected to diminish, shifting the equilibrium in the direction of the diaxial conformer.

The marked preference of the K-region *trans*-dihydrodiols for the diequatorial conformation contrasts with the earlier observation that 9-hydroxy-9,10-dihydrophenanthrene exists predominantly in the axial conformation in chloroform and other solvents.[19] This difference may reflect the importance of intramolecular hydrogen bonding in the vicinal dihydrodiols.

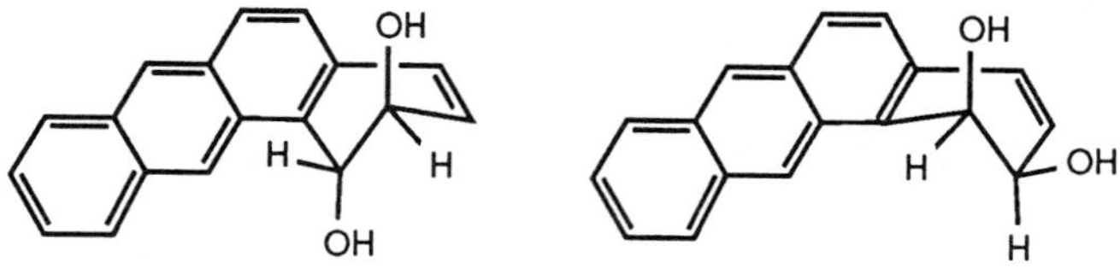

Figure 8.7 Conformations of the *trans*- and *cis*-1,2-dihydrodiols of benz[*a*]anthracene.

8.2.2 Bay-Region Dihydrodiols

The conformational properties of the bay-region dihydrodiols are largely determined by the steric crowding in this molecular region. Both the *cis*- and *trans*-dihydrodiols of this type tend to be locked into a conformation in which the benzylic hydroxyl group is oriented axially to relieve the potentially strong steric interaction with the proton of the adjacent aromatic ring (Figure 8.7). Thus, the hydroxyl groups of the *trans*-1,2-dihydrodiol of benz[*a*]anthracene were shown by X-ray crystallographic and NMR analysis to be diaxial in both the crystal lattice and in solution.[22] The observed coupling constant for the carbinol hydrogens of this dihydrodiol ($J_{1,2}$ = 1.7 Hz) was in good agreement with the calculated value ($J_{1,2}$ = 2.0 ± 0.1 Hz) using the average torsion angles derived from the X-ray data and the Karplus relationship as modified by Bothner-By.[23] Esterification of the *trans*-1,2-dihydrodiol had only minimal effect on $J_{1,2}$ (Table 8.2), showing that these diesters also exist exclusively in the diaxial conformation. NMR coupling constant data reported in the literature for other bay-region dihydrodiols and their diesters (Figure 8.8) is summarized in Table 8.2.[22] These data are generally consistent with the findings for the benz[*a*]anthracene 1,2-dihydrodiol, confirming the strong preference of bay-region *trans*-dihydrodiols for the diaxial conformation.

On the basis of the foregoing findings with the *trans*-dihydrodiols, it may be

Table 8.2 NMR Coupling Constants for the Carbinol Protons of PAH Bay-Region *trans*-Dihydrodiols and Diesters

Dihydrodiol	*J* (Hz)	Reference	Dihydrodiol	*J* (Hz)	Reference
4	2	24	**8**	2.3	28
4 diacetate	1.6	24	**9**	1.8	25
5	1.7	24	**10**	2	25
5 diacetate	1.8	24	**10** diacetate	2	25
5 dibenzoate	2	25	**11**	2	29
6	2	26	**11** dibenzoate	2	29
6 dibenzoate	1.8	26, 27			
7	1	25			
7 dibenzoate	2	25			

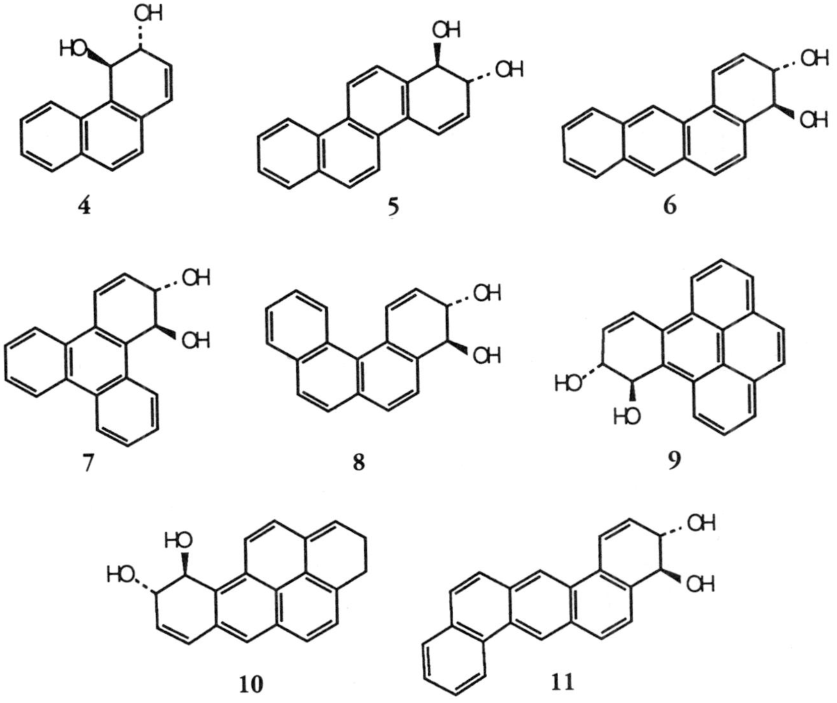

Figure 8.8 Structures of PAH bay-region *trans*-dihydrodiols.

assumed with reasonable confidence that the bay-region *cis*-dihydrodiols also exist preferentially in a conformation in which the benzylic hydroxyl group is axial. However, relatively few studies have been conducted with the bay region *cis*-dihydrodiols, and there is no available crystal structure data. NMR analysis is not useful in this regard, since the coupling constants for the carbinol axial–equatorial of both conformers are expected to be identical. The coupling constant for the carbinol protons of benz[*a*]anthracene *cis*-1,2-dihydrodiol and benzo[*a*]pyrene *cis*-9,10-dihydrodiol are $J_{1,2}$ = 5.4 Hz and $J_{9,10}$ = 5.2 Hz, respectively, which are in the range expected for axial–equatorial coupling.[30,31]

8.2.3 Dihydrodiols in Other Molecular Regions

From the standpoint of biological activity, the most important class of dihydrodiols are those that give rise to bay-region diol epoxides. Dihydrodiols of this type, such as the phenanthrene *trans*-1,2-dihydrodiol (**12**), the benz[*a*]anthracene *trans*-3,4-dihydrodiol (**14a**), and the benzo[*a*]pyrene *trans*-7,8-dihydrodiol (**17a**) (Figure 8.9), occur in the same benzo rings as the bay-region dihydrodiols, but in the less sterically restricted sites. These are termed

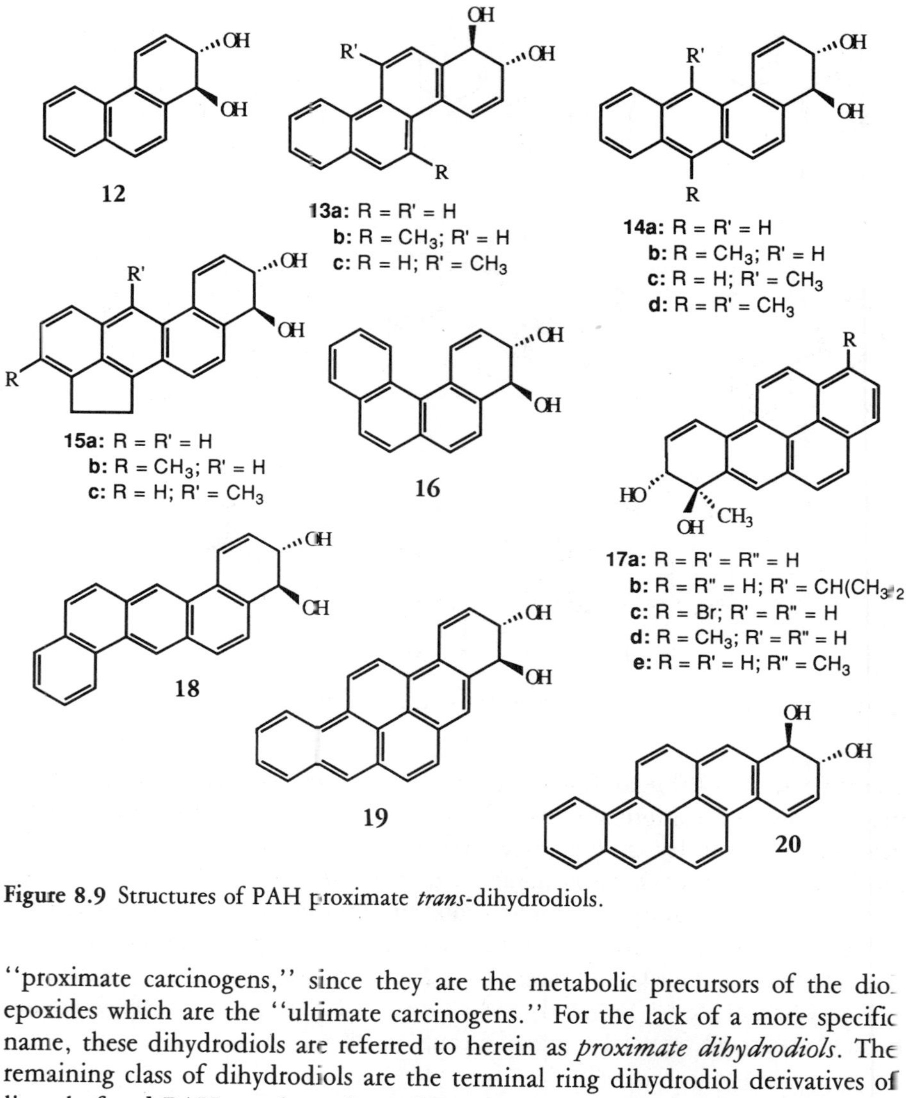

Figure 8.9 Structures of PAH proximate *trans*-dihydrodiols.

"proximate carcinogens," since they are the metabolic precursors of the diol epoxides which are the "ultimate carcinogens." For the lack of a more specific name, these dihydrodiols are referred to herein as *proximate dihydrodiols*. The remaining class of dihydrodiols are the terminal ring dihydrodiol derivatives of linearly fused PAHs, such as the anthracene *trans*-1,2-dihydrodiol (**22**) and the benz[*a*]anthracene *trans*-8,9- and *trans*-10,11-dihydrodiols (**23a** and **24**) (Figure 8.10).

In the absence of steric and other effects, the *trans*-dihydrodiols of both these classes exist preferentially in the diequatorial conformation. X-ray crystal structure analysis of **17a** and **24** shows that the aromatic ring systems of both dihydrodiols are essentially planar, while the hydroxyl groups are oriented diequatorially.[22,32] NMR spectral analysis of these compounds provides results in good agreement with the findings from the X-ray crystal analysis. The coupling

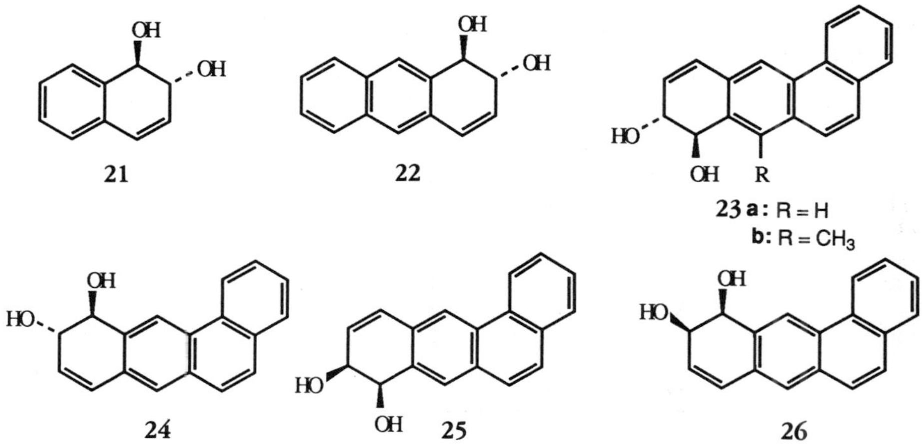

Figure 8.10 Structures of PAH terminal ring *cis*- and *trans*-dihydrodiols and diesters.

constant for the carbinol hydrogens of **24** was calculated to be $J_{10,11}$ = 12.7 ± 0.2 Hz using the average torsion angles from the X-ray data. The experimentally determined value was 9.5 Hz in dimethyl sulfoxide and 10.0 Hz in acetone. Since the observed coupling is less than anticipated for the diequatorial conformer alone, this dihydrodiol exists in solution as a mixture of conformers undergoing interconversion rapidly on the NMR time scale. On the assumption that the observed couplings represent a weighted average of the ratio of pure diequatorial and diaxial conformers in solution, the percentage of the diequatorial conformer of **24** is calculated to be 70–75% at room temperature. Similar calculations on **17a**, for which $J_{7,8}$ = 10 Hz is observed,[39] indicate that the equilibrium percentage of the diequatorial conformer in solution is 75%.

Esterification of **17a** and **24** resulted in a decrease in the couplings of the carbinol hydrogens, indicating a shift of the conformational equilibrium in favor of the diaxial structure. The value of $J_{10,11}$ for the dibenzoate ester of **24** was 6.3 Hz, while the diacetate ester of **17a** had $J_{7,8}$ = 6.9 Hz. The ratios of the diequatorial conformers in solution are calculated to be 30 and 35%, respectively. This shift toward the diaxial conformation results from the loss of intramolecular hydrogen bonding between the hydroxyl groups and steric interaction between the bulky ester groups.

The NMR spectral data reported in the literature for other structurally similar PAH dihydrodiols and their diesters are generally consistent with these findings. Thus, the couplings for the carbinol protons of a large series of proximate *trans*-dihydrodiols (Table 8.3) fall in the relatively narrow range of 10–12 Hz. The only exceptions are the dihydrodiols of 6-bromo- and 6-methylbenzo[*a*]pyrene (**17c** and **17d**) for which $J_{7,8}$ = 1.7–1.8 Hz. In these cases the diaxial conformation is favored due to the steric interaction of the hydroxyl groups with the peri substituents. Esterification of the proximate dihydrodiols, as for the K-region dihydrodiols, shifts the conformational equilibrium in the diaxial

Table 8.3 NMR Coupling Constants for the Carbinol Protons of PAH Proximate *trans*-Dihydrodiols

Dihydrodiol	*J* (Hz)	Reference	Dihydrodiol	*J* (Hz)	Reference
12	11.3	24, 33	**16**	10.8	28
13a	10.8	34	**17a**	10.0	39
13b	11.5	34	**17b**	10.6	40
13c	10.3	34	**17c**	1.8	39
14a	11.5	24	**17d**	1.7	39
14b	12	35	**18**	11	29
14c	11.3	36	**19**	10.5	41
14d	11.3	37	**20**	11	41
15b	10.7	38			
15c	11.3	36			

Table 8.4 NMR Coupling Constants for the Carbinol Protons of PAH Proximate *trans*-Dihydrodiol Diesters

Diesters	*J* (Hz)	Reference	Diesters	*J* (Hz)	Reference
12 diacetate	5.6	24	**17a** diacetate	6.9	39
13a diacetate	6.5	27	**17a** BMA	6.6	39
14a diacetate	6.1	24	**18** diacetate	5.8	27
14a BMA[a]	6	42	**18** dibenzoate	7	29
15a diacetate	5.7	36	**19** diacetate	6	41
15b diacetate	5.5	38	**20** diacetate	6	41
16 diacetate	6.1	28			

[a] BMA = bismenthoxyacetate.

Table 8.5 NMR Coupling Constants for the Carbinol Protons of PAH Terminal Ring *cis*- and *trans*-Dihydrodiols and Diesters

Dihydrodiol or Diester	*J* (Hz)	Reference	Dihydrodiol or Diester	*J* (Hz)	Reference
21	11.1	33	**24**	10.4	24
22	9.5	33	**24** dibenzoate	6.2	24
22 dibenzoate	7.0	24	**25**	4.8	30
23a	10.0	24	**25**	4.7	30
23a dibenzoate	7.0	24			
23b	1.8	39			

direction. Based on the observed couplings (Table 8.4), the calculated percentage of the diaxial conformers at equilibrium is 64–75%. The coupling constants for the terminal ring dihydrodiols and diesters (Table 8.5) closely resemble those for the proximate dihydrodiols and diesters. Therefore, it may be concluded that the free terminal ring dihydrodiols exist mainly in the diuequatorial conformation, while the related diesters favor the diaxial conformation.

The 7,8-dihydrodiol of 7-methylbenzo[*a*]pyrene (**17e**) is unusual in that it contains a methyl group in the benzylic carbinol position. Since this molecule lacks a $J_{7,8}$ proton–proton coupling, due to the presence of the methyl group, assignment of conformation on this basis is not possible. However, the small values of $J_{8,9}$ (2.2–2.5 Hz) of this dihydrodiol and its diacetate and dimenthoxyacetate esters in comparison with that of the parent dihydrodiol (**17a**), for which $J_{8,9}$ = 2.0 Hz, suggest that these compounds all preferentially adopt diequatorial conformations.[39] The close similarities in the CD Cotton effects of the 7-methylbenzo[*a*]pyrene 7*R*,8*R*-dihydrodiol to those of the corresponding enantiomer of **17a** provide further support for the diequatorial conformational assignment for **17e**.

8.3 STEREOCHEMISTRY OF DIOL EPOXIDES

Carcinogenesis research has focused primarily on the "bay-region diol epoxides," that is, diol epoxides containing an epoxide ring in a PAH bay region, since intermediates of this type are implicated as the principal active forms of carcinogenic PAHs.[1,4] The biologically active bay-region diol epoxides are formed metabolically by the addition of an atom of oxygen to the olefinic bond of a proximate dihydrodiol. Dependent on whether addition of oxygen is from the same or the opposite molecular face as the benzylic hydroxyl group, there is obtained a *syn*- or an *anti*-diol epoxide diastereomer. The best known examples are *anti*- and *syn*-BPDE (Figure 8.2). The conformational properties of the PAH diol epoxides may be expected to be closely related to those of the parent dihydrodiols, except that hydrogen bonding between the benzylic hydroxyl group and the epoxide oxygen of the *syn*-isomer may potentially contribute to the stability of the diaxial conformer.

Molecular orbital studies predict the diequatorial conformer of *anti*-BPDE to be more stable than the diaxial form.[43–45] CNDO/2 calculations on *syn*-BPDE indicate that transannular hydrogen bonding is absent in the semi-chair conformation of the tetrahydro ring but is energetically favored in an alternative more puckered conformation.[43] CNDO/2L and MINDO/3 calculations also support a hydrogen-bonded diaxial conformation for *syn*-BPDE.[44,45] However, it should be kept in mind that in the biological milieu the diol epoxide molecules are likely to be closely associated with water, decreasing the likelihood of intramolecular hydrogen bonding.

Contrary to theoretical prediction, X-ray crystallographic analysis of the structures of *anti*-BPDE and *syn*-BPDE show that both isomers exist exclusively as the diequatorial conformers in the crystal lattice.[46,47] However, the tetrahydrobenzo ring of the *syn*-isomer is more puckered than that of the *anti*-isomer as a consequence of severe steric interaction between the hydrogen atom on C-8 and the epoxide oxygen atom. In *anti*-BPDE the angle between the plane of the epoxide ring and the aromatic ring system is 83°, whereas in *syn*-BPDE this angle is splayed out to 118°.[47] Analysis of the molecular structure of the analogous *anti*-diol epoxide derivative of naphthalene by X-ray crystallographic techniques gave similar results.[48] The hydroxyl groups of the naphthalene derivative also occupy equatorial positions, and the angle between the plane of the epoxide ring and the aromatic ring system is closely similar (80°) to that of *anti*-BPDE.*

NMR analysis indicates that both diastereomers of BPDE exist in solution as rapidly equilibrating mixtures of conformers. The coupling of the carbinol hydrogens of *anti*-BPDE is in the range of $J_{7,8}$ = 8.25–9.0 Hz in dimethyl sulfoxide,[49,50] closely similar to that of its dihydrodiol precursor for which $J_{7,8}$ = 9.0 Hz. The percentage of the diequatorial conformer of *anti*-BPDE in solution is calculated to be 58–65%. *syn*-BPDE exhibits a somewhat smaller coupling constant ($J_{7,8}$ = 6.0 Hz), indicative of a slight excess of the diaxial conformer (~62%) in dimethyl sulfoxide at room temperature. The coupling constants for the carbinol hydrogens of the corresponding *anti*- and *syn*-diol epoxide derivatives of naphthalene were $J_{1,2}$ = 9.0 and 3.0 Hz, respectively. This finding indicates that the conformational preferences of these isomers are closely similar to those of *anti*- and *syn*-BPDE, except that the naphthalene *syn*-diol epoxide shows a markedly higher ratio of the diaxial conformer (~90%). It is interesting that the coupling constants for H_1 and H_2 of the dimethyl ether derivatives of the naphthalene *anti*- and *syn*-diol epoxide are essentially the same ($J_{1,2}$ = 8.8 and 2.8 Hz, respectively)[51] as those of the unmethylated compounds. This finding suggests that intermolecular hydrogen bonding may be less important in determining conformational preferences than is often assumed. It should also be kept in mind that the energy for conformational interconversion of 1,2-dihydroaromatic molecules is generally low in the absence of steric effects, and these conformational equilibria may be expected to shift readily with changes in the medium and other experimental conditions.

NMR data on the *anti* bay-region diol epoxide derivatives of other proximate dihydrodiols (Figure 8.11: **28–31**, **35–37**, **40**, **41**) show a relatively narrow range of values for the coupling constants of the carbinol protons (Table 8.6). In most cases these lie between 8 and 9.0 Hz, indicative of a small preference for existence of the *anti*-isomers in the diequatorial conformation. The most significant exceptions are **29** for which $J_{1,2}$ = 7.0 Hz and **37b** for which $J_{7,8}$ = 5.5 Hz. There is no apparent reason for the somewhat lower than normal value of the $J_{1,2}$ coupling for **29**, particularly in view of the normal values of $J_{1,2}$ for the closely related **28a,b**. On the other hand, the preference of **37b** for the diaxial conformation is readily understood as a consequence of the steric interaction between the benzylic hydroxyl group and the fluorine atom in the 6 position. The only *anti*-diol epoxide derivatives of terminal ring dihydrodiols investigated (**27**,**32**,**33**) show a

* Klein and Stevens[48] report this angle as 100°. In order to be consistent with the findings of Neidle et al.[46] for *anti*-BPDE we report this angle as its complement 80°.

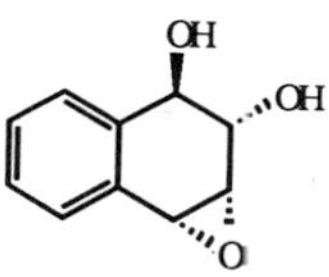

27

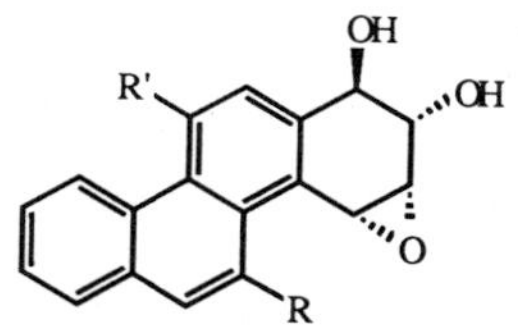

28a: R = R' = H
b: R = CH_3; R' = H
c: R = H; R' = CH_3

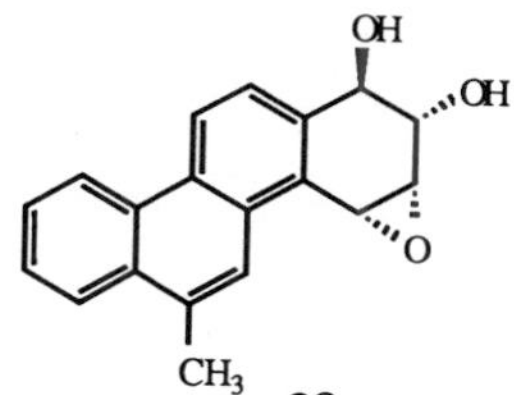

29

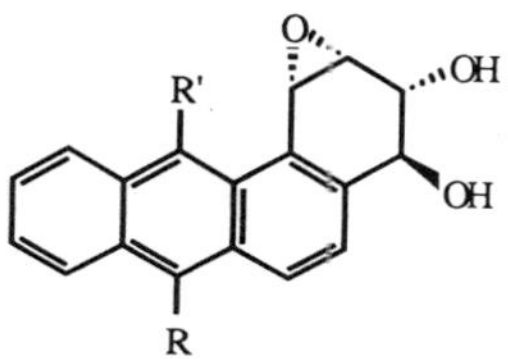

30a: R = R' = H
b: R = CH_3; R' = H
c: R = R' = CH_3

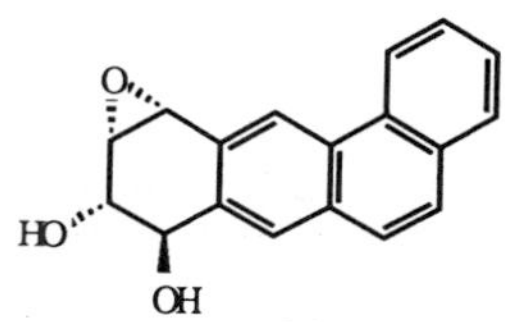

31

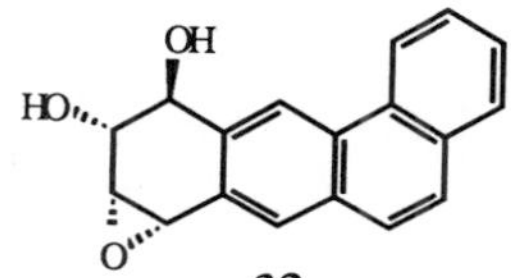

32

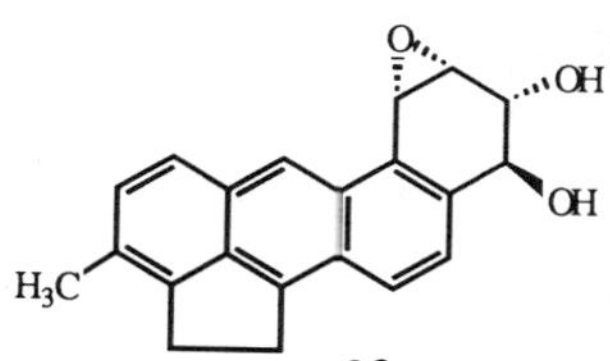

33

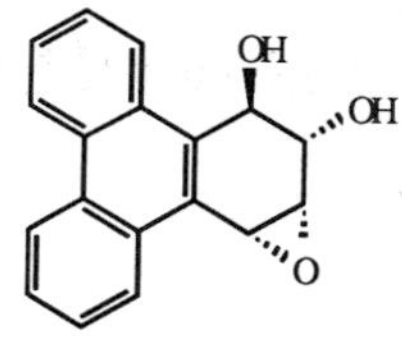

34

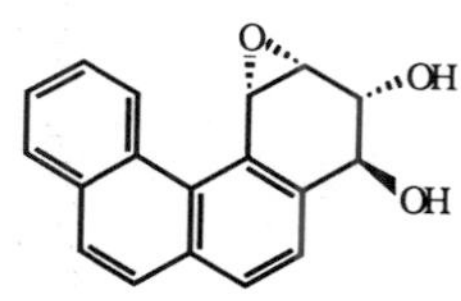

35

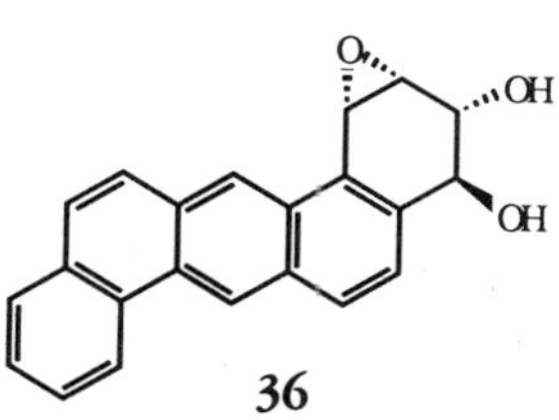

36

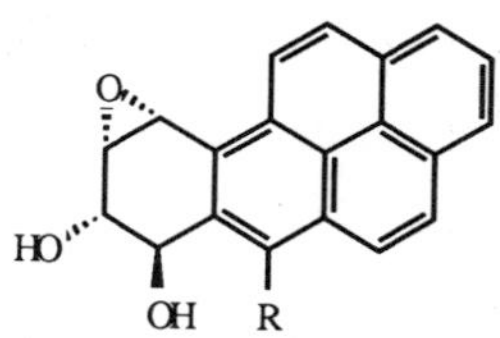

37a: R = H
b: R = F

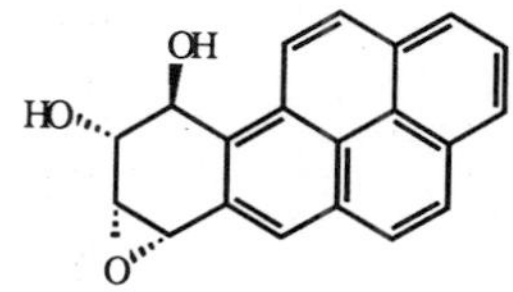

38

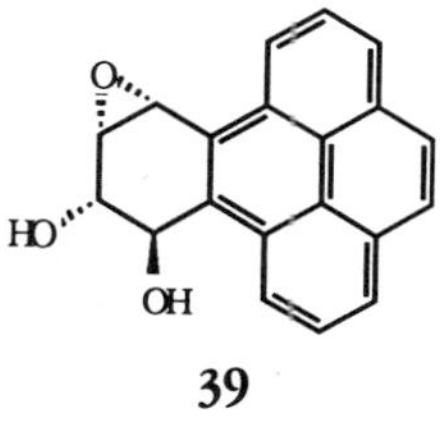

39

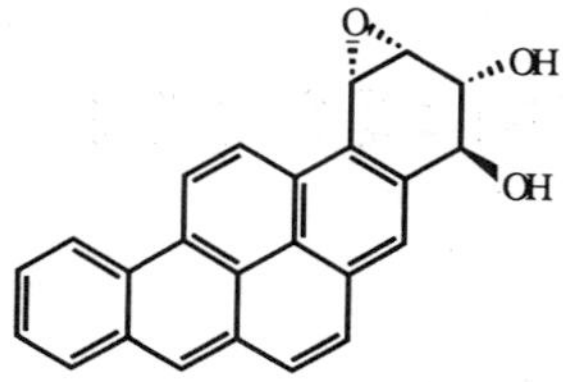

40

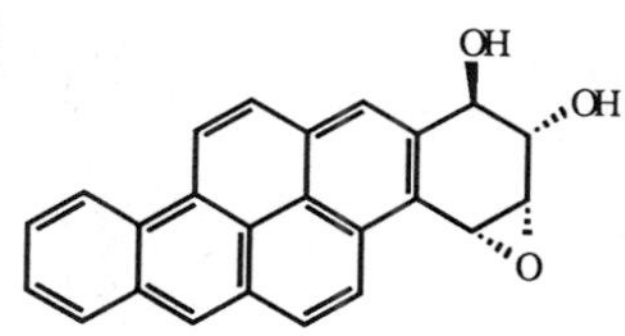

41

Figure 8.11 Molecular structures of the *anti*-diol epoxide derivatives of PAHs.

Table 8.6 NMR Coupling Constants for the Carbinol Protons of PAH *anti*- and *syn*-Diol Epoxides

Diol Epoxide	*Anti* J (Hz)	*Syn* J (Hz)	Reference	Diol Epoxide	*Anti* J (Hz)	*Syn* J (Hz)	Reference
27	9.0	3.0	49	**33**	9.0	3.0	53
28a	8.6	7.1	34	**34**	3.5	1.7	54
28b	8.6	9.6	34	**35**	8.0	9.0	55
28c	7.4	7.6	34	**36**	9.0		29
29	7.0		52	**37a**	8.25,9.0	6.0	49,50
30a	8.5	7.0	53	**37b**	5.5	2.5	56
30b	8.0		35	**38**	6.5	2.5	56
30c		8.1	37	**39**	3.5	2.5	54
31	7.1		38	**40**	8.75		41
32	9.0	3.2	53	**41**	9.0		41

similar preference for the diequatorial conformation, based on the coupling of their carbinol hydrogens. The only examples of *anti*-diol epoxides in which the hydroxyl groups reside in a bay region (**34**,**38**,**39**) exhibit small couplings, consistent with the diaxial conformation.

The *syn*-diol epoxide derivatives of proximate dihydrodiols also tend to exhibit a small preference for the diequatorial conformation (Table 8.6), while the *syn*-diol epoxides derived from terminal ring dihydrodiols (*syn*-isomers of **27** and **32**) and bay-region dihydrodiols (*syn*-isomers of **34**,**38**,**39**) show greater preference for the diaxial conformation. While the molecular basis of these conformational differences is not established with certainty, inspection of molecular models suggest that the diaxial preference of the terminal ring isomers, for example, *syn*-**27**, is largely dictated by the severe steric interaction between the hydrogen atom of the nonbenzylic carbinol position and the epoxide oxygen atom in the diequatorial conformer. This effect is also present in the *syn*-diol epoxide isomers in which the epoxide oxygen resides in a bay region, for example, *syn*-BPDE, but it is apparently partially counterbalanced by the additional steric interaction between the benzylic oxiranyl hydrogen atom and the aromatic proton in the adjacent bay-region site present in the diaxial conformers. This steric interference is relieved in the *syn*-diequatorial conformers.

8.4 STEREOCHEMISTRY OF ARENE OXIDES AND TETRAHYDRO EPOXIDES

The conformational properties of PAH arene oxides are relatively unexplored. This is hardly surprising in view of the chemical reactivity and thermal instability of most arene oxides. However, X-ray crystallographic studies throw some light on the steric factors that are operative for the relatively stable K-region oxides.[57] In general, the effect of epoxidation of an aromatic ring is a lengthening of the bonds in the ring to which the oxygen atom attaches and a slight shortening of the bonds in the adjacent rings. Phenanthrene 9,10-oxide and benzo[*a*]pyrene 4,5-oxide are both approximately planar with an angle of ~103° between the epoxide ring and the plane of the adjacent aromatic ring.[58,59] In contrast, 7,12-dimethylbenz[*a*]anthracene 5,6-oxide (**42**) deviates markedly from planarity, with an angle of 35° between the two most distant rings (Figure 8.12).[58] This nonplanarity is also found in the parent hydrocarbon and is a result of steric repulsion between the bay-region methyl group and the angular benzo ring.[57,60] The oxide rings of the phenanthrene and benzo[*a*]pyrene derivatives are both symmetrical, while the C—O bonds of 7,12-dimethylbenz[*a*]anthracene 5,6-oxide are asymmetric with the C_6—O bond being somewhat longer than the C_5—O bond. Although 7,12-dimethylbenz[*a*]anthracene 5,6-oxide is locked into a rigid conformation in the crystalline state, an alternative conformation may in principle

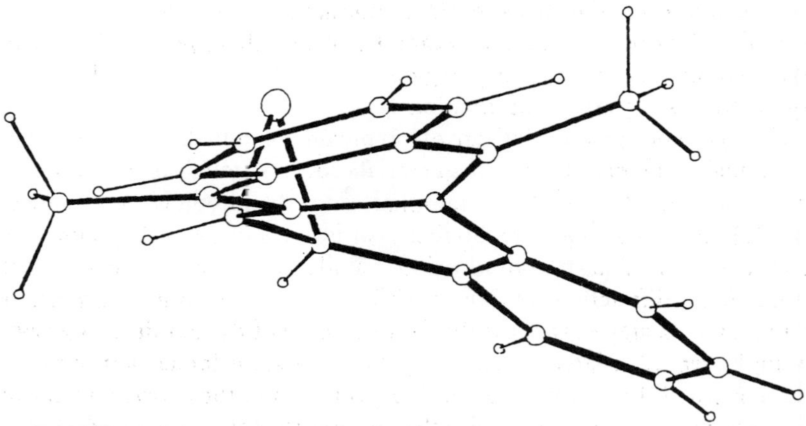

Figure 8.12 View of 7,12-dimethylbenz[*a*]anthracene 5,6-oxide (**42**) showing the marked deviation from planarity.[57]

arise in solution by inversion of the twisted aromatic ring system. However, there is currently no evidence for this alternative structure in the solid state or in solution.

Tetrahydro PAH epoxide derivatives, such as 9,10-epoxy-7,8,9,10-tetrahydrobenzo[*a*]pyrene (**43**) and 9,10-epoxy-7,8,9,10-tetrahydrobenzo[*e*]pyrene (**44**), are of considerable biological interest because of their exceptionally high mutagenicity.[61,62] Although direct evidence concerning their conformational properties is lacking, they may be regarded as analogs of diol epoxides without hydroxyl groups. Two conformers equivalent to the diequatorial and diaxial conformers of the corresponding diol epoxides are possible. In one of these (Figure 8.13, conformer I) the benzylic hydrogen atom on the same ring face as the epoxide oxygen atom is oriented axially, while in conformer II the same benzylic hydrogen atom is oriented equatorially. However, conformer II is anticipated to be energetically less favorable due to the severe steric interaction between the oxygen atom and the axial hydrogen atom adjacent to the epoxide ring. Therefore, in the absence of other effects, conformer I is expected to predominate in simple tetrahydro epoxides, such as anthracene 1,2-oxide (**45**). In more complex tetrahydro epoxides, such as **43** and **44**, in which the epoxide ring resides

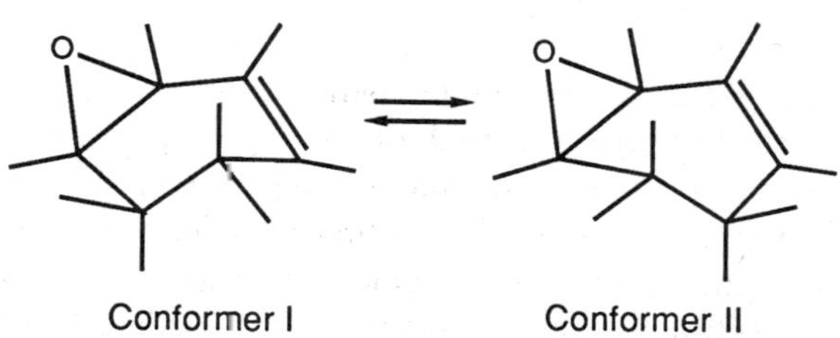

Figure 8.13 Conformational equilibrium of tetrahydro PAH epoxide derivatives.

in a bay region, steric repulsion between the benzylic oxiranyl hydrogen atom and the bay-region aromatic proton is expected to destabilize conformer I relative to conformer II. Unfortunately, experimental evidence pertinent to this question is unavailable.

8.5 STEREOCHEMISTRY OF PRODUCTS OF NUCLEOPHILIC ADDITION AND DNA BINDING

The reactions of PAH oxides and diol epoxides with nucleophiles are of considerable interest because of their critical role in the mechanism of carcinogenesis as well as in detoxification involving reaction with glutathione.

8.5.1 Adducts of PAH Oxides

Extensive physical chemical studies of the reactions of arene oxides with various nucleophiles have been carried out by Bruice and co-workers.[63] These studies show that nucleophilic attack by amines and oxygen bases on non-K-region arene oxides (phenanthrene 1,2-oxide and 3,4-oxides, benzene oxide, and naphthalene 1,2-oxide) is not sufficiently rapid to compete with the spontaneous aromatization reaction. However, thiols, because of their greater nucleophilic reactivity, add readily to the non-K-region arene oxides. The K-region arene oxide phenanthrene 9,10-oxide, as a result of its relatively slow rate of ring opening to a carbonium ion and the slow rate of rearrangement of the latter to phenol, undergoes reaction readily with a wide range of nucleophiles, including oxygen bases, amines, and thiols (including glutathione). NMR studies of the adducts show that nucleophilic attack is stereospecific and results in trans adducts. The conformation of these adducts is assumed to resemble that of the closely related K-region dihydrodiol derivatives. The coupling constants ($J_{9,10}$) for the benzylic hydrogens of the adducts formed with phenanthrene 9,10-oxide by various nucleophiles are given in Table 8.7. With the exception of the products formed from the addition of 2-mercaptoethanol and thiolacetic acid, the coupling constants fall in the range of 6.2–9.4 Hz, indicative of a mixture of diaxial and diequatorial conformers with a small preference for the diaxial conformer at equilibrium in most cases [47–70% diaxial calculated from Equation (8.1)]. Acetylation generally decreases the coupling constants and shifts the conformational equilibrium in favor of the diaxial conformer, due to elimination of internal hydrogen bonding and the increase in steric bulk.

Reactions of a more extensive series of K-region arene oxides with the model nucleophile *tert*-butylthiol were shown by Beland and Harvey[64] to afford the products of trans-stereospecific addition to the oxide ring. Product structures accorded with regioselective attack at the most electrophilic carbons atoms, in

Table 8.7 Coupling Constants for the Benzylic Protons of the Adducts Formed by Addition of Nucleophiles at the K-Region of Phenanthrene 9,10-oxide[a]

Nucleophile	$J_{9,10}$ Adduct (Hz)	$J_{9,10}$ Acetate of Adduct (Hz)
CH_3O^-	8.4	4.6
$HOCH_2CH_2S^-$	3.9	
$CH_3C(=O)S^-$	3.5	2.7
CH_3NH_2	7.2	~4
$(CH_3)_2NH$	6.2	3.4
tert-$BuNH_2$	9.4	3.6
n-$BuNH_2$	8.0	~3
N_3^-	7.2	5.2

[a] Data are from Ref. 63a.

agreement with MO theoretical prediction. The coupling constants for the benzylic protons of the products formed by the addition of *tert*-butylthiolate to the K oxides of phenanthrene, pyrene, benzo[*c*]phenanthrene, chrysene, benzo[*a*]pyrene, and benz[*a*]anthracene were all approximately $J = 8$ Hz in nonpolar solvents (CCl_4, $CDCl_3$), but decreased to $J = 3$ Hz in ionizing solvents (Me_2SO-d_6, acetone-d_6). These values confirm that all of these compounds exist in a preferred diaxial conformation. The larger values of J in less polar media are indicative of a shift of the dynamic equilibrium in the direction of the diequatorial conformation as a consequence of internal hydrogen bonding between the hydroxyl group and the sulfur atom. The acetate ester derivatives generally exhibited smaller couplings ($J = 3.0$–3.5 Hz), consistent with the expectation that these larger groups are capable of locking the conformation into the diaxial structure.

The presence of a methyl group in a position peri to the K region, as in 7,12-dimethylbenz[*a*]anthracene 5,6-oxide (Figure 8.12:**42**), shifted the conformational equilibrium of the *tert*-butyl sulfide adducts further toward the diaxial structure, as shown by the couplings for **46a** and **47a** which were $J_{5,6} = 3.6$ and 3.0 Hz, respectively, in $CDCl_3$. A similar effect was evident in the aniline adducts of **42** (**46b** and **47b**) for which $J_{5,6} = 3.0$ and 3.1 Hz, respectively, indicative of an equally strong preference for the diaxial conformation.[65]

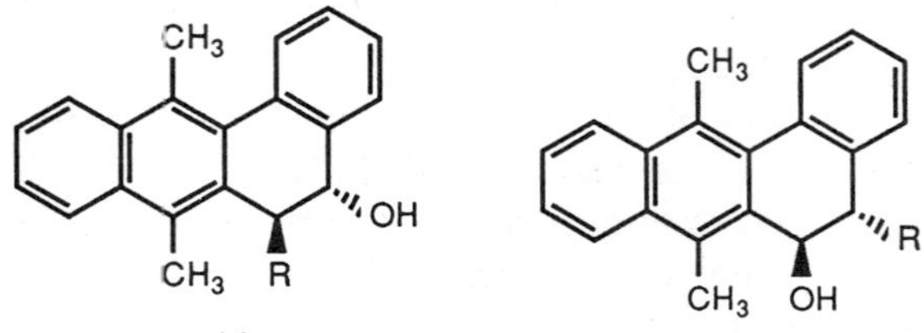

46 47

a: R = t-BuS; **b:** R = C_6H_5NH; **c:** R = Guanosine.

Similar reactions of the K-region oxides of phenanthrene, pyrene, and benzo[*a*]pyrene with glutathione (as its *N*-trifluoroacetyl dimethyl ester) furnished the corresponding products of trans-stereospecific addition.[66] The coupling constants of the benzylic protons of the benzoate ester derivatives of the adducts were quite small, ranging from 1.1 to 2.4 Hz, and indicative of a strong diaxial conformational preference.

Since K-region arene oxides, such as benzo[*a*]pyrene 4,5-oxide and **42**, are established mutagens[67] which have been shown to bind covalently to nucleic acids,[17,68] the stereochemistry of their interactions with DNA and RNA are of particular interest. It is expected, because of the steric bulk of nucleic acid molecule near the reactive site, that nucleophilic attack will occur from an axial direction to furnish an adduct in which the purine or pyrimidine base (dG in the major adduct)[69] is axially oriented. The only direct evidence bearing on this question comes from studies of the reaction of **42** with polyguanylic acid.[17] The modified polymer was degraded to the nucleoside level and the resulting PAH–nucleoside adducts were separated by HPLC to yield two pairs of diastereomers corresponding to **46c** and **47c**. All four of the adducts had $J_{5,6}$ coupling constants in the range of 3.2 ± 0.5 Hz. These small values are consistent with the tentative assignment of the *trans*-diaxial conformation.

PAH tetrahydro epoxides also enter into reactions with nucleophiles to afford the products of trans-stereospecific addition in which the entering group becomes covalently attached to the benzylic position. For example, reaction of 1,2-epoxy-1,2,3,4-tetrahydrobenz[*a*]anthracene (**48**) with diethyl phosphate yields the adduct **49**.[70] In the adducts of this type derived from bay-region

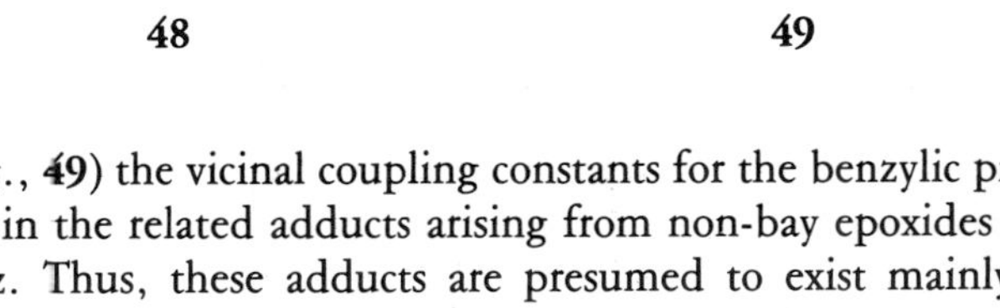

48 **49**

epoxides (e.g., **49**) the vicinal coupling constants for the benzylic protons are ~3 Hz, whereas in the related adducts arising from non-bay epoxides the couplings were ~7 Hz. Thus, these adducts are presumed to exist mainly in the axial conformation.

8.5.2 Adducts of PAH Diol Epoxides

The reactions of bay-region diol epoxides, such as *anti*-BPDE (Figure 8.2), with nucleophiles are also expected, as a consequence of the severe steric crowding in the bay region, to afford adducts in which the covalently bound entering group is axially oriented in the most stable conformation. The reaction of *tert*-butylthiolate with *anti*- and *syn*-BPDE, the first such reaction to be investigated was initially employed to aid in the stereochemical assignments of these isomers.[50] The geometry of the saturated benzo ring in these adducts is closely related to that of cyclohexene, which can adopt either a half-chair or a flattened boat conformation (Figure 8.14). For cyclohexene, the half-chair form is favored by only 2.7 kcal/mol, so that conformational interconversion is facile.[71] For the *tert*-butylthiolate adducts of *anti*- and *syn*-BPDE following acetylation, the observed couplings of the bay-region benzylic protons were $J_{9,10}$ = 4.0 and 2.0 Hz, respectively. These small values are consistent with the *trans*-diaxial conformational assignment of the substituents in the 9,10 positions. For the trans-adduct of *anti*-BPDE, the observed large couplings for $J_{7,8}$ = 10 Hz and small coupling for $J_{8,9}$ = 2.8 Hz favor conformer A in which the C-7,8 substituents are diequatorial. For the trans-adduct of *syn*-BPDE, the relatively large couplings for $J_{7,8}$ = 8.25 Hz is not compatible with conformer E, which is also improbable because of its substantial steric crowding due to the four axial substituents. The flattened boat conformation G is most consistent with NMR data for this adduct.

The reactions of *anti*- and *syn*-BPDE with a variety of other nucleophiles have also been investigated (Table 8.8).[72] Regioselective attack of these reagents at the benzylic carbon atom was substantiated by the characteristic downfield shifts of the signal for the hydrogen atom at C-10. While these reactions were predominantly trans-stereospecific, the reactions with water and phenol afforded significant amounts of the cis-adducts. The conformational preference of these adducts (Figure 8.14) is expected to be influenced not only by steric interactions between the substituent groups on the saturated ring, but also by steric interactions with the aromatic hydrogens in the peri and bay-region positions (i.e., C-6 and C-11). For the trans-adducts of *anti*- and *syn*-BPDE, the coupling

Table 8.8 NMR Coupling Constants of the Acetates of the Products of cis and trans Addition to *anti*- and *syn*-BPDE[a]

	anti-BPDE			*syn*-BPDE		
R	$J_{7,8}$	$J_{8,9}$	$J_{9,10}$	$J_{7,8}$	$J_{8,9}$	$J_{9,10}$
trans-Adducts						
OH	8.8	2.5	3.6	8.0	5.0	3.5
OMe	9.0	2.5	3.6	8.8	3.8	3.8
$SC_6H_4NO_2$	9.0	2.5	3.5	8.0	2.8	2.0
NHC_6H_5	9.0	2.5	2.5	4.5	4.0	4.0
OC_6H_5				7.4	3.6	3.6
cis-Adducts						
OH	3.5	2.5	4.6	8.0	11.5	3.5
OC_6H_5	4.5	2.5	5.0	7.8	11.3	3.3

[a] Data are from Yagi et al.[71] Spectra were taken in $CDCl_3$ at 100 or 220 MHz, and coupling constants are given in hertz.

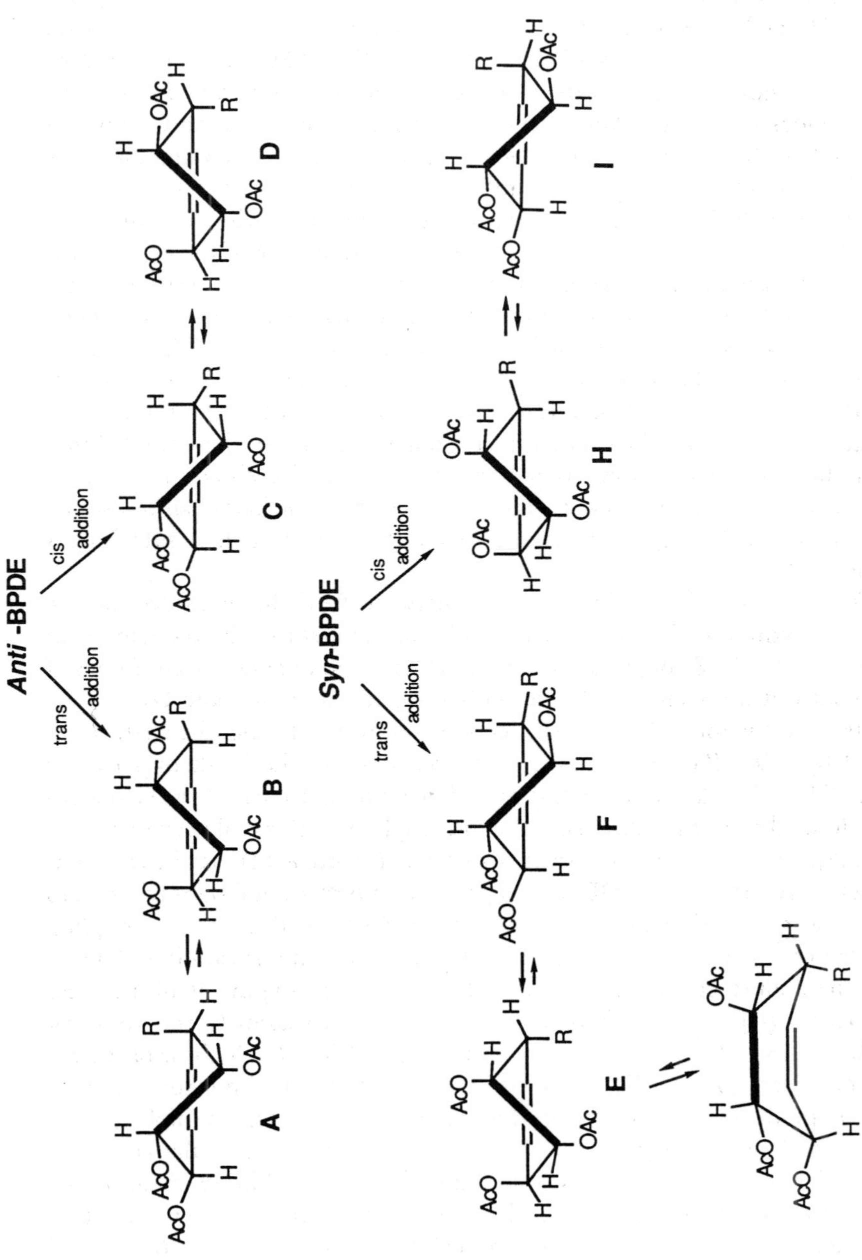

Figure 8.14 Possible conformations for the adducts formed by addition of nucleophiles to the diol *anti*- and *syn*-BPDE followed by acetylation.

constants for the 9,10 protons were generally small ($J_{9,10}$ = 2.5–3.6 and 2.0–4.0 Hz, respectively) and consistent with the expected axial orientation of the benzylic substituents. For the trans-adducts of *anti*-BPDE, the observed large couplings for $J_{7,8}$ = 9 Hz and small coupling for $J_{8,9}$ = 2.5 Hz are most compatible with conformer A. For the trans-adducts of *syn*-BPDE, like that of the *tert*-butylthiolate adduct, neither of the two chair forms is expected to be favored due to their serious steric interactions. The boat conformation G is most consistent with the relatively large values for $J_{7,8} \sim$ 8 Hz and the unusual coupling for $J_{8,9} \sim$ 2.8–5 Hz. The sole exception is the aniline adduct for which all three coupling constants are small. In this case, the data are most compatible with conformer E in which all four substituents are axial. Because of the steric bulk of the groups, it is likely that the actual conformation of the molecule is somewhat distorted from ideality. Hydrogen bonding between the amine hydrogen and the acetate carbonyl may be a stabilizing factor for this unusual conformation.[72] The adducts resulting from cis addition to *anti*-BPDE were assigned the half-chair conformation D in which the substituents at C-7,8,10 are axial and the acetoxy group at C-9 is equatorial. The small values for all the coupling constants in the saturated ring support this proposed conformational assignment. For the cis-adducts of *syn*-BPDE, the observed couplings were most consistent with the half-chair conformation I in which the bay-region group is axial and the remaining three substituents are equatorial.

The structures of the adducts formed between PAH diol epoxides and the purine and pyrimidine bases of nucleic acids are pertinent to the mechanism of PAH carcinogenesis. Evidence on this question has been obtained from studies of the reactions of *anti*- and *syn*-BPDE with polyguanylic acid (poly G).[8,73] The products were hydrolyzed to ribonucleosides, derivatized, and their structures analyzed by CD, NMR, and mass spectral techniques. The NMR data on the major *anti*-BPDE adduct (Figure 8.3) were closely similar to that of the analogous adducts formed by trans addition of other nucleophiles to this diol epoxide isomer (Table 8.8). Therefore, the conformer A structure (Figure 8.14) may be assigned. Analogous reaction of *syn*-BPDE yielded products of both cis and trans addition to the 2-amino group of guanine. For the trans-adduct, all three of the coupling constants were small ($J_{7,8}$ = 5.2, $J_{8,9}$ = 5.4, $J_{9,10}$ = 5.4 Hz), resembling those of the aniline adduct of *syn*-BPDE (Table 8.8), allowing assignment of a similar conformation (Figure 8.14: conformer E). For the cis-adduct, the coupling constants ($J_{7,8}$ = 8.0, $J_{8,9}$ = 12, $J_{9,10}$ = 4.0 Hz) paralleled closely those of the cis-adduct of phenol to *syn*-BPDE. These values are consistent with conformer I.

The products of reaction of the racemic anti-bay-region diol epoxide derivative of 5-methylchrysene **28b** with DNA have also been fully characterized.[10,74] The major adduct was shown to arise from trans addition of the amino group of deoxyguanosine to the benzylic position of the epoxide ring of the (+) enantiomer of **28b**. A minor product formed by analogous reaction of the (−) enantiomer was also identified. In addition, two diasteromeric products arising from trans addition of the amino group of adenosine to the epoxide ring and a third product formed by cis addition of adenosine to one enantiomer was also characterized. The NMR data for the acetylated derivatives of these adducts[10,74] is consistent in all cases with a conformation in which the covalently bound purine base is axially oriented. For the trans-adducts, the observed large couplings for $J_{1,2}$ = 9.0–9.4 Hz are most consistent with a half-chair conformation in which the C-1,2 substituents are diequatorial, analogous to conformer A (Figure 8.14). For the cis-adducts, the small values of all the coupling constants ($J_{1,2}$ = 6.0–6.7, $J_{2,3}$ ~ 2, $J_{3,4}$ ~ 4.0 Hz) paralleled those of the cis-adducts of *anti*-BPDE (Table 8.8), allowing assignment as conformer D in which the C-1,2,4 substituents are all axial.

8.6 CONFORMATION, DNA BINDING, AND BIOLOGICAL ACTIVITY

Knowledge of the conformational properties of oxidized PAH metabolites may potentially provide useful insight into their biological properties. There is now substantial evidence that the mechanism of PAH mutagenic and carcinogenic activities entails covalent interaction of reactive PAH metabolites with cellular macromolecules, most likely DNA.[1,3,74–76] While PAH epoxide derivatives are known also to react extensively with proteins, water, and other nucleophiles, these are believed to be secondary competing pathways that are not directly relevant to the biological activities of these molecules.

Current evidence suggests that the mechanism of covalent binding of PAH epoxide derivatives to nucleic acids in vivo involves prior intercalation of the fused aromatic ring system between the base pairs of the nucleic acids.[75,77] Kinetic studies of the reactions of *anti*-BPDE and other diol epoxides with DNA indicate that the mechanism entails protonation of the intercalated diol epoxide in the rate-determining step to furnish an intercalated triol carbonium ion intermediate which combines covalently with a nucleophilic site, principally the 2-amino group of deoxyguanosine, to yield products.[75,76,78] For covalent bonding to take place at the benzylic carbon atom of the epoxide ring, it is necessary for the reactive PAH intermediate to be rather precisely oriented in the intercalated reaction complex to allow sufficiently close approach of the appropriate nucleophilic site on DNA. Molecular modeling studies[79,80] indicate that the preferred orientation of *anti*-BPDE in the reaction complex is consistent with the experimental finding that reaction takes place preferentially on the 2-amino group of deoxyguanosine in the minor groove of DNA (Figure 8.15). The strong propensity for reaction to take place at the latter site is especially remarkable in view of the fact that theoretical MO calculations indicate that the N-7 atom is the most electron-rich center in the guanosine molecule.[81] Indeed, simple alkylating agents, such as methyl iodide, react preferentially at the N-7 atom of guanosine, in accord with this prediction.[82] Thus, it appears that the relatively rigid geometrical requirements imposed by the

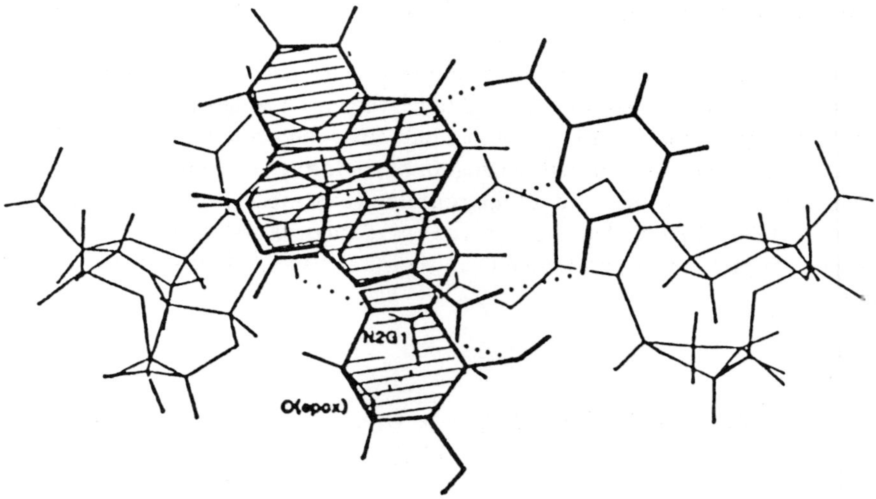

Figure 8.15 Minimum energy structure of a noncovalent adduct of *anti*-BPDE intercalated between dCpG base pairs calculated by Subbiah et al.[80] using computer modeling techniques.

intercalated intermediate complex are largely responsible for the observed regiospecificity of these reactions.

The relatively high trans-stereospecificity of the reactions of PAH epoxides with nucleic acids is also remarkable in view of the evidence from mechanistic studies for involvement of a triol carbonium ion intermediate as the active species.[78] The trans-adducts presumably arise from attack of the nucleophilic amino group on the benzylic carbon atom of the intercalated carbonium ion species on the molecular face opposite the hydroxyl group in the adjacent position (e.g. the 9 position of *anti*-BPDE). Attack is necessarily from the axial direction due to the relatively fixed geometry of the DNA base pairs at the reaction site. Thus, the trans-stereoselectivity of these reactions also appears to be largely a consequence of the stereochemical limitations of the intercalated reaction complex.

There is substantial evidence that existence in the diequatorial conformation is an essential requirement for the biological activity of PAH dihydrodiol and diol epoxide metabolites. The bay-region dihydrodiols and other dihydrodiols and diol epoxides, which are sterically constrained to adopt the diaxial conformation, are generally less active than their related diequatorial counterparts. The reason for this difference is unknown. However, it is likely that diaxially oriented hydroxyl groups will interfere sterically with insertion and proper orientation of the diol epoxide metabolites between the base pairs of the DNA helix. Thus, the bay-region diol epoxide derivative of benzo[*e*]pyrene **39** has been shown to bind less efficiently to DNA than the isomeric *anti*-BPDE.[83] The hydroxyl groups of the former are diaxial, while those of the latter are diequatorial.[25] Another difference between the properties of these molecules is in the structures of their covalent DNA-bound adducts.[83] Spectroscopic studies indicate that the principal adduct of *anti*-BPDE is covalently bound in the minor groove oriented approximately parallel to the DNA helix.[78,84] This conformation appears to be characteristic of the structures of the DNA-bound adducts of the diol epoxide derivatives of the most carcinogenic PAHs investigated to date.[75,76] In contrast, interaction of **39** with DNA affords at least two types of covalent adducts, the ratio of which is concentration dependent.[83] One of these exhibits properties characteristic of an external binding site, while the other is distinguished by strong interaction with the DNA bases, indicating that the aromatic ring system is oriented parallel to the bases and is at least partially intercalated. It is conceivable that the diaxial hydroxyl groups may sterically interfere with reorganization of the initially formed intercalated adducts into an externally bound structure or may cause substantial unwinding of the helix near the site of attachement. In any case, these differences in adduct structure are apparently related to the biological properties of these compounds. Further investigation will be required to establish their molecular basis.

While it is clear from the foregoing findings that the stereochemical properties of the epoxide derivatives of PAHs with nucleic acids are important to their binding to nucleic acids, little is known concerning the relation between adduct structure and DNA replication leading to mutation and ultimately to tumor induction. A more comprehensive account of current knowledge on the interaction of PAH diol epoxide metabolites with nucleic acids and the relation between adduct structure and bioactivity may be found in the recent review by Harvey and Geacintov.[76]

REFERENCES

1. Harvey, R. G. *Acc. Chem. Res.* **1981**, *14*, 218.
2. Yang, S. K.; Mushtaq, M.; Chiu, P. In *Polycyclic Hydrocarbons and Carcinogenesis*; Harvey, R. G., Ed.; Symposium Monograph 283; American Chemical Society: Washington, DC, 1985; pp. 19–34.
3. Sims, P.; Grover, P. L. In *Polycyclic Hydrocarbons and Cancer*, Vol. 3; Gelboin, H. V. and Ts'o, P. O. P., Eds.; Academic: New York, 1981; pp. 117–181.
4. Conney, A. H. *Cancer Res.* **1982**, *42*, 4875.
5. Estabrook, R. W.; Werringloer, J.; Capdevila, J.; Prough, R. A. In *Polycyclic Hydrocarbons and Cancer*, Vol. 1; Gelboin, H. V. and Ts'o, P. O. P., Eds.; Academic: New York, 1978; pp. 285–319.
6. Fu, P. P.; Harvey, R. G.; Beland, F. *Tetrahedron* **1978**, *34*, 857.
7. Nakanishi, K.; Kasai, H.; Cho, H.; Harvey, R. G.; Jeffrey, A. M.; Jennette, K. W.; Weinstein, I. B. *J. Am. Chem. Soc.* **1977**, *99*, 258.
8. Jeffrey, A. M.; Jennette,K. W.; Blobstein, S. H.; Weinstein, I. B.; Beland, F. A.; Harvey, R. G.; Kasai, H.; Muira, I.; Nakaniski, K. *J. Am. Chem. Soc.* **1976**, *98*, 5714.
9. Osborne, M. R.; Harvey, R. G.; Brookes, P. *Chem.-Biol. Interact.* **1978**, *20*, 123. Jeffrey, A. M.;

Grzeskowiak, K.; Weinstein, I. B.; Nakanishi, K.; Roller, P.; Harvey, R. G. *Science* **1979**, *206*, 1309. Baird, W. M.; Diamond, L. *Int. J. Cancer* **1978**, *22*, 189. Weinstein, I. B.: Jeffrey, I. M.; Leffler, S.; Pulkrabek, P.; Yamasaki, H.; Grunberger, D. In *Polycyclic Hydrocarbons and Cancer*, Vol. 2; Gelboin, H. V. and Tso's, P. O. P., Eds.; Academic: New York, 1978; pp. 1–36.

10. Melikian, A. A.; Amin, S.; Hecht, S. S.; Hoffman, D.; Pataki, J.; Harvey, R. G. *Cancer Res.* **1984**, *44*, 2524.
11. Dipple, A.; Pigott, M. A.; Agarwal, S. K.; Yagi, H.; Sayer, J. M.; Jerina, D. M. *Nature (London)* **1987**, *327*, 535.
12. Newbold, R. F.; Brookes, P. *Nature (London)* **1976**, *261*, 53. Huberman, E.; Sachs, L.; Yang, S. K.; Gelboin, H. V. *Proc. Natl. Acad. Sci. USA* **1976**, *73*, 607. Wood, A. W.; Wislocki, P. G.; Chang, R. L.; Levin, W.; Lu, A. Y.; Yagi, H.; Hernandez, O.; Jerina, D. M.; Conney, A. H. *Cancer Res.* **1976**, *36*, 3358.
13. Slaga, T. J.; Bracken, W. M.; Viaje, A.; Levin, W.; Yagi, H.; Jerina, D. M.; Conney, A. H. *Cancer Res.* **1977**, *37*, 4130. Buening, M. K.; Wislocki, P. G.; Levin, W.; Yagi, H.; Thakker, D. R.; Akagi, H.; Koreeda, M.; Jerina, D. M.; Conney, A. H. *Proc. Natl. Acad. Sci. USA* **1978**, *75*, 5358.
14. Pullman, A.; Pullman, B. *Adv. Cancer Res.* **1955**, *3*, 117.
15. Harvey, R. G. *Synthesis* **1986**, 605. Harvey, R. G. In *Polycyclic Hydrocarbons and Carcinogenesis*, Harvey, R. G., Ed.; Symposium Monograph 283; American Chemical Society: Washington, DC, 1985; pp. 35–62.
16. Zacharias, D. E.; Glusker, J. P.; Harvey, R. G.; Fu, P. P. *Cancer Res.* **1977**, *37*, 775.
17. Jeffrey, A. M.; Blobstein, S. H.; Weinstein, I. B.; Beland, F. A.; Harvey, R. G.; Kasai, H.; Nakanishi, K. *Proc. Natl. Acad. Sci. USA* **1976**, *73*, 2311.
18. Fu, P. P.; Evans, F. E.; Miller, D. W.; Chou, M. W.; Yang, S. K. *J. Chem. Res. (S)* **1983**, 158.
19. Harvey, R. G.; Fu, P. P.; Rabideau, P. W. *J. Org. Chem.* **1976**, *41*, 3722.
20. Harvey, R. G.; Goh, S. H.; Cortez, C. *J. Am. Chem. Soc.* **1975**, *97*, 3468.
21. Balani, S. K.; van Bladeren, P. J.; Shirai, N.; Jerina, D. M. *J. Org. Chem.* **1986**, *51*, 1773.
22. Zacharias, D. E.; Glusker, J. P.; Fu, P. P.; Harvey, R. G. *J. Am. Chem. Soc.* **1979**, *101*, 4043.
23. Bothner-By, A. A. *Adv. Magn. Reson.* **1965**, *1*, 195.
24. Lehr, R. E.; Schaefer-Ridder, M.; Jerina, D. M. *J. Org. Chem.* **1977**, *42*, 736.
25. Harvey, R. G.; Lee, H. M.; Shyamasundar, N. *J. Org. Chem.* **1979**, *44*, 78.
26. Fu, P. P.; Harvey, R. G. *J. Org. Chem.* **1979**, *44*, 3778.
27. Karle, J. M.; Mah, H. D.; Jerina, D. M.; Yagi, H. *Tetrahedron Lett.* **1977**, 4021.
28. Croisy-Delcey, M.; Ittah, Y.; Jerina, D. M. *Tetrahedron Lett.* **1979**, 2849.
29. Lee, H. M.; Harvey, R. G. *J. Org. Chem.* **1980**, *45*, 558.
30. Jerina, D. M.; van Bladern, P. J.; Yagi, H.; Gibson, D. T.; Mahadevan, V.; Neese, A. S.; Koreeda, M.; Sharma, N. D.; Boyd, D. R. *J. Org. Chem.* **1984**, *49*, 3621.
31. Gibson, D. T.; Jerina, D. M.; Yagi, H.; Yeh, H. J. *Science* **1975**, *189*, 295.
32. Neidle, S.; Subbiah, A.; Osborne, M. *Carcinogenesis* **1981**, *2*, 533.
33. Sukumaran, K. B.; Harvey, R. G. *J. Org. Chem.* **1980**, *45*, 4407.
34. Harvey, R. G.; Pataki, J.; Lee, H. *J. Org. Chem.* **1986**, *51*, 1407.
35. Lee, H. M.; Harvey, R. G. *J. Org. Chem.* **1979**, *44*, 4948.
36. Harvey, R. G.; Cortez, C.; Sugiyama, T.; Ito, Y.; Sawyer, T. W.; DiGiovanni, J. *J. Medic. Chem.*, **1988**, *31*, 154.
37. Lee, H.; Harvey, R. G. *J. Org. Chem.* **1986**, *51*, 3502.
38. Jacobs, S.; Cortez, C.; Harvey, R. G. *Carcinogenesis* **1983**, *4*, 519.
39. Chiu, P.; Fu, P. P.; Weems, H. B.; Yang, S. K. *Chem.-Biol. Interact.* **1985**, *52*, 265.
40. Pataki, J.; Harvey, R. G. *J. Org. Chem.* **1987**, *52*, 2226.

41. Lehr, R. E.; Kumar, S.; Cohenour, P. T.; Jerina, D. M. *Tetrahedron Lett.* **1979**, 3819.
42. Yagi, H.; Vyas, K. P.; Tada, M.; Thakker, D. R.; Jerina, D. M. *J. Org. Chem.* **1982**, *47*, 1110.
43. Yeh, C. Y.; Fu, P. P.; Beland, F. A.; Harvey, R. G. *Bioorg. Chem.* **1978**, *7*, 497.
44. Kikuchi, O.; Hopfinger, A. J.; Klopman, G. *Cancer Biochem. Biophys.* **1979**, *4*, 1.
45. Klopman, G.; Grinberg, H.; Hopfinger, A. J. *J. Theor. Biol.* **1979**, *79*, 355.
46. Neidle, S.; Subbiah, A.; Cooper, C. S.; Ribeiro, O. *Carcinogenesis* **1980**, *1*, 249.
47. Neidle, S.; Cutbush, S. D. *Carcinogenesis* **1983**, *4*, 415.
48. Klein, C. L.; Stevens, E. D. *Cancer Res.* **1984**, *44*, 1523.
49 Yagi, H.; Hernandez, O.; Jerina, D. M. *J. Am. Chem. Soc.* **1975**, *97*, 6881.
50. Beland, F. A.; Harvey, R. G. *J. Chem. Soc., Chem. Commun.* **1976**, 84.
51. Becker, A. R.; Janusz, J. M.; Bruice, T. C. *J. Am. Chem. Soc.* **1979**, *101*, 5679.
52. Amin, S.; Huie, K.; Hecht, S. S.; Harvey, R. G. *Carcinogenesis* **1986**, *7*, 2067.
53. Lehr, R. E.; Schaefer-Ridder, M.; Jerina, D. M. *Tetrahedron Lett.* **1977**, 539.
54. Yagi, H.; Thakker, D. R.; Lehr, R. E.; Jerina, D. M. *J. Org. Chem.* **1979**, *44*, 3439.
55. Sayer, J. M.; Yagi, H.; Croisy-Delsey, M.; Jerina, D. M. *J. Am. Chem. Soc.* **1981**, *103*, 4970.
56. Yagi, H.; Sayer, J. M.; Thakker, D. R.; Levin, W.; Jerina, D. M. *J. Am. Chem. Soc.* **1987**, *109*, 838.
57. Glusker, J. P. In *Polycyclic Hydrocarbons and Carcinogenesis*; Harvey, R. G., Ed.; Symposium Monograph 283; American Chemical Society: Washington, DC, 1985; pp. 125–185.
58. Glusker, J. P.; Carrell, H. L.; Zacharias, D. E.; Harvey, R. G. *Cancer Biochem. Biophys.* **1974**, *1*, 43.
59. Glusker, J. P.; Zacharias, D. E.; Carrell, H. L.; Fu, P. P.; Harvey, R. G. *Cancer Res.* **1976**, *36*, 3951.
60. Klein, C. L.; Stevens, E. D.; Zacharias, D. E.; Glusker, J. P. *Carcinogenesis* **1987**, *8*, 5.
61. Wood, A. W.; Chang, R. L.; Huang, M. T.; Levin, W.; Lehr, R. E.; Kumar, S.; Thakker, D. R.; Yagi, H.; Jerina, D. M.; Conney, A. H. *Cancer Res.* **1980**, *40*, 1985.
62. Wood, A. W.; Wislocki, P. G.; Chang, R. L.; Levin, W.; Lu, A. Y.; Yagi, H.; Hernandez, O.; Jerina, D. M.; Conney, A. H. *Cancer Res.* **1976**, *36*, 3358.
63. (a) Bruice, P. Y.; Bruice, T. C.; Yagi, H.; Jerina, D. M. *J. Am. Chem. Soc.* **1976**, *98*, 2973; (b) Bruice, T. C.; Bruice, P. Y. *Acc. Chem. Res.* **1976**, *9*, 378; (c) Becker, A.; Janusz, J. M.; Bruice, T. C. *J. Am. Chem. Soc.* **1979**, *101*, 5679.
64. Beland, F.; Harvey, R. G. *J. Am. Chem. Soc.* **1976**, *98*, 4963.
65. Posner, G.; Lever, J. R. *J. Org. Chem.* **1984**, *49*, 2029.
66. Hernandez, O.; Gopinathan, M. B. *J. Chem. Soc., Chem. Commun.* **1984**, 1491.
67. Maher, V. M. McCormick, J. J. In *Polycyclic Hydrocarbons and Cancer*, Vol. 2; Gelboin, H. V. and T'so, P. O. P., Eds.; Academic: New York, 1981; pp. 137–160.
68. Baird, W. M.; Harvey, R. G.; Brookes, P. *Cancer Res.* **1975**, *35*, 54.
69. Blobstein, S. H.; Weinstein, I. B.; Grunberger, D.; Weisgras, J.; Harvey, R. G. *Biochemistry* **1975**, *14*, 3451.
70. Di Raddo, P. Chan, T. H. *J. Chem. Soc., Chem. Commun.* **1983**, 16.
71. Beckett, C. W.; Freeman, N. K.; Pitzer, K. S. *J. Am. Chem. Soc.* **1958**, *80*, 1227.
72. Yagi, H.; Thakker, D. R.; Hernandez, O.; Koreeda, M.; Jerina, D. M. *J. Am. Chem. Soc.* **1977**, *99*, 1604.
73. Koreeda, M.; Moore, P. D.; Yagi, H.; Yeh, H. J.; Jerina, D. M. *J. Am. Chem. Soc.* **1976**, *98*, 6720.
74. Reardon, D. B.; Prakash, A. S.; Hilton, B. D.; Roman, J. M.; Pataki, J.; Harvey, R. G.; Dipple, A. *Carcinogenesis* **1987**, *8*, 1317.
75. Harvey, R. G In *Molecular Mechanisms of Carcinogenic and Antitumor Activity*; Chagas, C. and

Pullman, B., Eds.; Pontifical Academy of Sciences, Vatican Press: Vatican City, 1987; pp. 95–129.

76. Harvey, R. G.; Geacintov, N. E. *Acc. Chem. Res.*, **1988**, *21*, 66.

77. Harvey, R. G., Ed. *Polycyclic Hydrocarbons and Carcinogenesis*; Symposium Monograph 283; American Chemical Society: Washington, DC, 1985.

78. Geacintov, N. E. In *Polycyclic Hydrocarbons and Carcinogenesis*; Harvey, R. G., Ed.; Symposium Monograph 283; American Chemical Society: Washington, DC, 1985; pp. 107–124. Geacintov, N. E.; Hibshoosh, H.; Ibanez, V.; Benjamin, M. J.; Harvey, R. G. *Biophys. Chem.* **1984**, *20*, 121. Geacintov, N. E.; Yoshida, H.; Ibanez, V.; Harvey, R. G. *Biochemistry* **1982**, *21*, 1864.

79. Neidle, S.; Pearl, L.; Beveridge, A. In *Molecular Mechanisms of Carcinogenic and Antitumor Activity*; Chagas, C. and Pullman, B., Eds.; Pontifical Academy of Sciences, Vatican Press: Vatican City, 1987; pp. 67–93.

80. Subbiah, A.; Islam, S. A.; Neidle, S. *Carcinogenesis* **1983**, *4*, 211.

81. Pullman, B.; Pullman, A. *Quantum Biochemistry*; Interscience: New York, 1963.

82. Singer, B.; Grunberger, D. *Molecular Biology of Mutagens and Carcinogens*; Plenum: New York, 1983; pp. 45–74.

83. Gagliano, A. G.; Geacintov, N. E.; Ibanez, V.; Harvey, R. G.; Lee, H. M. *Carcinogenesis* **1982**, *3*, 969.

84. Geacintov, N. E.; Gagliano, A. G.; Ibanez, V.; Harvey, R. G. *Carcinogenesis* **1982**, *3*, 247. Geacintov, N. E.; Gagliano, A. G.; Ivanovic, V.; Weinstein, I. B. *Biochemistry* **1978**, *17*, 5256. Prusik, T.; Geacintov, N. E.; Tobiasz, C.; Ivanovic, V.; Weinstein, I. B. *Photochem. Photobiol.* **1979**, *29*, 233.

Appendix: Empirical Force-Field Method

Kenny B. Lipkowitz

A.1 EMPIRICAL FORCE FIELD

Much of the literature uses the term "empirical force-field (EFF) calculations;" this is synonymous with molecular mechanics (MM). Many of the terms and ideas of molecular mechanics have their origin in normal coordinate analysis. Vibrational spectroscopists are interested in the forces that hold molecules together, and the approach they have taken is to do a normal coordinate analysis. Briefly, vibrational energies, vibrational wavefunctions, and force constants are related by a secular determinant just like the electronic energy and electronic wavefunctions with which most of us are familiar. Both are eigenvalue problems.

$\|H-ES\|=0$	$\|FG-E\lambda\|=0$
electronic secular determinant	vibrational secular determinant

In the vibrational secular determinant F is the force constant matrix that tells us about the forces holding the molecule together, G is the so-called inverse kinetic energy matrix that depends on the mass and positions of the individual atoms in the molecule, E is a unit matrix, and λ are the characteristic roots (eigenvalues often called eigenfrequencies) related to the normal mode frequencies by $\lambda = 4\pi^2\nu^2$ where ν is the vibrational frequency. The basis set (counterpart to AOs in MO calculations) is the set of internal coordinates of the molecule. The normal modes are going to be some combination of this basis set of internal displacement coordinates, L. For water we might define the displacements as

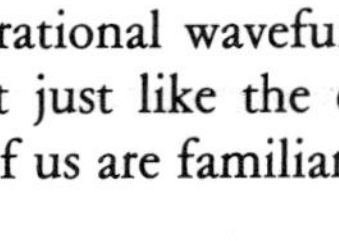

Thus, in the vibrational secular determinant, F_{11} is the force constant for O—H stretching along L_{11} and F_{22} is the force constant for stretching along L_{22}. F_{21} and F_{12} are off-diagonal elements of the force constant matrix referred to as

"interaction force constants" which describe how two isolated stretches interact with one another. For larger systems the interaction force constants may describe how isolated stretching motions, or bending motions, or stretch-bending motions interact.

Before we lose track of things, recall that the normal coordinate frequencies and the inverse kinetic energy matrix are known quantities. The spectroscopist then finds the force constants. This is a typical eigenvalue problem that is solved iteratively; one makes an initial guess at the values in the F matrix, computes the normal coordinate frequencies, and then with an optimization procedure readjusts the force constant matrix so that experiment and theory agree. This process is called the Wilson FG matrix formalism and need not be described further except to say that it allows us to find force constants.

The reason for mentioning this was to inform you that different treatments of the F matrix during a normal coordinate analysis give rise to different vibrational spectroscopic force fields. When all elements of the F matrix are considered in the analysis, that is, when all independent force constants are determined and no simplifications are made, we have the general harmonic force field (GHFF). The GHFF is good for very small molecules with high symmetry. For larger molecules we make the problem more tractable by developing systematic simplifications of the complete force field. These simplified force fields are referred to as "constrained force fields." In many of these force fields it is recognized that the off-diagonal elements of the F matrix are much smaller than the diagonal elements and can be set to zero; for example, a high-frequency stretch will not interact much with a low-frequency bend. Several of these simplified force fields are:

1. *Direct Constraints Force Field.* Fixes values of all interaction force constants from vibrations that are separated by at least two noncommon atoms.

2. *Valence Force Field.* All interaction force constants between stretching and bending of different bonds set equal to zero. No cross-terms are included. This model is so simple that it has no quantitative applicability. A slight relaxation of the VFF is to allow some stretch–stretch and stretch–bend terms.

3. *Central Force Field.* In its simplest form this model describes angle bending in terms of non bonded interaction distances. This force field is rarely used. Like the VFF, this model is too simple to explain molecular frequencies but a relaxation of some constraints resulting in the general central force field improves the situation somewhat by allowing interactions between interbond parameters.

4. *Urey–Bradley Force Field.* In addition to conventional stretching and bending force constants, all interactions that are stretching/stretch–bend coordinates are defined in terms of interactions between nonbonded atoms. Shortcomings of this model are well established; inadequacies are eliminated by allowing stretch–stretch and bend–bend interactions missing in the simple model. This enhanced force field is called the modified Urey–Bradley force field.

5. *Orbital Valence Force Field.* Like the UBFF in its treatment of interactions in terms of repulsions between nonbonded atoms but treats angle distortions in terms of maximizing orbital overlap.

6. Other force fields and "approximate force fields" exist.

A.2 MOLECULAR MECHANICS

Empirical force fields as incorporated in molecular mechanics are quite different from the vibrational spectroscopic force fields. For the spectroscopist the approach is to know the structure and vibrational frequencies of a molecule and then figure out the forces holding the molecule together. In molecular mechanics we assume we know the forces that hold the molecule together and try to find structures and vibrational frequencies. Several different empirical force fields have been used in molecular mechanics. These include the very successful modified Urey–Bradley and the central force fields, but by far the most popular have been the valence force fields. Because of their current popularity we focus on the valence force-field approach.

Molecular mechanics is an attempt to formulate as reliable a recipe as possible for reproducing the potential energy surface for the movement of atoms within a molecule. The basic philosophy of molecular mechanics rests on the fact that it is a computational model for describing the potential surface for all internal degrees of freedom in a molecule and, in many respects, is similar to hand-held mechanical models but is much more sophisticated. The treatment relies heavily on the availability of experimental data for parameterization; thus, the EFF methods are interpolations and sometimes extrapolations of existing data.

Historically, the application of empirical force fields has been directed toward a specific problem or even a specific class of molecules. During the past decade new force-field programs were developed that are applicable to a wide variety of structures. These programs often reflect the author's interests and abilities and are as different from one another as are the authors.

In general, there are no rules as to what functions are to be chosen or what parameters are to be used. Some authors want their models to reproduce structures and energies only (the idea being that if too much is asked from the force field it will perform inadequately), while others want to calculate structures, energies, and vibrational spectra (these are called consistent force fields, CFF).

All the force fields, however, consider a molecule as a collection of particles held together by some sort of elastic forces. These forces are defined in terms of potential energy functions of the internal coordinates of the molecules that constitute the molecular force field.

A.2.1 Potential Functions

The internal energy of a molecule can be expressed as a function V of the internal coordinates of a molecule. This function presumably has minima corresponding to stable equilibrium geometries of the molecule. The precise form of V is unknown but has been approximated as a sum of different types of energy contributions:

$$V_{\text{total}} \approx V_1 + V_\theta + V_\phi + V_{\text{nb}} + V_{\text{e}} + V_{\text{h}} + \cdots \quad [A.1]$$

The component terms contributing to the total molecular potential typically include V_1 for bond stretching or compression, V_θ for valence angle bending, V_ϕ for describing rotations around bonds, V_{nb} for nonbonded interactions, V_{e} for coulombic interactions, V_{h} for hydrogen bonding, and a variety of additional terms that may be needed to better reproduce experiment. Let me point out that this is a gross oversimplification that is justified only by the fact that it works! Let me also point out that authors partition the component terms contributing to V_{total} in vastly different ways. Some force fields will have 10–20 components that explicitly treat each conceivable kind of interaction while some may only have 5–10 terms; the missing terms are often incorporated into the force field by judicious parameterization. In any event this makes it difficult to compare force fields and leads up to this important warning. Do not become overzealous and rely too heavily on the information content of the component energies. They can be terribly misleading.

The numerical value of V has no inherent physical meaning. Its value depends on the potential functions and parameters. But, as with other computational methods, the differences between energies for rotational isomers (the underpinning of conformational analysis) usually is good. Therefore, even for its approximations and assumptions, molecular mechanics is well suited for conformational analysis.

The analytical form of the terms in Equation (A.1) is straight-forward and will not be critically reviewed here but will be commented upon.

1. V_ℓ, *Bond Stretching*. A general form of empirical potential function for stretching and compression of covalent bonds is

$$V_\ell = k_1^s(\ell - \ell_0) + \tfrac{1}{2}\, k_2^s(\ell - \ell_0)^2 + \tfrac{1}{6}\, k_3^s(\ell - \ell_0)^3 \quad [A.2]$$

where k^s is a stretching force constant, ℓ is the actual length of the bond, and ℓ_0 is the reference geometry corresponding to the "strain-free" state for that bond. Initially, $k_1^s = k_3^s = 0$ was assumed. The resulting harmonic approximation was quickly abandoned when it was realized that even modest deviations from ideal bond lengths required some sort of correction for the resulting anharmonicity. Generally, a cubic term is included. An alternative that is popular because it accounts for anharmonic behavior and the fact that bonds dissociate at finite energies is the Morse potential:

$$V_\ell = D(\{\exp[-\alpha(\ell - \ell_0)] - 1\}^2 - 1) \quad [A.3]$$

which includes two additional parameters, α and D, that are related to the force constant and bond dissociation energy, respectively. An important but somewhat subtle point often glossed over in seminars and research papers concerns the constant k^s, in Equation (A.2). These are not force constants in the vibrational sense; one cannot blindly adopt a spectroscopic force constant for the molecular mechanics force field. These constants are potential constants and differ in a very real way from force constants. The two terms are unfortunately used interchangeably.

2. V_θ, *Angle Bending*. The same general form of the empirical potential function used in bond stretching is used in bond bending in a valence force field:

$$V_\theta = k_1^b(\theta - \theta_0) + \tfrac{1}{2}\, k_2^b(\theta - \theta_0)^2 + \tfrac{1}{6}\, k_3^b(\theta - \theta_0)^3 \qquad [A.4]$$

where k^b are bending potential constants, θ is the bond angle, and θ_0 is the strain-free reference angle. Take note that a simple Hooke's law approach here too is inadequate. Hooke's law overestimates the bending energy. When a bond angle is severely compressed, the angle formed by the three nuclei are not the same as the angle formed by the bonding orbitals. Bent bonds have reduced orbital overlap resulting in an effective decrease in force constant. To correct for this, we reduce the potential constant at large angular deformations. In the Urey–Bradley force field, geminal interactions (1,3 interactions between atoms) are explicitly included as

$$V_{UB} = k_1^s(\ell - \ell_0) + \tfrac{1}{2}\, k_2^s(\ell - \ell_0)^2 \qquad (A.5)$$

Two points concerning the potential functions, one minor and the other major, can be made here. First, the potential constants for bending are smaller than those for stretching ($k^b < k^s$) usually by an order of magnitude. Thus, as a molecule is perturbed from its equilibrium geometry, the distortions show up more in angular deformations than in stretching or compression of covalent bonds. Second, the above equations assume that the natural bond lengths, bond angles, and the potential constants are transferable within the same class of compounds. Transferability means that standard bond lengths, standard bond angles, and potential constants in small molecules are the same as in larger molecules. Transferability of force constants in vibrational spectroscopy is a transaction that quickly fails but in molecular mechanics is a fine approximation.

3. V_ϕ, *Torsional Terms*. The torsional potential usually adopted in molecular mechanics calculation is a truncated Fourier series in which only the cosine terms are included:

$$V_\phi = \tfrac{1}{2} \sum_n k_n^\phi(1 + s \cos n\phi) \qquad [A.6]$$

where k^{ϕ} is the rotational barrier of the *n*th term and *s* is the multiplicity of the barrier. The terms in (A.6) depend on the local symmetries of the groups attached to the bond being rotated. Torsional potentials for rotation around bonds with two-fold symmetry as in double bonds can be represented by the single term

$$V_{\phi} = \frac{1}{2}\, k_2(1 - \cos 2\phi) \qquad [A.7]$$

Another commonly used function is

$$V_{\phi} = \frac{1}{2}\, k(\phi - \phi_0)^2 \qquad [A.8]$$

which because of its harmonic nature is expected to be applicable only for small distortions from equilibrium configurations. Rotating groups like CH_3 having C_{3v} symmetry as in ethane can be approximated by only a single term:

$$V_{\phi} = \frac{1}{2}\, k_3^{\phi}(1 - \cos 3\phi) \qquad [A.9]$$

Sixfold barriers originate from rotations of groups having C_{3v} and C_{2v} symmetries as in toluene and may be represented as

$$V_{\phi} = \frac{1}{2}\, k_6^{\phi}(1 - \cos 6\phi) \qquad [A.10]$$

Twelvefold barriers in organometal complexes due to rotation of groups with C_{4v} and C_{3v} symmetries are similarly formulated. Interestingly, this part of molecular mechanics has not been adequately treated and still remains one of the more controversial issues in conformational analysis.

4. V_{nb}, *Nonbonded Interactions*. For this term, vibrational spectroscopy uses potential functions that are quite different from those in molecular mechanics. Excluding short-range 1,3 interactions, spectroscopic force fields ignore nonbonded terms. These effects are indirectly accounted for when the force constants in the *F* matrix are determined. It is this indirect accounting of steric effects that does not permit transferability of force constants from molecule to molecule in vibrational spectroscopic force fields. (Even transferability from *n*-alkanes to branched alkanes fails.) This is also one of the major reasons why spectroscopic force constants are usually different from molecular mechanics potential constants. Molecular mechanics, in contrast, explicitly treats the nonbonded interactions. By treating long-range nonbonded terms we can more justifiably transfer parameters from one class of molecules to another.

The analytical form of the potential for nonbonded interactions varies from force field to force field. All of them divide nonbonded interactions into two parts: one attractive at long range and the other repulsive at short range. Two of the more popular analytical expressions used in molecular mechanics are the

Buckingham and Lennard-Jones potentials:

$$V_r = A \exp(-Br) - \frac{c}{r^6} \quad \text{Buckingham} \qquad [A.11]$$

$$V_r = \frac{A}{r^n} - \frac{c}{r^6} \quad \text{Lennard-Jones} \qquad [A.12]$$

Note that the last terms in both equations treat the attractive part of the potential the same way but the closed-shell repulsions between nonbonded atoms are treated differently. Other forms exist but these are the most common.

Several aspects of the molecular mechanics treatment of nonbonded interactions usually neglected in reviews that can provide you with additional insight about the method are:

(i) The potentials used to describe intramolecular effects are those developed to explain intermolecular interactions. The parameters in these equations are derived from crystal properties like heats of sublimation and interatomic distances or from gas-phase experiments like measurements of second virial coefficients and scattering experiments. The approximation thus made in molecular mechanics is that potential functions which describe intermolecular interactions will adequately describe intramolecular interactions. No justification is given but it seems to work.

(ii) The calculations as typically performed assume pairwise additivity of atom–atom interactions and neglect three- and higher-body terms. Thus, the long-range attraction of an atom with two otherwise equivalent atoms will be the same even if one of the atom–atom interactions has an intervening atom, bond, or molecule between them.

(iii) Atoms are treated as spherical species. This is not a bad assumption for some large polarizable atoms but is an insufficient assumption for smaller atoms like hydrogen. The fix-up for smaller atoms is to place the center of the sphere somewhere along the bond axis rather than on the nucleus. In some force fields nonspherical electron density is approximated by explicitly adding lone pairs, for example, on ether and alcohol oxygens.

The above assumptions and computational simplifications can be corrected for in a roundabout way by judicious selection of parameters. But you should be aware of this. A further complicating factor is the treatment of polar bonds or of atoms that are charged. For many years these terms were neglected. (Early work focused on hydrocarbon force fields where C—H and C—C bond moments are small.) Electronegativity differences of heteroatoms (and even carbon in different hybridization states) require an accounting of the uneven distribution of electron density between atom pairs. Thus, most molecular mechanics treatments add a coulombic term:

$$V_c = \frac{q_i q_j}{\epsilon r_{ij}} \qquad [A.13]$$

or a dipole–dipole term:

$$V_{\text{dipol}} = \frac{\mu_i \mu_j}{D r_{ij}^3} (\cos \chi - 3 \cos \alpha_i \cos \alpha_j) \qquad [\text{A.14}]$$

to supplement the nonbonded terms. These account for point–dipole or dipole–dipole interactions. Either way, one must come up with suitable partial charges or bond moments. A hot issue in this regard is how one obtains the partial atomic charges. Some people rely on simple schemes like Del Rey's method, others will use semiempirical (MNDO or CNDO) atomic charges, while some groups insist that charges should be selected to reproduce electrostatic contour maps of high-quality *ab initio* calculations for the molecule under consideration. There is no agreement in the scientific community as to which is best. Consequently, it is not uncommon to find two groups using the same potential functions but with bond moments reversed simply because their quantum calculations were different.

Another problem to think about is that the assigned charges remain fixed for all conformations of the molecule; that is, dipole-induced dipoles are not accounted for. In some instances this is acceptable, but in others it is untenable. Certainly the magnitude of the C—F bond moments change as one rotates around the C—C bond of 1,2-difluoroethane. Allinger recognized this type of dipolar coupling as an important feature of conformational analysis and has spent a considerable effort addressing how best to treat it. This and related work is summarized in an elegant review of intramolecular electrostatics in force field calculations by Meyer[24] and should be studied.

5. *Cross-Terms*. By their very nature, valence force fields are oversimplifications that often need cross-terms to give proper results. These are usually bilinear equations that couple stretching motions with bending motions, bending with torsions, stretches with stretches, and so on. An example of this is the stretch–bend term in Allinger's MM2 force field is

$$E_{\text{SB}} = k_{\text{SB}}(\ell - \ell_0)(\Theta - \Theta_0) \qquad [\text{A.15}]$$

A more exotic one is Lifson's early trilinear term is

$$V_{\text{BTB}} = k_{\text{BTB}}(\Theta - \Theta_0)(\phi - \phi_0)(\Theta' - \Theta_0') \qquad [\text{A.16}]$$

The coupling of bond elongation or shrinkage with angular deformation is depicted below. The structure on the left has standard bond angles and bond lengths. As the X—C—X angle is

```
   C             C
  / \          / Θ \
 X Θ X        X))  ((X

Θ = 109°      Θ = 90°
```

compressed, the repulsive interaction of the two X groups colliding into one another is offset by C—X bond elongation. Widening the X—C—X angle causes the opposite effect of bond shrinkage.

Similar types of torsion–bend, torsion–out-of-plane bend, and other interactions can be considered to make up for deficiencies in the valence force field. Some authors use only one or two cross-terms while others may use up to six:

Allinger	Lifson
Stretch–bend	Stretch–stretch
Torsion–bend	Stretch–bend (adjacent)
	Stretch–bend (opposite)
	Bend–bend (common central atom)
	Bend–torsion–bend
	Double bond oop bend–oop bend
	oop = out of plane

These terms are sometimes added into the force field at early stages of development and sometimes a posteriori; in the case of the Allinger force field the torsion–bend term was included primarily to ensure ring puckering in cyclobutane that could not be satisfactorily treated with torsion terms alone. Again you see how difficult it becomes to compare individual terms in the force field since different authors partition the total energy into different terms.

6. V_h, *Hydrogen Bonding*. The hydrogen bonding term is treated different ways by different authors. One form of the potential used successfully by Karplus is:

$$V_h = \left(\frac{A}{S^p} - \frac{B}{S^m}\right) \cos^\alpha(\phi_{D\text{-}H\text{-}A}) \cos^\beta(\phi_{AA\text{-}A\text{-}H}) \qquad [A.17]$$

S is the hydrogen bond length; A, B, p, and m are parameters that define the shape and depth of the hydrogen bond well; $\phi_{D\text{-}H\text{-}A}$ and $\phi_{AA\text{-}A\text{-}H}$ are donor-hydrogen acceptor and acceptor-antecedent-acceptor hydrogen bond angles with exponents $p = 12$, $m = 10$, $\alpha = 0$, $\beta = 0$ in one version and $p = 6$, $m = 4$, $\alpha = 4$, and $\beta = 4$ in another. Many of the force fields developed for peptides include a hydrogen bonding term explicitly because hydrogen bonding effects are so important. Other force fields developed for hydrocarbon analysis like Allinger's do not treat these terms properly although fix-ups have been published.

A.2.2 Parameterization

A.2.2.1 Optimization

Once the analytical forms of the potential energy function have been selected, it is necessary to obtain values of reference geometries (like ℓ_0, Θ_0), their corresponding potential constants (k^s, k^b), nonbonded parameters, bond moments, and so on. Initial, best-guess values are selected; geometries, energies, and sometimes spectra are computed and then compared with known quantities of a database. The parameters are then varied until the potential functions with those parameters reproduce the information content of the database. This is called parameter optimization and it is performed different ways by different groups.

The simplest technique is to optimize by inspection. Here one adjusts the parameters in a trial and error way until the deviation of computed and observed values in the database are satisfactorily small. This method works surprisingly well. The interdependence of parameters is nonetheless complex. Lifson and co-workers realized this and attempted a systematic way of optimizing parameters. Briefly (using their notation), the difference between a calculated and measured value of an observable in the database is

$$\Delta y = y^{\text{calc}} - y^{\text{meas}} \qquad [A.18]$$

The problem then is: given the parameters p, find the changes δ_p to make Δy as small as possible. To make a long story short, they end up with a nonlinear least-squares problem that is linearized by an expansion which is truncated after the first term. This requires δ_p to be small and extensive computing is required. More problematic is the fact that truncation of their Taylor series after the linear term requires the initial guess of parameters to be close to the optimum. Furthermore, since different kinds of parameters are involved, an elaborate weighting scheme involving inverse absolute uncertainties is used and many decisions (that are more opinion than fact) are made about how to weight things. In any event this method, albeit cumbersome, works and has been extensively reviewed.[29]

A.2.2.2 Databases

In the above discussion we mentioned that one attempts to optimize the parameters for a given potential energy function so that the calculated and measured observables in a database are in agreement. The question of what observables to calculate is left to the author of the force field; some want to reproduce structure and energies while others want vibrational frequencies as well. The databases used for fitting parameters are quite varied in scope and in completeness. AMBER, for example, has a relatively small databank (keep in mind that good experimental information is sometimes difficult to find, especially for highly functionalized fragments of molecules like DNA). Even with these shortcomings AMBER is one of the best protein force fields but I suspect it would do poorly on other classes of molecules like strained hydrocarbons.

In contrast, the consistent force field of Rasmussen has been optimized on five classes of observables. They are geometry, rotational constants, atomic charges, dipole moment, and internal frequencies of vibration. Other quantities like thermodynamic properties are being considered. The database also contains information about a large number of alkanes, cycloalkanes, linear and cyclic ethers, alcohols, esters, lactones, amides, amines, and so on.

The reason for mentioning Rasmussen's consistent force field is to point out that with his parameter optimization scheme he has been able to develop several different (albeit related) potential energy functions—each with its own unique set of parameter combinations. These are designated PEF (potential energy function) followed by a number. Thus, PEF 300, PEF 301, PEF 302, and PEF 303 exist along with the PEF 400 series. The latter differs from the former by using a Lennard-Jones rather than a Buckingham term and includes electrostatic monopole interactions. Both PEF 300 and PEF 400 parameters were developed by trial and error but were eventually optimized. Other groups have adopted a similar strategy of developing specific force fields for certain types of problems. Karplus, for instance, recently published (*J. Comput. Chem.* **1986**, 591) an improved empirical energy function, EF2, for nucleic acids in the absence of explicit solvent (this is done by scaling the shielding of charged groups in a way that approximates the effect of solvent).

This leads up to an important point. It is my opinion that one should not tamper with force fields. It is very common to find people adulterating the potential energy function without reoptimizing the parameters. This is especially rampant in the organic community with Allinger's force field. One cannot simply add a hydrogen bond function with associated parameters unless the rest of the parameters are adjusted for this new force. Even replacing MM2 bond moments with *ab initio* or semiempirical atomic changes gives, in my experience, completely different potential surfaces. Unfortunately, most of the software developers (vendors) who incorporate MM2 into their modeling systems will alter the force field by adding functions to make MM2 do calculations for which it was not intended. I am afraid there are many MM2 mutants running around and you should be cautious.

A.2.3 Energy Minimization

Given a potential energy function with suitable parameters we can calculate V for any arrangement of the component atoms that do not vary much from the reference geometry. The objective of energy minimization, sometimes called geometry optimization, is to find a set of coordinates (internal or cartesian) corresponding to minima on the molecular potential energy surface. Thus, all one needs to do is take the derivative of Equation (A.1) with respect to all degrees of freedom in the molecule and locate the place where each of those derivatives are simultaneously equal to zero. In practice, we often use first and second derivatives. The gradient vector **g** and the Hessian matrix **G** are, respectively,

$$\mathbf{g}_i = \frac{\partial V(\mathbf{x})}{\partial x_i} \qquad [A.19]$$

and

$$\mathbf{G}_{ij} = \frac{\partial^2 V(\mathbf{x})}{\partial x_i \partial x_j} \qquad [A.20]$$

where (**x**) represents parameters defining structural features of the molecule like bond lengths and bond angles. The gradient vector and the Hessian matrix are evaluated numerically by finite difference methods or, as most modern programs now do, analytically. (Examples of the analytical expressions for first and second derivatives are in Rasmussen's books.) The reason for computing the second derivatives is that the type of stationary point on a potential surface can be discerned, that is, one can determine whether the stationary point is a minimum or a maximum (e.g., transition structure). Additionally, the second derivative of the potential is the force constant and so the normal coordinate frequencies (i.e., spectra) can be obtained.

There exist a plethora of algorithms for performing function minimization. Although many of these procedures are adequate for minimizing 5–10 variables, most are not well suited for molecular mechanics calculations which typically involve over 100 variables. Those that have been used can be classified as (1) conjugate-direction methods and (2) Newton methods which may be subdivided into restricted-step methods or quasi-Newton methods. The mathematics of these minimization techniques is complicated and need not be further considered here. Review articles containing detailed information about these methods especially with regard to molecular mechanics exist[1,4–6,10,12,22,28,29] and should be read. Suffice it to say that each method has its advantages and disadvantages; some quickly converge to a minimum on a shallow surface while others flounder; some are CPU intensive while others are fast. Some of the more recent commercially available modeling programs will give you a choice of optimizers. Macromodel, for example, include steepest descent, variable metric, two types of conjugate gradient methods, block diagonal, and full matrix Newton–Raphson minimizers. Other programs are less ambitious.

All the optimization methods search for stationary points ($\partial V/\partial x = 0$) on the potential energy surface. All the methods are capable of finding minima on the surface but not all are capable of finding maxima. The full matrix Newton–Raphson method is used for searching both types of stationary point. Whether the molecular structure corresponds to a minimum or a maximum can be determined from the second derivatives. If all the eigenvalues of the F matrix are positive, we have a minimum. If a couple of the eigenvalues are negative we are at a hilltop while a saddlepoint (transition structure) is characterized by only one negative eigenvalue in the Hessian matrix. Finding transition states is much more difficult than finding minimum energy conformations and not all groups use this method.

An alternative approach for locating maxima on a conformational energy surface is to restrict certain torsional degrees of freedom. Usually, what is done is to incrementally rotate around a bond in small steps. At each step the torsion angle is kept rigid with an artificially high torsional potential constant (not included in the computed energies) and all other degrees of freedom are allowed to relax. This is sometimes referred to as the dihedral driver method and is quite popular for mapping conformational interconversions. These driver methods, as popular as they are, usually locate minima quite well but since reaction coordinates are rarely described by only one or two dihedral angles they do not accurately map out minimum energy pathways and they are notoriously poor for defining geometries and energies of transition states.

The take home message for energy minimization is that most methods can locate the minimum energy structures on a surface and assuming all minima are found, most force fields can give a reasonable Boltzmann population of conformers. Using anything but a full matrix Newton–Raphson minimizer does not accurately provide structures and energies of transition structures. Be aware of this as you read the literature; relatively few groups will report that transition structures are confirmed by one negative eigenvalue (an imaginary frequency corresponding to the transition vector) in the Hessian matrix.

A.2.4 Energies

The initial energy obtained from the minimization is the minimum potential energy or the ''steric energy.'' It is the energy relative to a hypothetical reference system. The steric energy itself has no real physical meaning; it corresponds to the energy of an isolated, gas-phase molecule in a fictitious motionless state (it is at the bottom of the potential well without zero-point vibrational energy). Furthermore, the magnitude of these energies have no significance because their numerical values depend on the potential energy functions and parameters.

These steric energies may be used directly to asses energy differences between stereoisomers and isologous molecules. However, if one wants to compare energies between molecules which differ in their bonding arrangement or in size, a more meaningful number is the heat of formation. Knowing the vibrational eigenvalues λ_i, we can obtain the vibrational enthalpy contributions from the statistical-mechanical relationship by

$$H_{\mathrm{vib}} = k_\beta T \sum_{i=1}^{3N-6} \frac{h\lambda_i^{1/2}}{k_\beta T}\left(\frac{1}{2}+\frac{1}{\exp\,(h\lambda_i^{1/2}/k_\beta T)-1}\right) \qquad \text{[A.21]}$$

Adding this to the steric energy along with the rotational and translational contribution of 3 *RT* gives the heat of formation, which can be used to compare isomers as well as conformers.

An alternative approach to computing heats of formation that works as well for small molecules and better for large molecules (it is difficult to accurately account for the low-frequency normal modes in large molecules) is simply to add bond increments or group increments to the steric energy along with 4 *RT* for translation, rotation, and *PV*. For many years chemists have made use of bond energy tables that allow one to compute heats of formation by summing numbers and types of bond in a molecule (this assumes transferability of information from one molecule to another) and is especially applicable for strainless molecules. All that is needed is to add the appropriate group or bond increment to the initial steric energy to approximate the heat of formation. Early on, Schleyer derived group increments based on least-squares optimization of the calculated steric energies with experimental heats of formation at 25 °C for a substantial number of hydrocarbons. Allinger treated the problem in a similar way but opted to use bond increment values instead. Both methods appear to work about equally well.

The energies calculated by molecular mechanics correspond to a molecule in a fictitious motionless state at 0 K. Corrections to the chemical binding energy, the vibrational zero-point energy, thermal energies of vibration, translation and rotation, and the enthalpy of mixing due to the various conformers must be made to convert these raw numbers into heats of formation in the gas phase at 25 °C. This process is completely empirical and, as with other approximations in molecular mechanics, assumes transferability of properties from molecule to molecule. A final point to be made here is that often you will see reaction coordinates with ordinates labeled in steric energy rather than heats of formation. This is acceptable because when comparing conformations many of the terms contributing to the heats of formation are the same and tend to cancel.

A.3 MOLECULAR MECHANICS IN PRACTICE

I have highlighted the main parts of any molecular mechanics program: the potential energy function and associated parameters, the minimization methods, and the energies. There are subtle nuances about these calculations that will not be mentioned here since they are thoroughly reviewed by Allinger.[22] Furthermore, there are extensions of molecular mechanics that include delocalized π systems, excited electronic states, heteroatoms, transition states involving bond-making–bondbreaking, and organometallics, of which you should be aware but they are not appropriate for this appendix.

My message for you is that molecular mechanics, albeit more sophisticated and more reliable than a mechanical model, is nonetheless a model. It relies extensively on good experimental data for parameterization but it allows one to accurately interpolate or extrapolate information about observables that have not yet been measured. Many of the simplifying approximations and assumptions described earlier have little theoretical justification but are acceptable because they seem to work.

Molecular mechanics when used properly is extraordinarily reliable. It is, to quote Ken Wiberg, "too important a theoretical tool to be left in the hands of the theoreticians." Unfortunately, the theoreticians seem better to understand what can and what cannot be done with the method than do organic chemists. There are major pitfalls that many of us have fallen into that seriously detract form the method's capability. Let me mention a few here:

1. *Minima*. There is no way of proving that the minimum you have located is the global minimum. You may be in a local minimum. Furthermore, depending on the optimization method employed and the convergence criteria selected to stop the calculation, you may have located a false minimum. This is more common than you would expect.

2. *Maxima*. Not all maxima quoted in the literature are truly transition states. These are sometimes difficult to locate and are confirmed only when a single negative eigenvalue in the Hessian matrix exists.

3. *Interconversion Pathways*. The minimum energy pathway is characterized by the requirement that all coordinates except the reaction coordinate are minimized. For most cases the reaction coordinate is composed of several torsion angles and/or other internal coordinates; it is hardly ever described by one or two torsion angles. Multiple dihedral drivers should be employed. Deficiencies of calculating conformational interconversions especially in cyclic compounds are discussed at the end of Chapter 3 in Allinger's book.[22]

4. *Solvent and Counterions*. These interactions are rarely treated explicitly. Karplus has developed a force field with solvent effects absorbed into the parameters but this is an exception. Most people treat solvation by changing the dielectric term in the electrostatic part of the potential and most groups neglect the gegenions altogether.

5. *Potential Functions*. Some groups adulterate programs by excising or adding to the original potential energy function. This could "unbalance" the force field by overestimating effects, which are already compensated for in the original terms. Several hydrogens bond potentials, for example, have been added to Allinger's programs but full parameter optimization has not been performed. Be careful when dealing with these types of mutant program.

6. *Parameters*. Most force fields are very limited in scope; they can all handle hydrocarbons but are not well suited for hetero atomics. A good rule of thumb is: the further away from a normal hydrocarbon you are, the less reliable

Table A.1 Computational Deficiencies in MM2

No.	Method	Deficiency
1.	MM2:	Underestimates high C—C rotational barriers in congested hydrocarbons.
2.	MM2:	Cl---CH_3 eclipsing in rotational barriers not well treated. MM1 gives better trends than MM2.
3.	MM2:	Overestimates close H---H nonbonded interactions.
4.	MM2:	Underestimates energies in strained propellanes.
5.	MM2:	Stretching is too stiff and not anharmonic enough.
6.	MM2:	Overestimates C—C and C—H bond lengths and underestimates C—O bond lengths in carbohydrates.
7.	MMP2:	Underestimates interaction energy for perpendicular benzene dimer; overestimates interaction energy of parallel benzene dimer.
8.	MM2:	Arguments for including $-V_2$ term in O—C—C—O to reproduce anomeric effect.
9.	MM2:	Allows for anomeric effect in torsion but not for anomeric effect on bond lengths. Also, O—C—O bond angles too large by 2–4°.
10.	MM2:	Overestimates energy difference between trans–gauche conformers for C—O bond rotation in O—C—C—O—C—C—O fragments.
11.	MMP2:	Improperly treats CH_3 rotation when methyl is attached to internal carbons of a polyene. CH_3 on the terminal carbons are treated properly.
12.	MM2:	Tends to make distorted alkynes, for example, cyclooctyne, etc. C—C≡C—C too flat as torsional contributions omitted from force field.
13.	MM2:	Underestimates alkene π---HO interactions by ~1.5 kcal/mol.
14.	MM2:	(See item 1). Three fold C—C—C—C torsion barrier too low. This underestimates eclipsing and gives unduly favorable stabilization of congested T.S.'s. Osawa's V_3 set at 0.600 may be too large. (The problem is partly in the entropy of activation, which is often nonzero in these compounds.)
15.	MM2:	Inadequately represents hydrogen bonds.

16. MM2: (See item 13). Underestimates —OH— phenyl π attraction.

17. MM2′: (Osawa's version of MM2). Does not properly treat expended O—C—O angles; makes them too small.

18. MM2: Radicals not accurately treated. Does not have torsion–stretch term that allows for hyperconjugative elongation of adjacent C—C bond when it is in plane of radical p A.O.

19. MM2′: (Osawa's version of MM2. See items 1 and 14). Overestimates barriers to C—C rotation under extremely congested situations.

20. MM2: Overestimates —C(=O)—O bond angles in medium size lactones because it lacks anomeric term which contracts this angle.

21. MM2: Bond shortening by attachment of electronegative substituents not allowed for. Improved in MM2(82), but heats of formation are erroneously raised somewhat by this correction (typically by only a few tenths of a kcal/mol).

22. MM2: Lack of a torsion–stretch interaction leads to eclipsed single bonds being too short in general. This shows up in cyclopentane rings in particular (C—C bonds too short by 0.010 Å or more).

23. MMP2: Conjugated bond lengths in most molecules can be calculated by a self-consistent field method, but some ([18]annulenes and probably other large cyclic polyenes) require an inclusion of electron correlation to give a proper result.

24. MMP2: The *f* factor used to modify out-of-plane bending of aromatic rings sometimes causes problems, allowing a bending that is too easy.

25. MM2: The parameters given for silicon versions of MM2 are obsolete. A new set, which is a big improvement, was derived.

Note: References to these problems and standard fix-ups can be found in the *Quantum Chemistry Program Exchange Bulletin*, **1987**, *7(1)*, 19–22 or directly from this author. Keep in mind that some of these problems are truly deficiencies of the force field while other examples may only pertain to the specific example cited and may not generally be a problem. Also be aware that some users may have transcended the areas of use based on the data used for parameterization; they may have simply overextended the method. No review is given and no judgement is passed (we leave that to you) but you may find this list of some value in your work.

will be the MM results. When parameters are not available for a particular force field, most people make ad hoc decisions about the missing values. These values may or may not be adequate. Certainly the validity of the results may be questioned, especially if no effort has been made to demonstrate that those ad hoc parameters can reproduce structures and energies of known systems. An even more devious problem with parameters exists, however. In current molecular modeling programs the system will revert to a set of default parameters if missing parameters are needed. Unfortunately, the user of these programs is not even aware this is happening. An example involves the original version of MODEL. If the MM2 force field is used, the program will write "MM2 Energy" on the top of the screen followed by the appropriate information. If the MM2 parameters are missing, a generalized parameter set is used and the program will write "MM Energy" followed by the appropriate information in the same format. Many of the organic chemists who use modeling software are oblivious to these goings on. Be careful!

7. *Component Analysis*. The total steric energy is a sum of component energies each of which measure the deviation from nonideal behavior. In most molecular mechanics programs the individual stretching energy, bending energy, torsion energies, and so on are printed along with the total steric energy. Be careful not to attribute too much meaning to the individual components since they may have some other effect inadvertently built in to them. Do not overinterpret your results. If several different force fields give the same result you may be on to something correct.

All these pitfalls are generic and are to be avoided for all empirical force fields. It would be nice to see a list of problems associated with a specific force field. To this end I have compiled in Table A.1 the deficiencies found in the MM2 force field. I selected this force field because many of you are already using it. The list is not comprehensive. Omitted are other problems like neglecting three- and four-body terms (recall the nonbonded interactions are treated pairwise only) the neglect of polarization effects (the bond moments or atomic charges will change as other dipoles or point charges come close) etc. The list also omits deficiencies associated with optimization methods, dihedral driver problems, and so on. Rather, we provide a list of problems relating to the potential function itself.

REFERENCES

1. Williams, J. E.; Stang, P. J.; Schleyer, P. v. R. *Annu. Rev. Phys. Chem.* **1968**, *19*, 531.
2. Bartell, L. S. *J. Chem. Educ.* **1968**, *45*, 754.
3. Engler, E. M.; Andose, J. D.; Schleyer, P. v. R. *J. Am. Chem. Soc.* **1973**, *95*, 8005.
4. Altona, C.; Faber, D. H. *Top. Current Chem.* **1974**, *45*, 1.
5. Allinger, N. L. *Adv. Phys. Org. Chem.* **1976**, *13*, 1.
6. Ermer, O. *Structure and Bonding* **1976**, *27*, 161.

7. Bentley, T. W. *International Review Science: Organic Chemistry,* Series Two, Vol. 5; D. Ginsburg, Ed., Butterworth: London, **1976**; pp. 342–346.

8. Warshel, A. *Semiempirical Methods of Electronic Structure Calculation; Modern Theoretical Chemistry, Vol. 7, Part A; G. A. Segal, Ed.; Plenum: New York, 1977*; pp. 133–171.

9. Niketic, S. R.; Rasmussen, K. *Lecture Notes in Chemistry, Vol. 3: The Consistent Force Field*; G. Berthier et al., Eds.; Springer: Berlin, **1977**.

10. Hursthouse, M. B.; Moss, G. P.; Sales, K. D. *Annu. Rep. Prog. Chem.* **1978**, *75B*, 23.

11. Beagley, B. *Molecular Structure by Diffraction Methods, Vol. 6, Specialist Periodical Reports*; The Chemical Society: London, *1978*; Chapter 3.

12. White, D. N. J. *Molecular Structure by Diffraction Methods, Vol. 6, Specialist Periodical Reports*; The Chemical Society: London, **1978**; Chapter 2.

13. Mislow, K.; Dougherty, D. A.; Hounshell, W. D. *Bull. Soc. Chim. Belg.* **1978**, *87*, 555.

14. Kitaigorodsky, A. I. *Chem. Soc. Rev.* **1978**, *7*, 133.

15. Duchamp, D. J. *In Computer-Assisted Drug Design*; Olson, E. C. and Christofferson, R. E., Eds.; ACS Symposium Series 112; American Chemical Society, Washington, DC, **1979**; Chapter 3.

16. Stuper, A. J.; Dyott, T. M.; Zander, G. S.; In *Computer-Assisted Drug Design*; Olson, E. C. and Christofferson, R. E., Eds.; ACS Symposium Series 112; American Chemical Society: Washington, DC, **1979**; Chapter 19.

17. Lifson, S. *In Supramolecular Structure and Function*, International Summer School in Biophysics (1981: Dubrovnik, Yugoslavia); Pifat, G. and Herak, J. N., Eds.; Plenum: New York, **1981**, pp. 1–44.

18. Ermer, O.; *Aspekte von Kraftfeldrechnungen*; Baur Verlag: Munchen, 1981.

19. Boyd, D. B.; Lipkowitz, K. B. *J. Chem. Educ.* **1982**, *59*, 269.

20. Bürgi, H. B. *Comput. Crystallogr., Pap. Int.*, Summer School (Meeting Date 1981); Sayre, D., Ed.; Oxford University Press: Oxford, **1982**; pp. 430–439.

21. Osawa, E.; Musso, H. *Top. Stereochem.* **1982**, *13*, 117.

22. Burkert, U.; Allinger, N. L. *Molecular Mechanics*; ACS Monograph 177, American Chemical Society: Washington, DC, **1982**.

23. Osawa, E.; Musso, H. *Angew. Chem. Int. Ed. Engl.* **1983**, *22*, 1.

24. Meyer, A. Y. In *Chemical Halides, Pseudo-Halides Azides*, Vol. 1; Patai, S. and Rappoport, Z., Eds.; Wiley: Chichester, **1983**; Chapter 1.

25. Hall, D.; Pavitt, N. *J. Comput. Chem.* **1984**, *5*, 441.

26. Brubaker, G. R.; Johnson, D. W. *Coordination Chem. Rev.* **1984**, *53*, 1.

27. Simonetta, M. *Int. Rev. Phys. Chem.* **1985**, *4*, 39.

28. Clark, T. *A Handbook of Computational Chemistry: A Practical Guide to Chemical Structure and Energy Calculations*; Wiley: New York, **1985**; Chapter 2.

29. Rasmussen, K. *Lecture Notes in Chemistry, Vol. 27: Potential Energy Functions in Conformational Analysis*; Berthier, G., et al., Eds.; Springer-Verlag: Berlin, **1985**.

30. Boeyens, J. C. A. *Structure and Bonding* **1985**, *63*, 67.

31. Wilson, S. *Chemistry by Computer. An Overview of the Applications of Computers in Chemistry*; Plenum: New York, **1986**; Chapter 5.

32. McCammon, J. A.; Harvey, S. C. *Dynamics of Proteins and Nucleic Acids*; Cambridge University Press: London, **1987**; Chapter 4.

Index